Mercury Handbook
Chemistry, Applications and Environmental Impact

Mercury Handbook
Chemistry, Applications and Environmental Impact

Leonid F Kozin and Steve Hansen
Email: kozin@ionc.kiev.ua, steve_hansen@adlt.com

RSC Publishing

Technical Editor: Cezary Guminski
Translated by Mark Kit

ISBN: 978-1-84973-409-7

A catalogue record for this book is available from the British Library

Published by The Royal Society of Chemistry,
Thomas Graham House, Science Park, Milton Road,
Cambridge CB4 0WF, UK

Registered Charity Number 207890

Visit our website at www.rsc.org/books

Preface

The Mercury Handbook attempts to cover all of the basic subject matter related to the condensed phase physics, chemistry, metallurgy, application and environmental aspects of mercury. The present book is derived from Leonid F. Kozin's book on the physical chemistry and metallurgy of high-purity mercury (*Fizikokhimiia i Metallurgiia Vysokochistoi Rtuti i ee Splavov*). The original Russian text was translated by Mark Kit of Language Interface in New York. Unfortunately a large percentage of the original work remains in Russian at the present time. Dr. Cezary Guminski assisted in the translation and technical editing of the book and also wrote a chapter on the use of mercury in small-scale gold mining.

Numerous important contributions were made to the book by others. Jason Gray of Nippon Instruments North America explained the practical operation of atomic fluorescence spectroscopy. A thorough discussion of the medical symptoms of mercury intoxication was generously providead by Bethlehem Apparatus, Inc. of Hellertown, Pennsylvania. The University of Illinois at Urbana-Champaign has provided invaluable access to its vast collection of online and print journals. Dr. Tim Brumleve of APL Engineered Materials, Inc. has assisted in the proof reading of key chapters of the present book.

The authors wish to thank their families for their extraordinary patience during the writing and editing of the manuscript. The staff of the Royal Society of Chemistry has endured more than necessary and is complimented for their professionalism and for their patience. Special mention should be given to Mrs. Janet Freshwater, Mrs. Alice Toby-Brant, Ms. Sarah Salter, Mrs. Katrina Harding, Mrs Rosalind Searle and others. They were exceptionally polite and patient throughout the entire writing process. Lastly, the staff of Strawberry

Mercury Handbook: Chemistry, Applications and Environmental Impact
By Leonid F Kozin and Steve Hansen

Published by the Royal Society of Chemistry, www.rsc.org

Fields in Urbana has provided encouragement and refreshment throughout the arduous task of writing this monograph. Additional information and updates to the *Mercury Handbook* can be found at www.mercuryhandbook.com.

Leonid F. Kozin
Kyiv, Ukraine

Steve C. Hansen
Urbana, Illinois

Contents

Mercury Handbook: Chemistry, Applications and Environmental Impact
By Leonid F Kozin and Steve Hansen

Published by the Royal Society of Chemistry, www.rsc.org

CHAPTER 1

Physicochemical Properties of Metallic Mercury

1.1 Atomic Properties

High-purity mercury is a dense, silvery white liquid with an extraordinarily low melting point of 234.321 K or –38.829 °C.[1] Mercury is a metal from Subgroup IIB of the Periodic Table and is related to zinc and cadmium. Mercury has the atomic number 80, atomic mass 200.59 and atomic volume $14.26 \times 10^{-6}\,m^3\,mol^{-1}$ at 298 K.[2] Its electronic configuration

$$1s^2 2s^2 2p^6 3s^2 3p^6 3d^{10} 4s^2 4p^6 4f^{14} 4d^{10} 5s^2 5p^6 5d^{10} 6s^2$$

(or more simply, Xe shell $4f^{14}5d^{10}6s^2$) qualifies it as a non-transition metal. It has valence states of +1 and +2. Natural mercury consists of seven stable isotopes[3] and 17 synthetic and radioactive isotopes with mass numbers 185–206. The natural mercury isotopes have the following mass numbers and abundances:

Isotope	Abundance (%)
196	0.146
198	10.02
199	16.84
200	23.13
201	13.22
202	29.80
204	6.85

The isotope with mass number 194 (^{194}Hg) has a half-life of 130 days, ^{203}Hg 47 days and ^{199}Hg 2.4×10^{-9} s. Isotopes of mercury can be obtained through the following reactions:

$$^{196}_{80}Hg + 1n \rightarrow\, ^{197}_{80}Hg + \gamma \tag{1.1}$$

Mercury Handbook: Chemistry, Applications and Environmental Impact
By Leonid F Kozin and Steve Hansen

Published by the Royal Society of Chemistry, www.rsc.org

$$^{204}_{80}Hg + ^{2}_{1}H_2 \rightarrow ^{205}_{80}Hg + ^{1}_{1}H \quad (1.2)$$

$$^{197}_{79}Hg + H \rightarrow ^{195}_{80}Hg + 3n \quad (1.3)$$

$$^{198}_{79}Au + 1p \rightarrow ^{199}_{80}Hg + \gamma \quad (1.4)$$

The thermal neutron capture cross-section for natural mercury is 380 ± 20 barn. Atomic and ionic ionization potentials (φ) and their radii are as follows:

	$Hg^0 \rightarrow Hg^+$	$Hg^+ \rightarrow Hg^{2+}$	$Hg^{2+} \rightarrow Hg^{3+}$
φ (eV)	10.438	18.756	34.2
Orbital radius (nm)	0.1126 (Hg)	0.1099 (Hg^+)	0.0605 (Hg^{2+})

The energy required for electron shell transfer from the basic state $6s^2$ (*i.e.* transfers $6s^2 \rightarrow 6s^1p^1$) is fairly large (524.26 kJ mol^{-1}) [4] and demonstrates the chemical inertness of metallic mercury. Moreover, the high $6s \rightarrow 6p$ transfer energy gives evidence that mercury tends to form two covalent bonds (and, as a result, further bonding of ligands is difficult). In contrast with the high energy of s,p transfer, $5d^{10} \rightarrow 5d^9s^1$ and $5d^{10} \rightarrow 5d^9p^1$ transfers in Hg^{2+} ions require a very low energy of 5.3 and 14.7 eV, respectively.[4] The electron affinity for α-mercury (α-Hg) is 1.54 eV, for β-Hg it is 1.37 eV and the electron work function is 4.52 eV.[5] The electronegativity of mercury, according to different authors, is given in Table 1.1.

The atomic, covalent and ionic radii of mercury[5] are given in Table 1.2.

Table 1.1 Electronegativity values of mercury.

Electronegativity (eV)	*Ref.*
1.9	6
2.0	7, 8
1.9	9
1.92	10
1.8	11

Table 1.2 Atomic radii of mercury.

Radius	*Distance (nm)*	*Coordination No.*
Atomic radius	0.155	
Covalent radius	0.149	
Ionic radius of Hg^{2+}	0.112	
Physical radius of Hg^+	0.111	3
	0.133	6
Physical radius of Hg^{2+}	0.083	2
	0.110	4
	0.116	6
	0.128	8

1.2 Crystallography

Solid mercury has a rhombohedral structure (α-Hg) with the lattice parameters $a = 0.29925$ nm (2.9925 Å), $\beta = 70° 44' 6''$. Each atom of mercury is surrounded by six neighboring atoms at a distance of 0.300 nm and six other atoms at a distance of 0.347 nm.

1.2.1 *P–T* Diagram

The *P–T* diagram for mercury[12] is given in Figure 1.1.

α-Hg: Mercury is a liquid under ambient conditions but crystallizes into the α-Hg structure at room temperature upon compression to 1.2 GPa.[13]

β-Hg: The α-Hg structure transforms to the β-Hg structure at 3.4 GPa at room temperature.[13] The β-Hg lattice, formed at temperatures below 79 K, is a body-centered tetragonal structure with lattice parameters $a = 0.3995$ nm, $c = 0.2825$ nm.[14] The space group of β-Hg is *I*4/*mmm*.[13] Swenson[15,16] determined the enthalpy of the solid-state transition α-Hg→β-Hg to be $\Delta H^{\alpha\rightarrow\beta} = 122$ J mol^{-1}, the volume change to be $\Delta V^{\alpha\rightarrow\beta} = -0.21$ cm^3 mol^{-1} and the entropy of transformation to be $\Delta S^{\alpha\rightarrow\beta} = -1.54$ J K^{-1} mol^{-1} at –194 °C and 0.101 MPa.

γ-Hg: According to Takemura *et al.*,[17] the structure of γ-Hg is monoclinic, *C*2/*m*. It forms at 12 GPa,[12] with six atoms in the unit cell. Each mercury atom is coordinated by 10–11 atoms.

δ-Hg: The structure of hexagonal close-packed (hcp) δ-Hg forms at pressures above 37 GPa[18] and is reported to be stable up to 193 GPa. At 193 GPa, the lattice parameters are $a = 0.2612$ nm and $c = 0.4284$ nm, which give $c/a = 1.640$. The c/a ratio of mercury under high pressure decreases from 1.75 at 50 GPa to 1.64 at about 200 GPa.[12,18] Yan *et al.*[19] also reported results on the high-pressure behavior of mercury.

Abell and King[20] performed mechanical tests on solid mercury below 77 K. Solid mercury turns white and becomes ductile at low temperatures and was

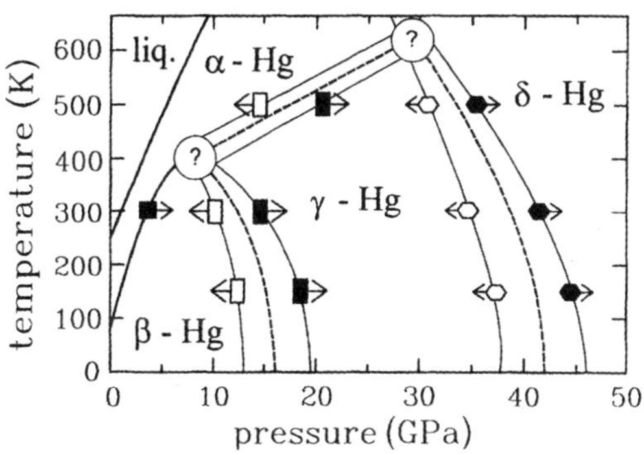

Figure 1.1 *P–T* phase diagram of mercury according to Schulte and Holzapfel.[12] Closed and open symbols represent forward and backward transitions, respectively. Reprinted with permission from O. Schulte and W. B. Holzapfel, *Phys. Rev. B*, 1993, **48**, 14009 [http://jpsj.ipap.jp/link?JPSJ/76/023601/]. Copyright (c) 1993 by the American Physical Society. Ref. 12.

found to recrystallize at ~160 K. Slip in single crystals of mercury was studied by Rider and Heckscher.[21] At 90 K, solid mercury deforms by slip and twinning. Plastic deformation of Hg single crystals has also been studied.[22,23]

At the melting temperature, the structure of liquid mercury is close to that of solid mercury. Each atom is surrounded by six other atoms at a distance of 0.307 nm. Thus, at the melting temperature, the coordination number in the liquid is 6, the same as for solid mercury. With increase in temperature, the mercury coordination number increases as follows:[24]

T(K)	Coordination No.
234.9	6.0
296	8.2–8.3
301	10–11

1.3 Melting Point

Mercury is the lowest melting point metal. Its melting point, measured by different groups, is given in Table 1.3. The data indicate the high purity of the samples studied. With increase in pressure, the melting point of mercury shifts towards higher temperature, $dT/dP = 42.44–49.84\ K\ GPa^{-1}$. Between 1.01325 and 6.0795 GPa, the melting point of mercury increases from 286.9 ± 0.3 to 515 K.[31,33]

1.4 Heat of Fusion

The heat of fusion (ΔH^{fusion}) of mercury, according to different sources, is given in Table 1.3. The heat of fusion increases with increase in pressure. At a pressure of 1.01325 MPa the heat of fusion is $2623 \pm 83.7\ J\ mol^{-1}$ and at 2.0265 GPa it is $2958 \pm 83.7\ J\ mol^{-1}$.[33]

1.5 Heat Capacity

The heat capacity of mercury has been studied over a broad range of temperatures.[30,36,37] The dependence of the specific heat capacity of mercury on

Table 1.3 Melting point and heat of fusion of mercury at 1 atm (101.325 kPa).

T *(K)*	T *(°C)*	ΔH^{fusion} *($J\ mol^{-1}$)*	*Ref.*
234.45	−38.700	2320	25
234.31	−38.840	2295	26
234.40	−38.750	2301	27
		2301.2 ± 20.9	28
		2310 ± 10	29
		2295 ± 40	30
		2343	31
		2308	32
234.288	−38.862 ± 0.003		33
234.314	−38.836		34
234.34	−38.810		27
234.33 ± 0.01	−38.82		35

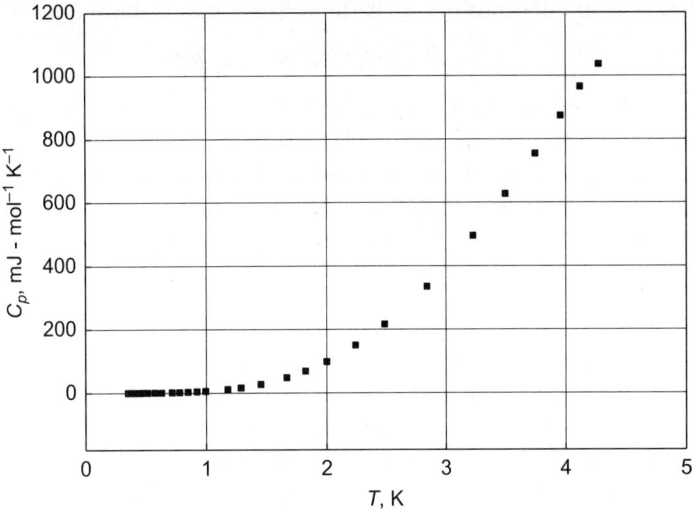

Figure 1.2 Low temperature heat capacity of mercury.[40]

temperature is shown in Figure 1.2. The heat capacity of solid mercury was determined by Regel and Glazov[36] with 257 experimental points and by Busey and Giaque.[26] In the temperature range 150.90–233.79 K, the heat capacity curve is represented by two temperature ranges:[36]

$$171.03 < T < 215.54\ K$$

$$213.20 < T < 233.58\ K$$

In the first range, the molar heat capacity of solid mercury is described by the equation[36]

$$C_p = C_{\text{vibr}} + C_{\text{el}} + C_{\text{an}} + C_{\text{vac}} \tag{1.5}$$

where C_{vibr} is the lattice vibration contribution, C_{el} is the electronic contribution, C_{an} is the anharmonic contribution and C_{vac} is the vacancy contribution. The sum of the lattice vibration contribution is calculated using the equation

$$C_{\text{vibr}} = 3R\left[1 - 0.05(\Theta_{\text{D}}/T)^2\right] \tag{1.6}$$

where Θ_{D} is the Debye temperature, which for α-Hg is 79 K.[38] C_{el} is the molar electronic contribution:

$$C_{\text{el}} = \gamma T \tag{1.7}$$

where γ is a constant equal to 1.81 mJ mol^{-1} K^{-2}.[39] C_{an} is the anharmonic component of C_p:

$$C_{\text{el}} + C_{\text{an}} = BT + DT^2 \tag{1.8}$$

where B and D are constants. C_{vac} is the vacancy contribution:

$$C_{vac} = (LU_0^2/RT^2)\exp(-U_0/RT) \tag{1.9}$$

where U_0 is the vacancy formation energy. The constants B, D and L in eqns (1.8) and (1.9) are found through the least-squares analysis of the U_0–T relationship in a given range of values. Experimentally obtained values of molar heat capacity of solid and liquid mercury are given in Tables 1.4 and 1.5. Constant-pressure heat capacity values at very low temperatures, below 20 K, were measured by van der Hoeven and Keesom[40] and others.[41–43] Van der Hoeven and Keesom measured an electronic specific heat coefficient of 1.79 ± 0.02 mJ mol^{-1} K^{-2}.

Analysis of the data in Table 1.5 reveals that in the temperature range 140–234 K, when approaching the melting temperature, the heat capacity of mercury increases non-linearly with increase in temperature.[30] The heat capacity of mercury at high temperatures [Figure 1.3] does not differ much from the classical value ($C_p/3R = 1.13$), which is due to the small effect of the anharmonic and electronic contributions.[44]

1.6 Thermal Conductivity

The thermal conductivity of solid mercury is anisotropic. The thermal conductivity of mercury single crystals on the trigonal axis ($\lambda_{\parallel}$) and perpendicular to it ($\lambda_{\perp}$), in the temperature range 80–234.288 K, is described by eqns (1.10) and (1.11), respectively.[45]

$$\lambda_{\parallel} = (44.8 - 0.0237T)\,\mathrm{W\,m^{-1}K^{-1}} \tag{1.10}$$

$$\lambda_{\perp} = (31.4 - 0.0279T)\,\mathrm{W\,m^{-1}K^{-1}} \tag{1.11}$$

Table 1.4 Specific heat of mercury at temperatures below 20 K. Data from Ref. 40.

T (K)	C_p (mJ mol^{-1} K^{-1})	T (K)	C_p (mJ mol^{-1} K^{-1})
0.3522	0.233	1.178	11.33
0.3669	0.263	1.286	16.62
0.3968	0.336	1.451	27.33
0.4243	0.406	1.665	47.93
0.4529	0.488	1.822	68.71
0.4809	0.587	2.000	98.15
0.5173	0.729	2.241	150.3
0.5766	1.001	2.485	216.0
0.6328	1.323	2.842	335.6
0.7189	1.968	3.230	495.6
0.7790	2.535	3.499	626.8
0.8480	3.383	3.746	754.8
0.9228	4.542	3.956	874.7
0.9947	5.943	4.121	965.6
		4.273	1036.7

Table 1.5 Experimental values of molar heat capacity of solid and liquid mercury. Data from Ref. 30.

T *(K)*	C_p *(J mol^{-1} K^{-1})*	T *(K)*	C_p *(J mol^{-1} K^{-1})*	T *(K)*	C_p *(J mol^{-1} K^{-1})*	T *(K)*	C_p *(J mol^{-1} K^{-1})*
5.19	1.834	29.97	14.78	89.66	23.74	193.50	27.09
5.62	2.068	31.46	15.30	92.44	23.88	203.46	27.40
6.12	2.361	32.90	15.81	95.17	24.02	209.10	27.58
6.72	2.658	34.32	16.27	97.86	24.14	214.70	27.77
7.36	2.994	35.82	16.71	100.71	24.27	217.44	27.87
7.97	3.304	37.34	17.18	103.72	24.40	219.60	27.95
8.64	3.669	38.88	17.58	106.69	24.51	222.18	28.05
9.33	4.076	40.44	17.98	109.62	24.62	224.75	28.14
9.97	4.439	42.09	18.37	113.03	24.75	227.30	28.24
10.80	4.946	43.89	18.77	117.09	24.89	229.84	28.34
11.89	5.650	46.94	19.39	121.76	25.04	232.37	28.45
12.93	6.351	49.11	19.79	126.88	25.21	237.90	28.50
13.87	6.998	51.22	20.15	130.94	25.37	239.14	28.49
14.77	7.542	56.58	20.97	136.00	25.48	240.63	28.48
15.66	8.097	58.98	21.29	140.99	25.62	242.64	28.43
16.80	8.646	61.32	21.57	145.92	25.76	245.98	28.41
17.76	9.183	63.97	21.85	150.78	25.90	250.56	28.35
18.77	9.719	66.93	22.15	155.60	26.03	256.35	28.29
19.86	10.28	69.90	22.41	160.12	26.16	263.46	28.22
20.98	10.85	72.89	22.66	164.60	26.28	271.09	28.15
22.11	11.41	75.81	22.89	169.28	26.41	278.15	28.10
23.27	11.97	78.67	23.08	174.10	26.55	285.19	28.02
24.66	12.61	82.25	23.61	179.02	26.68	292.20	27.98
26.17	13.27	84.25	23.43	183.88	26.81	297.19	27.93
27.74	13.90	86.94	23.59	188.71	26.95	299.05	27.89

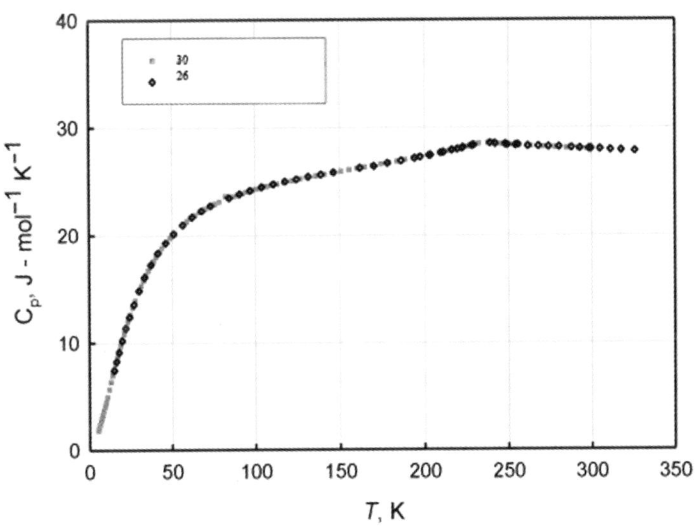

Figure 1.3 Specific heat capacity of mercury *versus* temperature.[26,30]

The thermal conductivity of liquid mercury, shown in Figure 1.4, has been extensively studied. The main contribution to the thermal conductivity of liquid mercury is made by conduction electrons. Therefore, the main heat flux in metallic mercury is transmitted, as in other metals, by conduction electrons.

The ratio of thermal conductivity, λ, to electrical conductivity, σ, at a given temperature is called the Lorentz number, L:

$$L = \frac{\lambda}{\sigma T} \tag{1.12}$$

Lorentz numbers calculated for the main axes of mercury single crystals agree within 3%.[45] Table 1.6 gives the values of the Lorentz number at different temperatures.

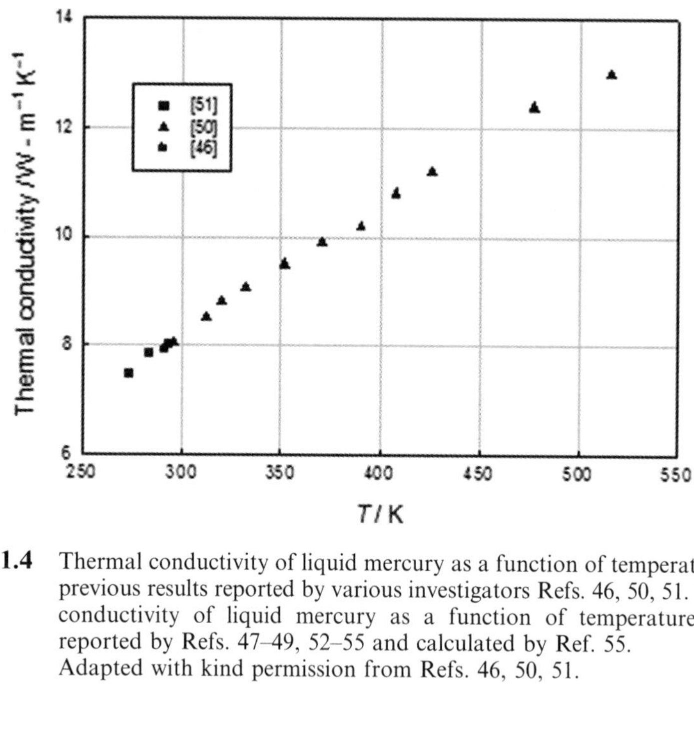

Figure 1.4 Thermal conductivity of liquid mercury as a function of temperature with previous results reported by various investigators Refs. 46, 50, 51. Thermal conductivity of liquid mercury as a function of temperature is also reported by Refs. 47–49, 52–55 and calculated by Ref. 55. Adapted with kind permission from Refs. 46, 50, 51.

Table 1.6 Lorentz number of mercury (W Ω K^{-2}).

T *(K)*	T *(°C)*	L× *10^{-8} (W Ω K^{-2})*	*Ref.*
303	30	2.75	56
348	75	2.50	57
373	100	2.47	57
423	150	2.45	57
473	200	2.45	57

1.7 Emissivity

The emission coefficient, ε_λ, of mercury from a smooth non-oxidized surface is 0.10–0.12. However, the reflectivity of polished solid mercury and a liquid surface, χ, for light flux of wavelength λ is as follows:[2]

Form	Wavelength, λ (μm)	Reflectivity, χ (%)
Solid	0.45–0.70	72.3–72.8
Liquid	0.75–1.00	77.3–77.9

1.8 Boiling Point, Heat and Entropy of Vaporization

The boiling point of mercury (T^{boil}) has been reported in the literature with an accuracy of 0.01–0.08 °C. Experimental results along with the heat of vaporization are given in Table 1.7.

According to Hultgren *et al.*,[68] mercury vapor is best described as a non-ideal monomer.

Values for the heat of evaporation (ΔH^{evap}) and entropy of vaporization (ΔS^{evap}) also depend on pressure. Table 1.8 summarizes ΔH^{evap} and ΔS^{evap} values[33] at different pressures. Thermodynamic values for

Table 1.7 Boiling point and heat of vaporization of mercury.

T^{boil} (K)	T^{boil} (°C)	Heat of vaporization, ΔH^{vap} (kJ mol^{-1})	Ref.
629.814	356.664		58
629.7653	356.6153		59
629.7683	356.6183		60
		59.10	26
		61.42	61
		61.41	62
		61.40	63
		61.44	64
		61.02	65
		61.29	66
		61.76	67

Table 1.8 Enthalpy and entropy of evaporation of mercury.[33]

P (Pa)	ΔH^{evap} (kJ mol^{-1})	ΔS^{evap} (J mol^{-1} K^{-1})
3.07×10^{-4}	61.883 ± 0.0628	264.136
2.62×10^{-1}	61.404 ± 0.0628	205.936
1.013×10^{-5}	59.228 ± 0.0628	94.0563

the sublimation of mercury at 234.288 K are $\Delta H^{subl} = 64.1784 \pm 0.06276$ kJ mol^{-1} and $\Delta S^{subl} = 273.926$ J mol^{-1} K^{-1}.[33]

1.9 Vapor Pressure

Studies of the temperature dependence of mercury vapor pressure were summarized by Huber *et al.*[69,70] Diatomic molecules of Hg_2 were found in mercury vapor.[71] Hg_2 molecules oscillate at ~ 36 cm^{-1}, their internuclear distance is 3.34×10^{-10} m and their dissociation energy is 7.53 ± 2.09 kJ mol^{-1}.[72] Values of the heat of sublimation of monatomic and diatomic molecules are 61.304 ± 0.063 and 103.64 kJ mol^{-1}, respectively. The heat of dimerization of mercury is 8.008 kJ mol^{-1}.[73] A small energy of dissociation of Hg_2 molecules in the vapor causes gaseous mercury to be virtually monatomic and to have significant vapor pressure even at low temperatures. The thermodynamic properties of Hg_2 molecules were also studied by Hilpert.[74]

1.9.1 Solid Mercury

Measurements of the vapor pressure over solid mercury are relatively scarce. Values obtained by Poindexter[67] are given in Table 1.9.

The vapor pressure over solid mercury[75] is given by

$$\log P\,(\mathrm{Pa}) = 5.00572 + \left(9.453 - 0.2011 \times \log T - 6.558 \times 10^{-4} T - \frac{3379}{T}\right) \quad (1.13)$$

The saturated vapor pressures over a broad range of temperatures for solid and liquid mercury are fairly consistent. Analysis of experimental data in coordinates of $\ln P_{Hg}$–$1/T$[70] demonstrated good agreement between data from different authors. The most accurate results are shown in Figure 1.5. The $\ln P_{Hg}$–T curve in Figure 1.5 also shows the triple point, boiling point and critical temperature.[70]

Table 1.9 Vapor pressure above solid mercury.[67]

T *(K)*	T *(°C)*	P *(kPa)*
193.58	–79.57	4.00×10^{-10}
203.25	–69.90	8.00×10^{-10}
206.30	–66.85	5.33×10^{-9}
209.47	–63.68	1.69×10^{-9}
215.43	–57.72	6.47×10^{-9}
216.31	–56.84	6.91×10^{-9}
223.30	–49.85	3.16×10^{-8}
229.75	–43.40	9.75×10^{-8}
230.38	–42.77	1.00×10^{-7}
231.39	–41.76	1.14×10^{-7}

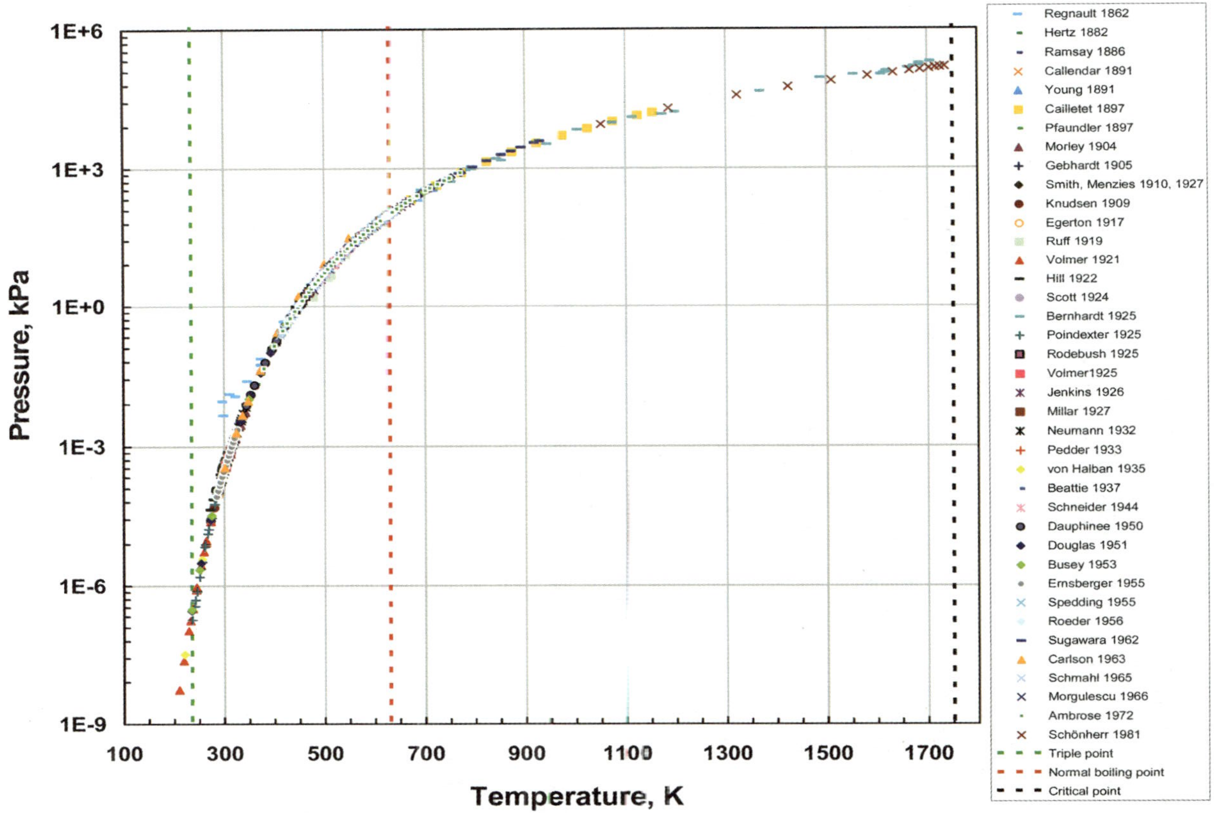

Figure 1.5 Mercury vapor pressure *versus* temperature. For the references that appear in the key, please refer to the original reference.[70] Official contribution of the National Institute of Standards and Technology; not subject to copyright in the United States.

1.9.2 Liquid Mercury

Based on a great amount of data, Nesmeyanov[76] suggested the following relationship between vapor pressure and temperature for liquid mercury up to 673 K (400 °C):

$$\log P\ (\mathrm{Pa}) = 5.00572 + \left(216.70428 - \frac{9078.658}{T} + 0.05481736T - 82.87205 \times \log T\right) \tag{1.14}$$

The vapor pressure of mercury according to the literature[26,77] in the temperature range 298–629.810 K (T^{boil}) can be determined using the equation

$$\log P\ (\mathrm{Pa}) = 5.00572 + \left(10.355 - 0.795 \times \log T - \frac{3305}{T}\right) \tag{1.15}$$

A more elaborate vapor pressure equation for mercury was suggested by Huber *et al.*[69,70] Very accurate experimental measurements of the vapor pressure of mercury were performed by Beattie *et al.*[59] from 623 to 636 K and by Spedding and Dye[78] from 534 to 630 K. The normal boiling point of mercury was determined by Beattie *et al.* as 629.7653 ± 0.0016 K on the ITS-90 international temperature scale.[79] Table 1.10 gives the vapor pressure of mercury up to its normal boiling point and Table 1.11 above the normal boiling point.

At very high temperatures, a small but noticeable change in slope on the vapor pressure curve occurs. A metal–non-metal transition[80] occurs at ~1360 K (1087 °C).

1.9.3 Triple Point

When analyzing the thermodynamic parameters of mercury, it was found that the triple point is located at 234.3156 K with a vapor pressure of

Table 1.10 Vapor pressure of saturated mercury up to the boiling point.[78]

T *(K)*	T *(°C)*	P *(kPa)*
533.825	260.675	13.06
549.811	276.661	19.337
558.948	285.798	23.954
564.721	291.571	27.351
565.743	292.593	27.964
573.610	300.460	33.293
586.013	312.863	43.390
594.741	321.591	51.918
597.253	324.103	54.588
604.288	331.138	62.792
613.886	340.736	75.568
620.254	347.104	85.144
630.244	357.094	102.22

Table 1.11 Vapor pressure of mercury above the boiling point.[58]

T *(K)*	T *(°C)*	P *(kPa)*	P *(mmHg)*
629.814	356.664	101.325	760
670.06	396.91	200	1500
733.41	460.26	500	3750
790.2	517.05	1000	7500
856.7	583.55	2000	15 000
964.8	691.65	5000	375×10^4
1068	794.85	10 000	7.5×10^4
1197	923.85	20 000	1.5×10^5
1425	1151.85	50 000	3.75×10^5
1639	1365.85	100 000	7.5×10^5
1765	1491.85	151 000	1.13×10^6
1769 ± 0.042	1495.85	153 000	1.15×10^6

Table 1.12 Triple point of mercury.

T *(K)*	T *(°C)*	*Temperature scale*	*Ref.*
234.3156	–38.8344	ITS-90	81
234.3159	–38.834	ITS-90	82
234.316	–38.834		29
234.314	–38.836	IPTS-68	34
234.3083	–38.842	IPTS-68	83
234.3086	–38.841	IPTS-68	84
234.306	–38.844		85

Table 1.13 Critical temperature and pressure of mercury. Adapted from Ref. 70.

T_c *(K)*	T_c *(°C)*	P_c *(MPa)*	ρ_c *(g cm^{-3})*	*Ref.*
1750	1477	172		86
1751 ± 1	1478 ± 1	167.3	5.8	87
1751 ± 1	1478 ± 1	167.3 ± 0.2	5.77	88
1753	1480	152 ± 1		89
1763.15 ± 15	1490 ± 15	151 ± 3	4.2 ± 0.4	90, 91
1764 ± 1	1491 ± 1	167 ± 3		92
1768 ± 8	1495	167.5 ± 2.5		93

$P_{Hg}=0.157$ Pa (1.55×10^{-6} Torr).[1] This value is close to the values calculated by means of eqns (1.18) and (1.19), namely 3.33×10^{-4} and 3.35×10^{-4} Pa (2.510×10^{-6} and 2.514×10^{-6} Torr), respectively. Table 1.12 gives experimentally determined values of the triple point of mercury.

1.9.4 Critical Temperature and Pressure

Table 1.13 gives measurements of the critical temperature and pressure of mercury.[70] The coexistence curve of liquid and gaseous mercury is shown in Figure 1.6.

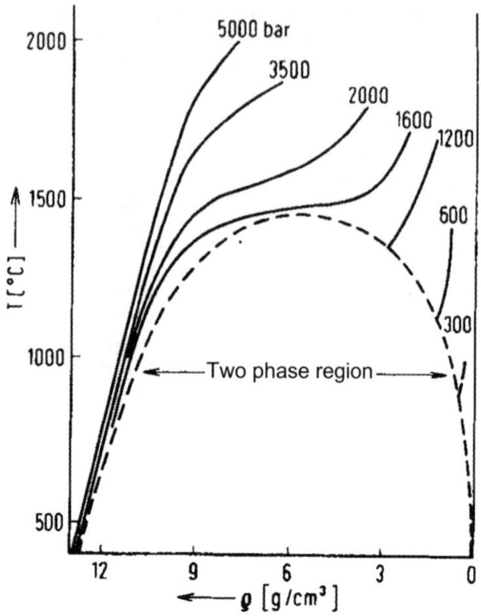

Figure 1.6 Phase diagram of liquid and gaseous mercury. The dashed line indicates the liquid–vapor coexistence curve. With kind permission from Taylor & Francis Ltd. D. R. Postill, R. G. Ross and N. E. Cusack, 'Equation of state and electrical resistivity of liquid mercury at elevated temperatures and pressures', *Adv. Phys.*, 1967, **16**, 493.[95]

1.10 Density

The density of mercury has been extensively studied. Literature values of the density of solid and liquid mercury are listed in Table 1.14 and plotted in Figure 1.7.

Solid mercury at 234.25 K has a density of 14.193 g cm^{-3},[96] and liquid mercury at 234.288 K has a density of 13.691 g cm^{-3}. The change in density at the liquid–solid transition is +3.5–3.7% (compared with the density of solid mercury). The volume change upon solidification is given in Table 1.15. Further studies on the density of mercury have been reported.[105–113]

Analysis of the density of liquid mercury as a function of temperature has shown that it may be expressed by a linear equation. Equation (1.16) gives the density of solid mercury *versus* temperature in the range 50–234.288 K:

$$d = d_0 + (T_0 - T)\left(\frac{\mathrm{d}\rho}{\mathrm{d}T}\right) \tag{1.16}$$

where T_0 is the melting temperature of mercury, d_0 is the density of liquid mercury at the melting point T_0, equal to 13.690 g cm^{-3}, and $\mathrm{d}\rho/\mathrm{d}T$ is the temperature coefficient, -2.4386 ± 0.01744 g cm^{-3} K^{-1}.

Experimental and calculated (straight line) values for the density of solid and liquid mercury are presented in Figure 1.7, showing good agreement of the

Table 1.14 Density of solid and liquid mercury.

T (K)	*T (°C)*	*ρ (g cm^{-3})*	*Ref.*
77	–196.15	14.70	15
78.15	–195	14.49	97
82.15	–191	14.469	98
194.15	–79	14.29	97
234.321	–38.829	14.182	31
234.321	–38.829	13.690	99
293.15	20	13.5460	95,100
293.15	20	13.545892	101
293.15	20	13.545884	102
298.15	25	13.5338	95,100
303.15	30	13.5213	103
313.15	40	13.4969	103
323.15	50	13.4725	103
333.15	60	13.4482	103
343.15	70	13.4239	103
353.15	80	13.3997	103
363.15	90	13.3755	103
373.15	100	13.3514	103
383.15	110	13.3273	103
393.15	120	13.3033	103
403.15	130	13.2792	103
413.15	140	13.2553	103
423.15	150	13.2314	103
425	151.85	13.23	104
471	197.85	13.12	104
519	245.85	13.00	104
571	297.85	12.88	104
633	359.85	12.73	104
681	407.85	12.61	104
691	417.85	12.60	95,100
693	419.85	12.60	95,100
715	441.85	12.59	95,100
758	484.85	12.45	95,100
787	513.85	12.35	95,100
838	564.85	12.25	95,100
900	626.85	12.11	95,100
930	656.85	12.02	95,100
979	705.85	11.86	95,100

data. According to Duval,[115] the equation for calculating the density of mercury at a selected temperature (t, °C) is more complex:

$$d = \frac{13.595080}{1+1.814401\times10^{-4}t+7.016\times10^{-9}t^2+2.8625\times10^{-11}t^3+2.617\times10^{-14}t^4}\ \text{g cm}^{-3} \tag{1.17}$$

Values for the density of liquid and gaseous mercury along the saturation line have been determined,[95,100,116] and Table 1.16 reports values for the density of gaseous mercury from 968 to 1523 K.[117]

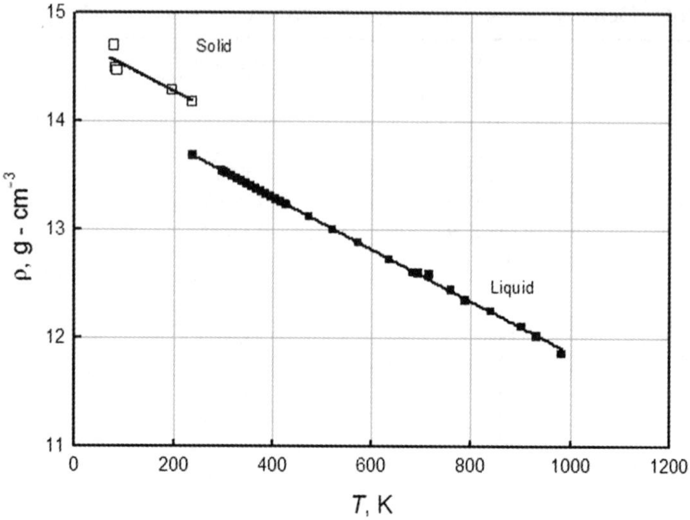

Figure 1.7 Density of solid and liquid mercury *versus* temperature.[15,31,95,97–104]

Table 1.15 Change in the volume of mercury during solidification ($\Delta V = V_{liquid} - V_{solid}$).

ΔV *(%)*	*Ref.*
3.59	31
3.66	114
3.67	99

Table 1.16 Density of gaseous mercury.[117]

T *(K)*	T *(°C)*	ρ *(g cm^{-3})*
968	694.85	0.121
1040	766.85	0.185
1096	822.85	0.25
1163	889.85	0.335
1263	989.85	0.53
1346	1072.85	0.85
1445	1171.85	1.05
1523	1249.85	1.40

The temperature, pressure, density and structure of liquid mercury have been measured at elevated temperatures and pressures.[118–120] The results are presented in Table 1.17.

A graphical representation of the density from room temperature to the critical point is given in Figure 1.8.[121]

The influence of pressure (up to 800 MPa) on the density of mercury in the temperature ranges 303.15–423.15 K,[103] 400–1833 K[116] and up to 1723 K[124] has been studied The isotherms of mercury density *versus* pressure are shown in

Table 1.17 Temperature, pressure and density of liquid mercury.

T *(K)*	T *(°C)*	P *(bar)*	ρ *(g cm^{-3})*	*Ref.*
293	20	5	13.55	118
523	250	49	12.98	118
773	500	55	12.42	118
1073	800	157	11.56	118
1273	1000	405	10.98	118
1373	1100	610	10.67	118
1473	1200	830	10.26	118
1573	1300	1137	9.81	119
1623	1350	1325	9.53	119
1673	1400	1570	9.25	119
1723	1450	1750	8.78	119
1773	1500	1980	8.26	119
1803	1530	1970	6.5	119

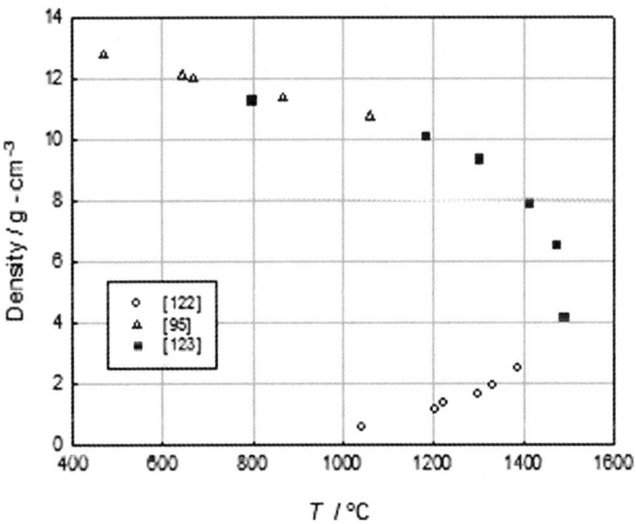

Figure 1.8 Density (ρ) of liquid and gaseous mercury as a function of temperature up to the critical point (C.P.).
Adapted from Refs. 95, 122, 123.

Figure 1.9. It can be seen that the course of the isotherms becomes more complicated at high pressures. These effects are due to changes in the physicochemical properties of metallic mercury.[116,124]

1.11 Surface Tension

Nizhenko and Floka[125] and Wilkinson[126] summarized the large amount of experimental data on the surface tension of mercury (σ_{Hg}) and amalgams.[125] Depending on the experimental conditions (purity of mercury, temperature, gaseous atmosphere, measurement method, *etc.*), the surface tension

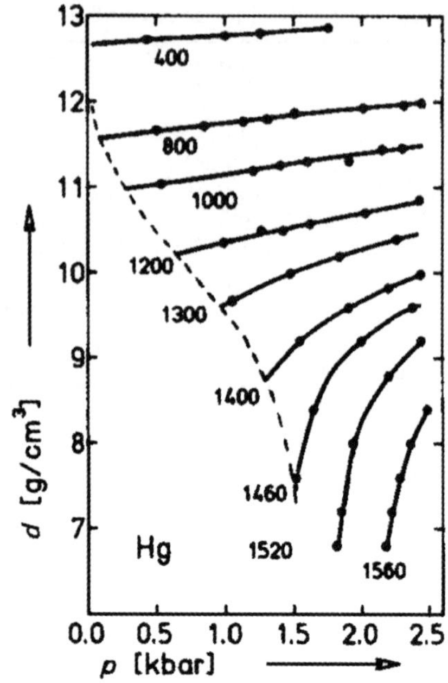

Figure 1.9 Density of pure mercury *versus* pressure at selected temperatures. Reproduced with kind permission from Deutsche Bunsen-Gesellschaft, G. Schonherr and F. Hensel, *Ber. Bunsenges Phys. Chem.*, 1981, **85**(5), 361–367. Ref. 116.

of mercury changes from 470 to 497 dyn cm^{-1} (mN m^{-1}). According to Smithells,[127] at the melting point of mercury $\sigma_{Hg} = 498$ dyn cm^{-1} and the temperature coefficient is $d\sigma/dT = -0.20$ dyn cm^{-1} K^{-1}. Nizhenko and Floka[125] recommended values of $\sigma_{Hg} = 497$ dyn cm^{-1} and $d\sigma/dT = -0.281$ dyn cm^{-1} K^{-1}. It follows from the data presented by Wilkinson[126] that the conditions of the experiment have a critical impact on the value of the surface tension. Analysis that involved almost 200 independent measurements showed that the surface tension is a normal distribution with a mean close to 466.3 ± 33 dyn cm^{-1}. The experimental values of σ_{Hg} ranged from 359 to 563 dyn cm^{-1}. In vacuum and dry air, the surface tension of mercury at 298.15 K was found to be 475.5 ± 10 dyn cm^{-1}.[126] The relationship between surface tension and the atomic and thermodynamic properties of metals, including mercury and amalgams, has been discussed.[128,129]

1.12 Viscosity

The dynamic viscosity of mercury, η, is 1.56 ± 0.015 mPa s at 293 K.[106,130] Its dependence on temperature can be calculated by means of an Arrhenius-type equation such as

$$\eta = B\exp(-E/RT) \tag{1.18}$$

where $B = 0.560541$ mPa s and $E = 2483.137$ J mol^{-1}. According to Vukalovich *et al.*,[45] the temperature dependence of dynamic viscosity in the range 623–898 K is described by more sophisticated equations.

Kinematic viscosity is calculated based on the dynamic viscosity data by means of the equation

$$\nu = \frac{\eta}{\rho} \tag{1.19}$$

where ρ is the density of mercury at the selected temperature. According to Kozin *et al.*,[131] kinematic viscosity depends linearly on the reciprocal of temperature.

The viscosity of liquid and gaseous mercury at high temperature (up to 1520 K) and high pressure was studied by Tippelskirch *et al.*[117] using a viscometer with an oscillating cylinder. The amplitude of the cylinder oscillations was measured with a helium–neon laser. The viscosity of liquid and gaseous mercury was measured along the saturation line up to 1520 K over a range of pressures. Data from the experiments of Tippelskirch *et al.*[117] and from other studies[131–135] are compared with the theoretical curve using the modified Enskog equations describing the overall saturation zone. Good agreement between the calculated curves and the experimentally obtained points was observed. The upper curve represents the viscosity of metallic mercury and the lower curve the viscosity of gaseous mercury. The viscosity around the critical temperature is shown with dashed lines. Tippelskirch *et al.*[117] correlated the calculated curves of the temperature dependence of viscosity for mercury, sodium and lead with experimental results and observed good agreement between the data. The viscosity of liquid mercury lies between the values for the other two metals ($\eta_{Na} < \eta_{Hg} < \eta_{Pb}$).

Viscosity values for mercury are given in Table 1.18. Further experimental investigations of the viscosity have been reported elsewhere.[136,137]

1.13 Isothermal Compressibility

The effect of pressure on the compressibility of mercury at high temperatures was studied by Schönherr and Hensel.[116] Figure 1.10 shows isothermal compressibility coefficients at different temperatures and constant pressure. At temperatures above 1400 K, the compressibility increases abruptly according to the equation

$$\chi = \frac{1}{d(\partial\rho/\partial P)_T} \times 10^{10}\ \mathrm{Pa}^{-1} \tag{1.20}$$

At high temperatures (>1400 K) and pressures of ~9 kg m^{-3}, a metal–non-metal transition is observed. This transition alters the interatomic forces and increases the isothermal compressibility coefficient. Isothermal compressibility values for mercury were summarized by Holman and ten Seldam.[138]

Table 1.18 Viscosity of mercury at different temperatures.[131,132]

T (K)	T (°C)	Kinematic viscosity (cSt)	ρ (g cm^{-3})	Viscosity (cP)	Ref.
623	350	0.070			131
573	300	0.072			131
523	250	0.076			131
513	240	0.0783	13.02	1.02	132
503	230	0.0791	13.04	1.03	132
493	220	0.0798	13.07	1.04	132
483	210	0.0805	13.09	1.05	132
473	200	0.080			131
473	200	0.0812	13.11	1.06	132
463	190	0.0821	13.14	1.07	132
453	180	0.0829	13.16	1.09	132
443	170	0.0840	13.18	1.10	132
433	160	0.0851	13.21	1.12	132
423	150	0.086			131
423	150	0.0863	13.23	1.14	132
413	140	0.0877	13.26	1.16	132
403	130	0.0891	13.28	1.18	132
393	120	0.0906	13.30	1.20	132
383	110	0.0922	13.33	1.22	132
373	100	0.094			131
373	100	0.0939	13.35	1.25	132
363	90	0.0957	13.38	1.28	132
353	80	0.0976	13.40	1.30	132
348	75	0.099			131
343	70	0.0998	13.42	1.33	132
333	60	0.102	13.45	1.37	132
323	50	0.104			131
323	50	0.105	13.47	1.41	132
313	40	0.108	13.50	1.45	132
303	30	0.111	13.52	1.50	132
298	25	0.110			131
293	20	0.115	13.55	1.55	132
283	10	0.119	13.57	1.61	132
273	0	0.123	13.60	1.68	132
263	–10	0.129	13.62	1.76	132
253	–20	0.135	13.64	1.84	132
243	–30	0.141	13.67	1.93	132
238	–35	0.145	13.68	1.98	132
235	–38	0.147	13.69	2.02	132

1.14 Thermal Expansion Coefficient

An equation for the isobaric volume thermal expansion coefficient, $\bar{\alpha}$, taken from Beattie *et al.*,[139] refitted by Ambrose[140] for temperatures in °C on the ITS-90 temperature scale, and later refitted for temperatures in K on the ITS-90,[138] is

$$\bar{\alpha} = 182.3887 \times 10^{-6} - 1.01689 \times 10^{-8} T + 2.2231 \times 10^{-11} T^2 + 1.5558 \times 10^{-14} T^3 \tag{1.21}$$

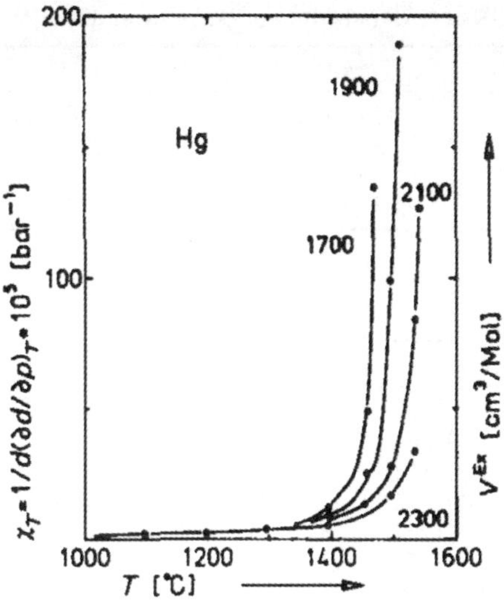

Figure 1.10 Isothermal compressibility, χ_T, of pure mercury. Reproduced with kind permission from Deutsche Bunsen-Gesellschaft, G. Schonherr and F. Hensel, *Ber. Bunsenges Phys. Chem.*, 1981, **85**(5), 361–367. Ref. 116.

where T is in K and $\bar{\alpha}$ is in K^{-1}. This equation is valid only at ambient pressure for the temperature range 253–573 K. Liquid mercury expands considerably more than solid mercury with increase in temperature, its expansion being described by

$$v_{T_2} = v_{T_1}(1 + \beta\Delta T) \qquad (1.22)$$

where v_{T_1} and v_{T_2} are the volume of liquid mercury at T_1 and T_2, respectively, $\Delta T = T_2 - T_1$ is the temperature difference and β is the coefficient of volumetric expansion, equal to $0.181 \times 10^{-3}\,K^{-1}$.

1.15 Self-diffusion

Coefficients of mercury self-diffusion and diffusion of metals in mercury can be calculated from Arrhenius equations similar to

$$D = D_0 \exp\left(\frac{E_D}{RT}\right) \text{cm}^2\,\text{s}^{-1} \qquad (1.23)$$

where D_0 is the pre-exponential factor, which has a constant value for the metal being studied, and E_D is the activation energy. Different authors have reported different values for E_D and D_0, as shown in Table 1.19.

These data suggest that values of E_D are determined with an accuracy of only 25–38% and values of D_0 with the accuracy of only 14–48%. The lack of accuracy of D_0 is caused by both the specifics of the experiment in a broad range of temperatures, under conditions that rule out convective diffusion, and

Table 1.19 Self-diffusion constants for pure mercury.

E_D (kJ mol^{-1})	D_0 (cm^2 s^{-1})	T (K)	Ref.
5.104	1.40×10^{-4}	238–533	141
6.820	1.63×10^{-4}	273–568	142
4.853	1.26×10^{-4}	275.5–364.2	143
4.205	0.85×10^{-4}	273–371	144

Table 1.20 Self-diffusion of liquid mercury.

T (K)	1/T (K^{-1})	D × 10^5 (cm^2 s^{-1})	Ref.
373	0.002681	2.62	145
423	0.002364	3.31	145
473	0.002114	4.08	145
523	0.001912	4.91	145
566	0.001767	5.68	145
273.2	0.003660	1.50	142
273.2	0.003660	1.51	142
315.4	0.003171	1.88	142
315.4	0.003171	1.88	142
315.4	0.003171	1.89	142
352.6	0.002836	2.44	142
352.6	0.002836	2.33	142
381.6	0.002621	2.69	142
381.6	0.002621	2.71	142
381.6	0.002621	2.72	142
381.6	0.002621	2.77	142
414.2	0.002414	3.26	142
415.1	0.002409	3.23	142
452.9	0.002208	3.80	142
497.6	0.002010	4.42	142
512.5	0.001951	4.70	142
283	0.003534	1.436	147
298	0.003356	1.59	147
313	0.003195	1.745	147
333	0.003003	1.949	147
308	0.003247	1.86–1.89	146

the need for extrapolating D_0 from the plot of $\ln D$ *versus* $1/T$ to $1/T \rightarrow 0$. Data relating the effect of temperature on the self-diffusion coefficient of mercury have been published.[142,145–147] Table 1.20 gives values for D_{Hg} in the temperature range 273–566 K.

A least-squares analysis of experimental results showed that the self-diffusion coefficient of mercury is described by Equation (1.24) with $D_0 = 1.8601\times10^{-4}$ cm^2 s^{-1} and $E_D = 6005.1$ J mol^{-1}. Efforts were made to calculate E_D based on correlations between E_D and T, E_D and ΔH^{melt} and E_D and ΔH^{evap}.[148–152] The values of E_D obtained agree with the experiments.[149,151]

To calculate diffusion coefficients of metals in mercury, the well-known equation of the hydrodynamic mass transfer theory is used, where the motion of a macroscopic ball with a radius $r \to \infty$ is traced in a non-compressive medium having viscosity η:[148,153]

$$D = \frac{kT}{6\pi\eta r} \tag{1.24}$$

where η is the dynamic viscosity of the medium and k is the Boltzmann constant. This equation is known as the Stokes–Einstein relation and is applicable, strictly speaking, to ideal solutions where no strong interaction of diffusing particles and the medium is observed. Mercury shows a strong affinity to many metals and forms intermetallic compounds (see Appendix I). Hence the size of the diffusing particle will depend on the nature of the metal. It has been shown[149,151,153,154] that when the particle diameter $d \to \infty$, the experimental data are better described by the equation

$$D = \frac{kT}{4\pi\eta r} = \frac{RT}{N_{\mathrm{Me}_i} \times 4\pi\eta r} \tag{1.25}$$

However, this equation requires knowledge of the nature of diffusing particles (atom, ion, associate, intermetallic molecule) and the sizes of the particles. Methods for modification of eqns (1.30) and (1.31) depending on the ratio between sizes of the diffusing particle and atoms of the solvent, and also other factors, have been discussed.[149,155] It was shown that the experimental data on self-diffusion coefficients are fairly consistent with the data[155–157] obtained using the equation

$$D_\mathrm{c} = \left(\frac{RT}{4\sqrt{2}\eta}\right) d^{\frac{1}{3}} M^{-\frac{1}{3}} N^{-\frac{1}{3}} \,\mathrm{cm^2\,s^{-1}} \tag{1.26}$$

where M is the atomic mass and N is Avogadro's number. When the properties of mercury are taken into account, eqn (1.32) is converted to the form[157]

$$D_\mathrm{c} = 4.743 \times 10^3 d_\mathrm{eff}{}^{-1} \,\mathrm{cm^2\,s^{-1}} \tag{1.27}$$

where d_eff (cm) is the effective diameter of diffusing particle, $d_\mathrm{eff} = (v_\mathrm{Me}/N)^{\frac{1}{3}}$, and v_Me is the molar volume of the diffusing metal at $T = 298$ K. According to Vukalovich *et al.*,[45] the self-diffusion coefficient of a metal up to a temperature of 1073 K (800 °C) is expressed by the equation

$$D_\mathrm{c} = 0.9264 \times 10^{-10} \frac{T}{\eta} \,\mathrm{cm^2\,s} \tag{1.28}$$

where η is the dynamic viscosity, described by

$$\eta = 0.310 \times 10^{-3} T^{0.07939} \exp\left(\frac{341.13}{T}\right) \mathrm{ns\,m^{-2}} \tag{1.29}$$

Figure 1.11 shows experimental values of mercury self-diffusion coefficients from Table 1.20. The values of v_Me were taken from the literature[2,68,158] when calculating d_eff. The linear dependence of $\ln D_c$ on $1/T$ can be clearly seen.

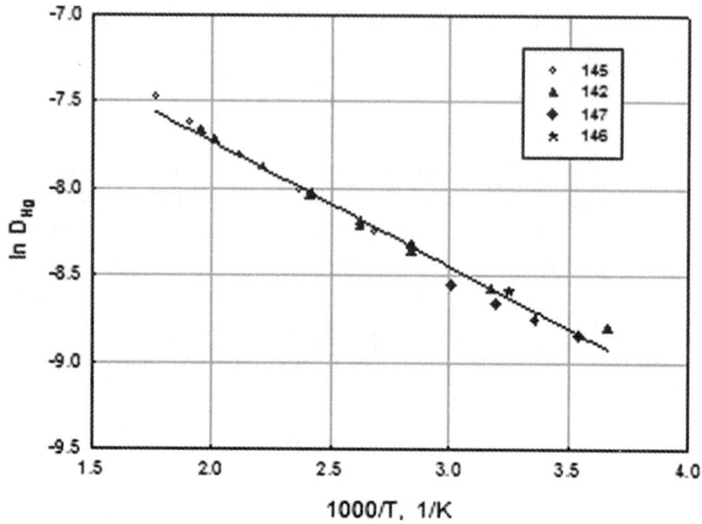

Figure 1.11 Self-diffusion of mercury.[142,145–147]

Equation (1.29) allows one to calculate the diffusion coefficients of metals in mercury for given values of D_0 and E_D. Values of D_0 and E_D for many metals have been published.[2,148,159] A more detailed discussion of the diffusion coefficients of metals in mercury is given in Chapter 4.

1.16 Electrical and Magnetic Properties

The specific resistance (resistivity) of mercury (ρ) depends on its physical state and structure. The anisotropy of the resistivity of solid mercury can be up to 30% lower along the trigonal axis ($\rho_{\parallel}$) and up to 30% higher perpendicular to that axis ($\rho_{\perp}$). The temperature dependence of the resistivity of mercury at $T = 80$–234.321 K is described by the equations[45]

$$\rho_{II} = 1.315 + 35.79 \times 10^{-3} T + 0.1588 \times 10^{-3} T^2 \ \mu\Omega \ \mathrm{cm} \quad (1.30)$$

and

$$\rho_{\perp} = 1.288 + 54.13 \times 10^{-3} T + 0.1899 \times 10^{-3} T^2 \ \mu\Omega \ \mathrm{cm} \quad (1.31)$$

with an accuracy $\leq 0.5\%$. The temperature coefficient of resistance is 0.92×10^{-3} K^{-1}. Results of resistivity measurements on solid mercury are given in Table 1.21.

Resistivity values for liquid mercury at different temperatures[44,124,160–162] are given in Table 1.22. At the melting point, a significant change in resistivity is observed, as shown in Figure 1.12. For liquid mercury at the melting point it is $\rho_{liq} = 90.96\ \mu\Omega$ cm.[32] Ratio values are $\rho_{liq}/\rho_{\parallel} = 4.94$[163] and $\rho_{liq}/\rho_{\perp} = 3.73$,[164] and at the melting point $\rho_{liq}/\rho_{solid} = 4.2$ for polycrystalline samples.

The resistivity ratio ($\rho_{\perp}/\rho_{\parallel}$) of mercury at 100 and 200 K is $8.60/6.48 = 1.327$ and $19.71/14.82 = 1.330$, respectively.[44] The resistivity of liquid mercury was measured

Table 1.21 Resistivity of solid mercury *versus* temperature.

T *(K)*	T *(°C)*	ρ *(μΩ cm)*	*Ref.*
15	–258.150	0.0188	115
77	–196.150	5.8	115
89.5	–183.650	6.97	160
100	–173.150	7.89	44
200	–73.150	18.08	44
223	–50.150	12.3	115
227.5	–45.650	21.2	115
233.8	–39.350	25.5	115
234	–39.150	22.0	44

Table 1.22 Resistivity of liquid mercury *versus* temperature.

T *(K)*	T *(°C)*	ρ *(μΩ cm)*	*Ref.*
234.288	–38.862	90.96	124
234.29	–38.860	94.8	44
253	–20.150	91	161
273	–0.150	94.7	160
293	19.850	95.8	160
298	24.850	95	162
300	26.850	102.0	160
323	49.850	98.5	160
373	99.850	103.25	160
400	126.850	113.0	44
473	199.850	114.27	160
500	226.850	126.0	160
573	299.850	127.0	160
600	326.850	137.0	160
623	349.850	135.5	160
700	426.850	150.0	44

with high accuracy over a broad range of temperatures [234.321–629.88 K at a pressure of 1 atm (101.325 kPa) and up to 1273 K following the saturation curve]. Liquid resistivity (ρ_{liq}) is approximated with a polynomial function:[45]

$$\frac{\rho}{\rho_0} = 1 + 0.88857\times10^{-3}(T-273) + 1.0075\times10^{-6}\times(T-273)^2 - 0.105\times10^{-9}(T-273)^3 + 0.2702\times10^{-12}(T-273)^4 + 1.199\times10^{-15}(T-273)^5 \quad (1.32)$$

where ρ_0 is the density of mercury at $T = 273$ K and $P = 101.325$ kPa and is 94.12×10^{-8} Ω m.

The temperature coefficient of resistivity of mercury (α) at 15 and 273 K is equal to 2×10^{-6} and $(0.89–0.92)\times10^{-3}$ K^{-1}, respectively.[160] The resistivity of mercury is higher than that of any other metal except bismuth. Therefore, it has been used to define the international ohm standard. The effect of pressure

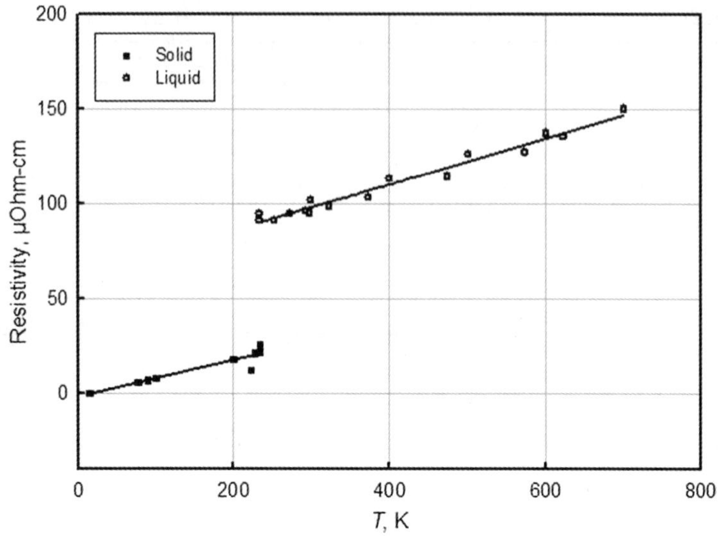

Figure 1.12 Temperature dependence of the resistivity of solid[44,115,160] and liquid mercury.[44,160]

on the resistivity of mercury was studied by Schönherr and Hensel.[116] It was found that at a density of 13.5–11.0 g cm^{-3}, the resistivity of pure mercury, κ_{Hg}, decreased from 1×10^4 to $3\times10^3\,\Omega^{-1}\,cm^{-1}$. In that case, the electron mean free path, L, is greater than the smallest distance between atoms of mercury, a, and the resistivity decreases abruptly. At a density of mercury of 11.0–9.0 g cm^{-3}, L is close to the lattice parameter a and electron transfer is then obstructed. It was shown in a number of studies that the decrease in the density of mercury results in changes of properties such as electrical conductivity, χ_{Hg}, thermal electromotive force and Hall coefficient (the Knight shift is an exception from this rule).

Mercury is diamagnetic; its magnetic susceptibility (χ) is a function of temperature and its physical state:[160]

T(K)	80.0	293.0	295.5	560.5
Physical state	Solid	Liquid	Liquid	Liquid
$\chi\times10^9$	–0.118	–0.167	–0.1681	–0.1637

It has been found that the magnetic susceptibility also depends on the crystal structure of mercury. The Knight shift for liquid mercury is 2.45%.[165]

1.17 Hall Coefficient

The Hall coefficient changes insignificantly with temperature. Values of the Hall coefficient for liquid mercury in the temperature range from –30 to 210 °C are given in Table 1.23.

Table 1.23 Measured Hall coefficients of liquid mercury.

T *(K)*	T *(°C)*	R × 10^{-5} *(cm^3 °C^{-1})*	*Ref.*
303–483	30–210	–7.6	166
243–373	–30–100	–7.46	167,168
293–573	20–300	–7.3	169
293–473	20–200	–9.3	170
293	20	–8.0	171

Table 1.24 Superconducting properties of α- and β-Hg.

Property	*α-Hg*	*β-Hg*	*Ref.*
T_c (K)	4.153 ± 0.001	3.949 ± 0.001	38
	4.1540 ± 0.0010		39
	4.16		40
H_0 (G)	412 ± 1	339.1 ± 1	38
	410		172
	415.40 ± 0.12		39[a]
	380 ± 60		40
γ (mJ mol^{-1} K^{-1})	1.91 ± 0.05	1.37 ± 0.04	38
	2.04 ± 0.03		175
	2.1 ± 0.1		176
	1.809 ± 0.012		39[a]
	1.86		177

[a]These values supersede those of Finnemore *et al.*[173]

1.18 Superconductivity

The superconductivity of solid mercury has been extensively studied.[38,172,173] Mercury is an example of a superconductor that exists in two crystallographic modifications. The two polymorphs of mercury exhibit nearly ideal superconducting behavior.[38] Schirber and Swenson[38] measured critical field quantities, T_c, H_0 and $(\partial H/\partial P)_T$, as a function of temperature for α- and β-Hg. These results and those of other investigators are given in Table 1.24. In addition, the coefficient of the electronic specific heat in the normal state, γ, was also measured by several researchers and results are reported in Table 1.24. Schirber and Swenson also estimated the Grüneisen constant of solid mercury to be ~2.00.[38]

1.19 Excited-state Properties

The UV emission and corresponding wavelength values for excited electronic states are as follows:

$6s^1s_0 \rightarrow 6pp_1$ $\lambda = 184.957$ nm

$6s^1s_0 \rightarrow 6p^3p_1$ $\lambda = 253.652$ nm

$6p^3p_1 \rightarrow 7s^3s_1$ $\lambda = 435.835$ nm

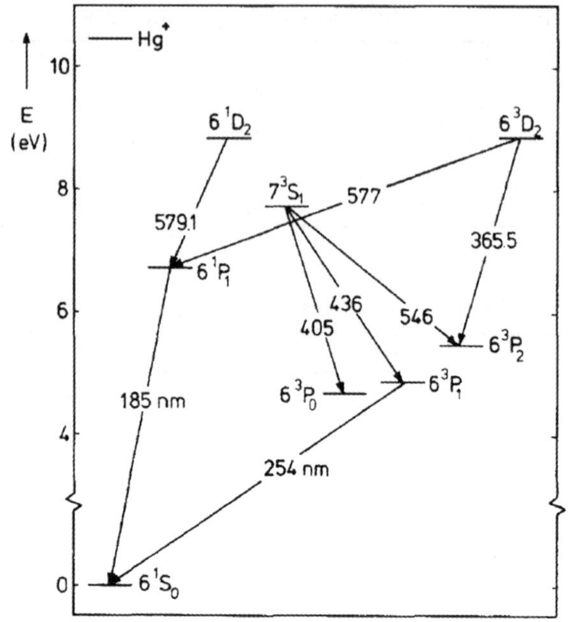

Figure 1.13 Partial energy level diagram of mercury.
Reproduced with permission from Ref. 178.

Excitation energies for reactions of metallic mercury and its ions are

Hg	$ns^2 \rightarrow ns^1p^1$	4.67 eV
Hg^{2+}	$nd^{10} \rightarrow nd^9\ (n+1)s^1$	5.3 eV
Hg^{2+}	$nd^{10} \rightarrow nd^9\ (n+1)p_1$	14.7 eV

Figure 1.13 shows a simplified energy level diagram for atomic mercury.

References

1. C. Guminski, *J. Phase Equilib.*, 1992, **13**, 339.
2. G. V. Samsonov (ed.), *Properties of Elements*, Metallurgiya, Moscow, 1976.
3. Yu. M. Buleev, N. P. Grudinkina and H. G. Pevzner, *Studies in Physico-Chemical Measurements, Proceedings of the Metrological Institutes of the USSR*, 1968, **96**, pp. 176, 180.
4. H. L. Roberts, *Adv. Inorg. Chem. Radiochem.*, 1968, **11**, 309.
5. V. K. Vainshtein, V. M. Fridkin and V. L. Indenbom, *Modern Crystallography*, Nauka, Moscow, 1979, vol. 2.
6. L. Pauling, *The Nature of the Chemical Bond*, Cornell University Press, Ithaca, NY, 3rd edn, 1960, p. 93.

7. A. Allred and E. Rochow, *J. Inorg. Nucl. Chem.*, 1958, **5**, 264.
8. A. L. Allred, *J. Inorg. Nucl. Chem.*, 1961, **17**, 215.
9. R. Mulliken, *J. Chem. Phys.*, 1955, **46**, 497.
10. R. T. Sanderson, *Chemical Periodicity*, Reinhold, New York, 1960, p. 34.
11. S. S. Batsanov, *Zh. Strukt. Khim.*, 1964, **5**, 293.
12. O. Schulte and W. B. Holzapfel, *Phys. Rev. B*, 1993, **48**, 14009.
13. M. I. McMahon and R. J. Nelmes, *Chem. Soc. Rev.*, 2006, **35**, 943.
14. M. Atoji, J. E. Schirber and C. A. Swenson, *J. Chem. Phys.*, 1959, **31**, 1628.
15. C. A. Swenson, *Phys. Rev.*, 1958, **111**, 81.
16. C. Guminski, *J. Phase Equilib.*, 1992, **13**, 657.
17. K. Takemura, H. Fujihisa, Y. Nakamoto, S. Nakano and Y. Ohishi, *J. Phys. Soc. Jpn.*, 2007, **76**, 023601.
18. K. Takemura, S. Nakano and Y. Ohisi, presented at the Joint 20th AIRART–43rd EHPRG, Karlsruhe, 2005.
19. J. Yan, B. Chen, S. Raju, J. Knight and B. K. Godwal, *High Pressure Res.*, 2011, **31**, 555.
20. J. S. Abell and H. W. King, *Cryogenics*, 1970, **10**, 119.
21. J. G. Rider and F. Heckscher, *Philos. Mag.*, 1966, **13**, 687.
22. V. V. Pustovalov, V. V. Vershinina, S. V. Tsivinsky and B. N. Aleksandrov, *Phys. Met. Metallogr.*, 1970, **30**, 95.
23. J. S. Abell, *Acta Metall.*, 1976, **24**, 11.
24. L. F. Kozin, *Physicochemistry and Metallurgy of High-Purity Mercury and Its Alloys*, Naukova Dumka, Kiev, 1992.
25. F. Pollitzer, *Z. Elektrochem.*, 1911, **17**, 5.
26. R. H. Busey and W. F. Giaque, *J. Am. Chem. Soc.*, 1953, **75**, 806.
27. J. E. Callanan, K. M. McDermott and E. F. Westrum Jr, *J. Chem. Thermodyn.*, 1990, **22**, 225.
28. A. Ya. Mitus and A. Z. Patashinskiy, *Zh. Eksp. Teor. Fiz.*, 1981, **80**, 1554.
29. E. Méndez-Lango and J. Ancsin, *Thermochim. Acta*, 1996, **287**, 183.
30. E. B. Amitin, E. G. Lebedeva and I. E. Paukov, *Zh. Fiz. Khim.*, 1979, **53**, 2666.
31. P. W. Bridgman, *Proc. Natl. Acad. Sci. U. S. A.*, 1911, **47**, 347.
32. L. van der Putten, J. A. Schouten and N. J. Trappeniers, *High Temp. High Press.*, 1984, **16**, 281.
33. V. P. Glushko (ed.), *Thermal Constants of Substances*, VINITI, Moscow, 1972.
34. National Bureau of Standards, *Standard Reference Material Certificate, SRM 743, Mercury*, NBS, Gaithersburg, MD, 1976.
35. O. Kubaschewski, *Z. Eletrochem.*, 1950, **54**, 275.
36. A. R. Regel and V. M. Glazov, *Physical Properties of Electronic Melts*, Nauka, Moscow, 1980.
37. E. B. Amitin, Yu. F. Minenkov, O. A. Nabutovskaya and I. E. Paukov, *Zh. Eksp. Teor. Fiz.*, 1985, **89**, 2092.
38. J. E. Schirber and C. A. Swenson, *Phys. Rev.*, 1961, **123**, 1115.
39. D. K. Finnemore and T. Mapother, *Phys. Rev.*, 1965, **140**, A507.

40. B. J. C. van der Hoeven Jr and P. H. Keesom, *Phys. Rev.*, 1964, **135**, A631.
41. P. L. Smith and N. M. Wolcott, *Philos. Mag.*, 1956, **1**, 854.
42. F. Simon, *Z. Phys. Chem.*, 1923, **107**, 279.
43. F. Simon, *Ann. Phys.*, 1922, **68**, 241.
44. V. E. Zinoviev, *Thermophysical Properties of Metals at High Temperatures*, Metallurgiya, Moscow, 1989.
45. M. P. Vukalovich, A. I. Ivanov, L. R. Fokin and T. A. Yakovlev, *Thermophysical Properties of Liquids*, Nauka, Moscow, 1970.
46. H. Fukuyama, T. Yoshimura, H. Yasuda and H. Ohta, *Int. J. Thermophys.*, 2006, **27**, 1760.
47. S. Nakamura, T. Hibiya, T. Yokota and F. Yamamoto, *Int. J. Heat Mass Transfer*, 1990, **33**, 2609.
48. S. Nakamura, T. Hibiya and F. Yamamoto, *J. Appl. Phys.*, 1990, **68**, 5125.
49. S. Nakamura and T. Hibiya, *Rev. Sci. Instrum.*, 1988, **59**, 2600.
50. M. V. Peralta, M. Dix, M. Lesemann and W. A. Wakeham, *High Temp. High Press.*, 2002, **34**, 35.
51. E. Yamasue, M. Susa, H. Fukuyama and K. Nagata, *Metall. Mater. Trans. A*, 1999, **30A**, 1971.
52. R. F. Brooks, B. J. Monaghan, A. J. Barnicoat, A. Mccabe, K. C. Mills and P. N. Quested, *Int. J. Thermophys.*, 1996, **19**, 1151.
53. J. T. Schriempf, *High Temp. High Press.*, 1972, **4**, 411.
54. M. J. Duggin, in *Proceedings of the 8th Thermal Conductivity Conference*, Purdue University, Plenum Press, New York, 1969, p. 727.
55. B. Sundqvist, *High Temp. High Press.*, 1986, **18**, 655.
56. R. W. Powell and R. P. Tye, *ASME*, 1961, **103**, 856.
57. M. J. Duggin, *Phys. Lett.*, 1968, **27A**, 257.
58. D. Ambrose and C. H. S. Sprake, *J. Chem. Thermodyn.*, 1972, **4**, 603.
59. J. A. Beattie, B. E. Blaisdell and J. Kaminsky, *Proc. Am. Acad. Arts Sci.*, 1937, **71**, 375.
60. K. N. Marsh (ed.), *Recommended Reference Materials for the Realization of Physicochemical Properties*, Blackwell Scientific, Oxford, 1987.
61. C. G. De Kruif, C. H. D. van Ginkel and A. Langenberg, *Recl. Trav. Chim. Pays-Bas*, 1973, **92**, 599.
62. J. D. Burlingame, PhD thesis, University of Pennsylvania, 1968.
63. W. H. Rodebush and A. L. Dixon, *Phys. Rev.*, 1925, **26**, 851.
64. H. Mayer, *Z. Phys.*, 1931, **67**, 240.
65. M. Merlin, *Tec. Ital.*, 1948, **3**, 248.
66. I. Filosofo, M. Merlin and A. Rostagni, *Nuovo Cimento*, 1950, **7**, 69.
67. F. E. Poindexter, *Phys. Rev.*, 1925, **26**, 859.
68. R. R. Hultgren, P. D. Desai, D. T. Hawkins, M. Gleiser, K. K. Kelley and D. D. Wagman, *Selected Values of Thermodynamic Properties of Metal and Alloys*, Wiley, New York, 1963.
69. M. L. Huber, A. Laesecke and D. G. Friend, *Ind. Eng. Chem. Res.*, 2006, **45**, 7351.

70. M. L. Huber, A. Laesecke and D. G. Friend, *The Vapor Pressure of Mercury*, NISTIR 6643, NIST, Boulder, CO, 2006 and references therein.
71. K. D. Carlson and K. R. Kuschnir, *J. Phys. Chem.*, 1964, **68**, 1566.
72. V. N. Kondratev (ed.), *Chemical Bonds Rupture Energy, Ionization Potentials and Electron Affinities*, Nauka, Moscow, 1974.
73. R. E. Honig and D. A. Kramer, *RCA Rev.*, 1969, **30**, 103.
74. K. Hilpert, *J. Chem. Phys.*, 1982, **77**, 1425.
75. S. Dushman, *Scientific Foundations of Vacuum Technology*, Wiley, New York, 1962.
76. A. N. Nesmeyanov, *Vapor Pressure of Chemical Elements*, USSR Academy of Sciences Publishing House, Moscow, 1961.
77. T. M. Dauphinee, *J. Chem. Phys.*, 1951, **19**, 389.
78. F. H. Spedding and J. L. Dye, *J. Phys. Chem.*, 1955, **59**, 581.
79. H. Preston-Thomas, *Metrologia*, 1990, **27**, 3.
80. S. R. Hubbard and R. G. Ross, *Nature*, 1982, **295**, 6683.
81. G. F. Strouse and J. Lippiatt, in *TEMPMEKO 2001: 8th International Symposium on Temperature and Thermal Measurements in Industry and Science (Berlin, Germany, 19–21 June, 2001)*, ed. B Fellmuth *et al.*, VDE Verlag, Berlin, 2002, p. 453.
82. K. D. Hill, *Metrologia*, 1994, **31**, 39.
83. G. T. Furukawa and W. R. Bigge, *Bureau International des Poids et Mesures, Paris*, 1976, 11, Annexe T14, T138–T144.
84. M. V. Chattle and J. Butler, *Cells for the Realisation of the Triple Point of Mercury*, NPL Report QU 79, National Physical Laboratory, Teddington, 1988.
85. S. L. Knina, A. A. Nechai, A. A. Semenov and V. A. Petrushina, *Izmer. Tekh.*, 1989, **no. 8**, 41.
86. S. R. Hubbard and R. G. Ross, *J. Phys. C*, 1983, **16**, 6921.
87. W. Götzlaff, G. Schönherr and F. Hensel, *Z. Phys. Chem., NF*, 1988, **156**, 219.
88. U. Brusius, J. Popielawski and H. Uchtmann, *Z. Phys. Chem.*, 1994, **184**, 107.
89. I. K. Kikoin and A. P. Senchenkov, *Phys. Met. Metallogr.*, 1967, **24**, 74.
90. F. Hensel and E. U. Franck, *Ber. Bunsenges. Phys. Chem.*, 1966, **70**, 1154.
91. E. U. Franck and F. Hensel, *Phys. Rev.*, 1966, **147**, 109.
92. V. Kozhevnikov, D. Arnold, E. Grodzinskii and S. Naurzakov, *Fluid Phase Equilib.*, 1996, **125**, 149.
93. F. E. Neal and N. E. Cusack, *J. Phys. F*, 1979, **9**, 85.
94. F. Hensel, *Angew. Chem.*, 1974, **86**, 459.
95. D. R. Postill, R. G. Ross and N. E. Cusack, *Adv. Phys.*, 1967, **16**, 493.
96. J. W. Mallett, *Proc. R. Soc. London*, 1877, **26**, 71.
97. A. Sapper and W. Biltz, *Z. Anorg. Allg. Chem.*, 1931, **198**, 184.
98. E. Grüneisen and O. Sckell, *Ann. Phys.*, 1934, **411**, 387.
99. W. E. Ayrton and J. Perry, *Philos. Mag.*, 1886, **22**, 325.
100. D. R. Postill, R. G. Ross and N. E. Cusack, *Philos. Mag.*, 1968, **18**, 519.

101. A. H. Cook and N. W. B. Stone, *Philos. Trans. R. Soc. London Ser. A*, 1957, **250**, 35.
102. A. H. Cook, *Philos. Trans. R. Soc. London Ser. A*, 1961, **254**, 125.
103. T. Grindley and J. E. Lind, *J. Chem. Phys.*, 1971, **54**, 3983.
104. Kh. Ibragimov and V. S. Savvin, *Inorg. Mater.*, 1996, **32**, 963.
105. M. Fürtig, *Exp. Techn. Phys.*, 1973, **21**, 521.
106. H. Bettin and H. Fehlauer, *Metrologia*, 2004, **41**, S16.
107. H. Adametz and M. Wloka, *Metrologia*, 1991, **28**, 333.
108. P. H. Bigg, *Br. J. Appl. Phys.*, 1964, **15**, 1111.
109. J. B. Patterson and D. B. Prowse, *Metrologia*, 1985, **21**, 107.
110. A. H. Cook, *Br. J. Appl. Phys.*, 1956, **7**, 285.
111. L. N. Kusmenkov, *Trudy Inst. VNIIM*, 1968, **96**, 92.
112. H. L. Callendar and H. Moss, *Philos. Trans. R. Soc. London Ser. A*, 1912, **211**, 1.
113. F. J. Harlow, *Proc. Phys. Soc. London*, 1913, **26**, 85.
114. A. Eucken, *Angew. Chem.*, 1942, **55**, 163.
115. C. Duval, *Le Mercure*, Presses Universitaires de France, Paris, 1968.
116. G. Schönherr and F. Hensel, *Ber. Bunsenges Phys. Chem.*, 1981, **85**(5), 361–367.
117. H. V. Tippelskirch, E. V. Franck, F. Hensel and J. Kestin, *Ber. Bunsenges. Phys. Chem.*, 1975, **79**, 889.
118. S. Hosokawa, T. Matsuoka and K. Tamura, *J. Phys. Condens. Matter*, 1991, **3**, 4443.
119. K. Tamura and S. Hosokawa, *Phys. Rev. B*, 1998, **58**, 9030.
120. M. Inui, X. Hong and K. Tamura, *Phys. Rev. B*, 2003, **68**, 094108.
121. F. Hensel and E. U. Franck, *Rev. Mod. Phys.*, 1968, **40**, 697.
122. J. Bender, *Phys. Z.*, 1918, **19**, 410.
123. F. Hensel and E. U. Franck, *Ber. Bunsenges, Phys. Chem.*, 1966, **70**, 1154.
124. F. E. Neale, N. E. Cusack and R. D. Johnson, *J. Phys. F*, 1979, **9**, 113.
125. V. I. Nizhenko and L. I. Floka, *Surface Tension in Liquid Metals and Alloys*, Metallurgiya, Moscow, 1981.
126. M. C. Wilkinson, *Chem. Rev.*, 1972, **72**, 575.
127. C. J. Smithells, *Metals Reference Handbook*, 2nd edn, Butterworths, London, 1955, vol. 1 .
128. V. K. Semenchenko, *Surface Phenomena in Metals and Alloys*, Metallurgiya, Moscow, 1963.
129. S. N. Zadumkin, *Zh. Eksp. Teor. Fiz.*, 1953, **24**, 615.
130. Physikalisch-Technische Bundesanstalt, *PTB-Stoffdatenblätter, Quecksilber, Mercury*, PTB, Braunschweig, 1995.
131. L. F. Kozin, R. Sh., V. F. Nigmetova, Kiselev and I. I. Churkin, *Proc. Inst. Org. Catal. Electrochem. Acad. Sci. Kazakh. SSR*, 1976, **12**, 20.
132. T. Iida, A. Kasama, Z. Morita, I. Okamoto and S. Tokumoto, *J. Jpn. Inst. Met.*, 1973, **37**, 841.
133. C. Khalilov, *Zh. Tekhn. Fiz.*, 1938, **8**, 1249.
134. V. R. Suhrmann and E. O. Winter, *Z. Naturforsch.*, 1955, **10A**, 985.
135. S. Koch, *Wied. Ann*, 1881, **14**, 1.

136. W. Menz and F. Sauerwald, *Acta Metall.*, 1966, **14**, 1617.
137. J. M. Grouvel, J. Kestin, H. Khalifa, E. U. Franck and F. Hensel, *Ber. Bunsenges. Phys. Chem.*, 1977, **81**, 339.
138. G. J. F. Holman and C. A. ten Seldam, *J. Phys. Chem. Ref. Data*, 1994, **23**, 807.
139. J. A. Beattie, B. E. Blaisdell, J. Kaye, H. T. Gerry and C. A. Johnson, *Proc. Am. Acad. Arts Sci.*, 1941, **71**, 371.
140. D. Ambrose, *Metrologia*, 1990, **27**, 245.
141. E. F. Broome and H. A. Walls, *Trans. Met. Soc. AIME*, 1968, **242**, 2177.
142. R. E. Meyer, *J. Phys. Chem.*, 1961, **65**, 567.
143. R. E. Hoffman, *J. Chem. Phys.*, 1952, **20**, 1567.
144. R. E. Norman, N. H. Nachtrieb and J. Petit, *J. Chem. Phys.*, 1956, **24**, 746.
145. R. P. Chhabra, *Met. Trans.*, 1986, **17A**, 355.
146. Z. Galus, *Pure Appl. Chem.*, 1984, **56**, 635.
147. V. M. M. Lobo and R. Mills, *Electrochim. Acta*, 1982, **27**, 969.
148. J. R. Wilson, *Metall. Rev.*, 1965, **10**, 381.
149. B. S. Bakshtein, *Diffusion in Metals*, Metallurgiya, Moscow, 1978.
150. D. K. Belashchenko, *Transport Phenomena in Liquid Metals and Semiconductors*, Atomizdat,Moscow, 1970.
151. V. Zayt, *Diffusion in Metals*, Foreign Literature Publishing, Moscow, 1958.
152. I. B. Borovsky, K. P. Gurov, I. D. Marchuk and Yu. E. Uchaste, *Interdiffusion Processes in Alloys*, Nauka, Moscow, 1973.
153. A. G. Stromberg and D. P. Semenchenko, *Physical Chemistry*, Vysshaya Shkola, Moscow, 1988.
154. H. A. Walls and W. R. Upthegrove, *Acta Met.*, 1964, **42**, 461.
155. B. Ottar, *Acta Chem. Scand.*, 1955, **9**, 344.
156. G. M. Panchenkov, N. N. Borisenko and V. V. Erienkov, *Zh. Fiz. Khim.*, 1971, **45**, 2338.
157. V. N. Korshunov, *Amalgam Systems, Structure and Electrochemical Properties*, Moscow University Publishing, Moscow, 1990.
158. D. Stull and G. Sinke, *Thermodynamic Properties of the Elements*, American Chemical Society, Washington, DC, 1956.
159. C. Gumiński, in *Liquid Metal Systems*, ed. H. U. Borgstedt, Plenum Press, New York, 1995.
160. M. E. Drits (ed.), *Properties of Elements*, Metallurgiya, Moscow, 1985.
161. N. W. Ashcroft and J. Lekner, *Phys. Rev.*, 1966, **145**, 83.
162. R. Evans, D. A. Greenwood, P. Lloyd and J. M. Ziman, *Phys. Lett.*, 1969, **30A**, 313.
163. O. Sckell, *Ann. Phys. Leipzig*, 1930, **6**, 932.
164. W. Jaeger and H. Steinwehr, *Ann. Phys. Leipzig*, 1914, **45**, 1089.
165. E. F. Broome and H. A. Walls, *Trans. Met. Soc. AIME*, 1968, **242**, 2177.
166. A. J. Greenfield, *Phys. Rev.*, 1964, **135**, A1589.
167. N. Cusack and P. Kendall, *Philos. Mag.*, 1960, **5**, 100.
168. N. Cusack and P. Kendall, *Philos. Mag.*, 1961, **6**, 419.

169. Y. Tièche, *Helv. Phys. Acta*, 1960, **33**, 963.
170. E. G. Wilson, *Philos. Mag.*, 1962, **7**, 989.
171. J. Enderby, *Proc. Phys. Soc. (London)*, 1963, **81**, 772.
172. I. E. Maxwell and O. S. Lutes, *Phys. Rev.*, 1954, **95**, 333.
173. D. K. Finnemore, T. Mapother and R. W. Shaw, *Phys. Rev.*, 1960, **118**, 127.
174. S. J. C. van der Hoeven Jr and P. H. Keesom, *Phys. Rev.*, 1964, **135**, A631.
175. R. G. Chambers and J. G. Park, in *Proceedings of the Seventh International Conference on Low Temperature Physics*, Toronto, 1960, ed. M. Graham and A. C. Hollis, University of Toronto Press, Toronto, 1960.
176. B. B. Goodman, Proceedings of Kammerlingh-Onnes Memorial Conference on Low-Temperature Physics, Leiden, 1958, *Physica*, 1958, **24**, S149.
177. N. E. Phillips, M. H. Lambert and W. R. Gardner, *Rev. Mod. Phys.*, 1964, **36**, 132.
178. P. van de Weijer and R. M. M. Cremers, *J. Appl. Phys.*, 1983, **54**, 2835.
179. N. W. Ashcroft and N. D. Mermin, *Solid State Physics*, Holt, Reinhart and Wilson, New York, 1976.
180. F. Bernhardt, *Phys. Z.*, 1925, **26**, 265.
181. L. Cailletet, E. Colardeau and C. Rivière, *C. R. Acad. Sci.*, 1900, **130**, 1585.
182. H. L. Callendar and E. H. Griffiths, *Philos. Trans. R. Soc. London A*, 1891, **182**, 119.
183. K. D. Carlson, P. W. Gilles and R. J. Thorn, *J. Chem. Phys.*, 1963, **38**, 2725.
184. T. M. Dauphinee, *J. Chem. Phys.*, 1951, **19**, 389.
185. T. B. Douglas, A. F. Ball and D. C. Ginnings, *J. Res. Natl. Bur. Stand.*, 1951, **46**, 334.
186. A. C. Egerton, *Philos. Mag.*, 1917, **33**, 33.
187. F. M. Ernsberger and H. W. Pitman, *Rev. Sci. Instrum.*, 1955, **26**, 584.
188. A. Gebhardt, *Ber. Dtsch. Phys. Ges.*, 1905, **7**, 184.
189. H. H. von Halban, *Helv. Phys. Acta*, 1935, **7**, 856.
190. H. Hertz, *Ann. Phys. Chem.*, 1882, **17**, 193.
191. C. F. Hill, *Phys. Rev.*, 1922, **20**, 259.
192. C. H. M. Jenkins, *Proc. R. Soc. London A*, 1926, **110**, 456.
193. M. Knudsen, *Ann. Phys.*, 1909, **29**, 179.
194. A. W. C. Menzies, *Z. Phys. Chem.*, 1927, **130**, 90.
195. R. W. Millar, *J. Am. Chem. Soc.*, 1927, **49**, 3003.
196. E. W. Morley, *Philos. Mag.*, 1904, **7**, 662.
197. I. G. Murgulescu and L. Topor, *Rev. Roum. Chim.*, 1966, **11**, 1353.
198. K. Neumann and E. Völker, *Z. Phys. Chem.*, 1932, **161**, 33.
199. J. S. Pedder and S. Barratt, *J. Chem. Soc.*, 1933, 537.
200. L. Pfaundler, *Ann. Phys. Chem.*, 1897, **63**, 36.
201. W. Ramsay and S. Young, *J. Chem. Soc.*, 1886, **49**, 37.
202. V. Regnault, *Mem. Acad. Sci. Inst. Fr.*, 1862, **26**, 506.
203. A. Roeder and W. Morawietz, *Z. Elektrochem.*, 1956, **60**, 431.

204. O. Ruff and B. Bergdahl, *Z. Anorg. Allg. Chem.*, 1919, **106**, 76.
205. G. Schönherr and F. Hensel, *Ber. Bunsenges. Phys. Chem.*, 1981, **85**, 361.
206. D. H. Scott, *Philos. Mag.*, 1924, **47**, 32.
207. N. G. Schmahl, J. Barthel and H. Kaloff, *Z. Phys. Chem.*, 1965, **46**, 183.
208. A. Schneider and K. Z. Schupp, *Elektrochem. Angew. Phys. Chem.*, 1944, **50**, 163.
209. A. Smith and A. W. C. Menzies, *J. Am. Chem. Soc.*, 1910, **32**, 1434.
210. S. Sugawara, T. Sato and T. Minamiyama, *Bull. JSME*, 1962, **5**, 711.
211. M. Volmer and P. Kirchhoff, *Z. Phys. Chem.*, 1925, **115**, 233.
212. S. Young, *J. Chem. Soc.*, 1891, **59**, 629.

CHAPTER 2

Amalgam Solubility

2.1 Solubility of Metals in Mercury

The properties of mercury and its alloys are essential information for the theory and application of electrochemistry,[1,2] synthesis of semiconducting materials,[3,4] metallurgical processes using amalgams to obtain high-purity and super-purity metals, lighting, the chlor-alkali process, application as a heat-transfer medium, liquid electrical contacts, *etc.*[5–9] Knowledge of the solubility of metals in mercury and the temperature dependence of their solubility is of the utmost importance for these applications. This chapter discusses predictive models for solubility and presents extensive data on the measured solubility of metals in mercury.

Various researchers[6–16] have contributed critical analyses of studies dedicated to the solubility of metals in mercury. Currently there are experimental data confirming the solubility of 75 metals[6–8,17–26] in mercury. Therefore, several authors have developed mathematical relationships between the solubility of metals in mercury and the physicochemical properties of pure metals. The goal is to calculate the solubility of a metal in mercury based on the physical parameters of the pure metal.[6–8,14] These efforts have met with a fair degree of success. Appendix V gives experimental values for the solubility of metals in mercury, and Figure 2.1 shows the solubility of metals in mercury in graphical form.

The possibility of calculating the solubility of metals in mercury is of great practical interest for high-purity mercury production processes. The chemical potential of a metal in mercury obeying an ideal solution is expressed by the equation

$$\mu_{Me}^{ideal} = \mu_0 + RT\ln x_1 \tag{2.1}$$

where μ_0 is the standard chemical potential of the metal in mercury and x_1 is the mole fraction of the metal dissolved in mercury.

Mercury Handbook: Chemistry, Applications and Environmental Impact
By Leonid F Kozin and Steve Hansen

Published by the Royal Society of Chemistry, www.rsc.org

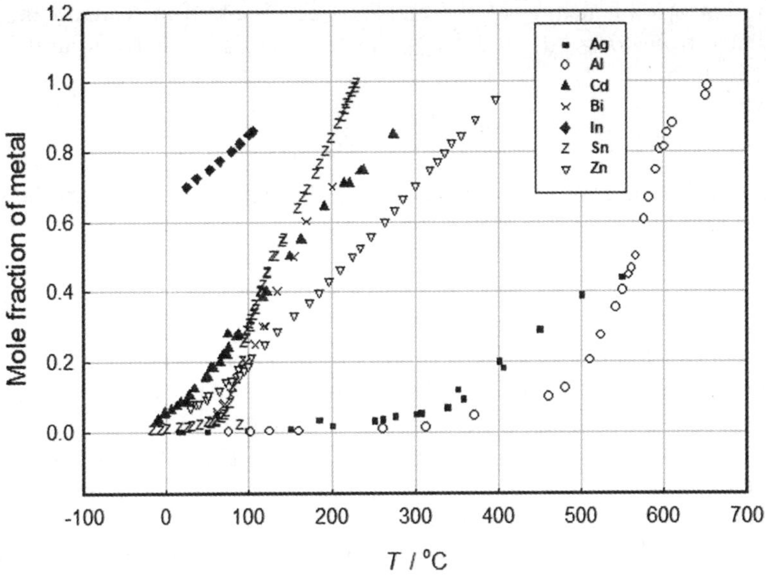

Figure 2.1 Solubility of different metals in mercury. References are in Appendix V. Adapted from sources.[29,73–89]

For thermodynamic equilibrium (at p, T = constant), it is essential that there is equality of the chemical potentials of metals contained in a saturated mercury solution (amalgam) $\mu_{\mathrm{Me}}^{*\mathrm{saturated}}$ and the solid metal $\mu_{\mathrm{Me}}^{\mathrm{solid}}$:

$$\mu_{\mathrm{Me}}^{\mathrm{ideal}} = \mu_{\mathrm{Me}}^{*\mathrm{saturated}} = \mu_0 + RT\ln x_1^{*\mathrm{saturated}} \tag{2.2}$$

For a pure solid metal, $\mu_{\mathrm{Me}}^{\mathrm{solid}} = \mu_0$. In the case of a saturated amalgam that obeys an ideal solution, after substitution into eqn (2.2), the solubility of the metal in mercury may be represented, considering that $\Delta\mu = \mu_{Me}^{*\mathrm{saturated}} - \mu_0 = \Delta G$, *via* a change of Gibbs free energy ($\Delta G_1^{*\mathrm{saturated}}$):

$$\ln x_1^{*\mathrm{saturated}} = \frac{\mu_{Me}^{*\mathrm{saturated}} - \mu_0}{RT} = \Delta\bar{G}_1^{*\mathrm{saturated}} \tag{2.3}$$

In the case of formation of ideal solutions (at $x_1 < x_1^{*\mathrm{saturated}}$) and considering that

$$\Delta\bar{G}_1 = \Delta H_{\mathrm{melt.Me_1}} - T\Delta S_{\mathrm{melt.Me_1}} \tag{2.4}$$

eqn (2.3) will appear as

$$\ln x_1 = \frac{\Delta H_{\mathrm{melt.Me_1}}}{RT} - \frac{\Delta S_{\mathrm{melt.Me_1}}}{R} \tag{2.5}$$

where $\Delta H_{\mathrm{melt.Me_1}}$ is the heat of fusion of the metal and $\Delta S_{\mathrm{melt.Me_1}}$ is the entropy of fusion of metal $\mathrm{Me_1}$ to be dissolved in mercury.

The entropy of melting of the metal to be dissolved in mercury may be calculated from $\Delta H_{\mathrm{melt.Me_1}}$ and $T_{\mathrm{melt.Me_1}}$, which are known fairly accurately for many metals:

$$\Delta S_{\mathrm{melt.Me_1}} = \frac{\Delta H_{\mathrm{melt.Me_1}}}{T_{\mathrm{melt.Me_1}}} \tag{2.6}$$

By substituting eqn (2.6) into eqn (2.5) and simplifying, we obtain the well-known Schroder equation:

$$\ln x_1 = \frac{\Delta H_{\mathrm{melt.Me_1}}}{R}\left(\frac{1}{T_{\mathrm{melt.Me_1}}} - \frac{1}{T}\right) \tag{2.7}$$

from which it follows that, given an ideal solubility of metals in mercury, the logarithm of solubility is a linear function when plotted against the reciprocal of temperature. When $\ln x_1$ is plotted *versus* $1/T$, the result should be a straight line with slope $\Delta H_{\mathrm{melt.Me_1}}/R$. Ideal curves are linear. The difference between the actual and ideal curves of solubility of metals in mercury lies in their behavior at low temperatures ($T = 0.30$–$0.5T_{\mathrm{melt}}$).

Kozin [7] analyzed the experimental data on the solubility of metals in mercury at different temperatures in terms of compliance with the Schroder equation by plotting $\ln x_1$ against $1/T$. The following systems were analyzed: In–Hg, Tl–Hg, Cu–Hg, Pb–Hg, Ag–Hg, Au–Hg, Bi–Hg and Sn–Hg. It was demonstrated that the curves describing the actual solubility of metals in mercury deviate strongly from the ideal solubility curves, obtained from eqn (2.7). It was concluded that eqn (2.7) cannot be used to calculate the solubility of metals in actual mercury (amalgam) systems. This is because, in an actual solution, the chemical potential of the metal dissolved in mercury is

$$\mu_{\mathrm{Me}} = \mu_0 + RT\ln a_1 \tag{2.8}$$

where a_1 = activity of metal $\mathrm{Me_1}$ in mercury. As reported,[6–8,28,29] the numerical value of a metal's activity in an amalgam is determined by the choice of the standard state. To facilitate the analysis of the deviation of the behavior of a mercury solution from ideal behavior, it is best to choose the pure metal for a standard state. The activity coefficient, γ, is used for a quantitative assessment of how much an amalgam deviates from ideal solution parameters:

$$\gamma_1 = \frac{a_1}{x_1} \tag{2.9}$$

or

$$a_1 = \gamma_1 x_1 \tag{2.10}$$

By substituting eqn (2.10) in eqn (2.8), we obtain

$$\mu_{\mathrm{Me_1}} = \mu_0 + (RT\ln\gamma_1)x_1 \tag{2.11}$$

By subtracting eqn (2.1) from eqn (2.11), we obtain

$$\mu_{\mathrm{Me_1}} - \mu_{\mathrm{Me}}^{\mathrm{ideal}} = RT\ln\gamma_1 \tag{2.12}$$

Considering that the activity coefficient γ_1 is functionally dependent on the partial excess Gibbs free energy:

$$\bar{G}^{\text{excess}} = RT\ln\gamma_1 \tag{2.13}$$

and taking into account the equation

$$\bar{G}^{\text{excess}} = \Delta\bar{H}_{\text{mix}} - T\Delta S_{\text{mix}}^{\text{excess}} \tag{2.14}$$

we may use equations (2.12)–(2.14) to deduce

$$RT\ln\gamma_1 = \Delta\bar{H}_{\text{mix}} - T\Delta S_{\text{mix}}^{\text{excess}} \tag{2.15}$$

The atomic interaction between metals and mercury in concentrated amalgams may produce compounds MeHg_n or solid solutions Me_xHg_y. In this case, eqn (2.11) may be represented as

$$\mu_{\text{Me}_1}^{*\text{saturated}} = \mu_0 + RT\ln\gamma_1 x_1 \tag{2.16}$$

The change in chemical potential, $\Delta\mu$, for a saturated amalgam is $\Delta\mu = \mu_{\text{Me}_1}^{*\text{saturated}} - \mu_0$. Keeping in mind eqns (2.3)–(2.16), we obtain

$$\ln x_1 - \left(\frac{\Delta H_{\text{mix,Me}_1}}{RT} - \frac{\Delta S_{\text{melt,Me}_1}}{T}\right) - \left(\frac{\Delta\bar{H}_{\text{min}} - T\Delta\bar{S}_{\text{mix}}^{\text{melt}}}{RT}\right) \tag{2.17}$$

From classical thermodynamics,

$$\Delta\bar{S}^{\text{excess}} = \Delta\bar{S}_{\text{mix}} - \Delta\bar{S}_{\text{ideal}} \tag{2.18}$$

and

$$\Delta\bar{S}_{\text{ideal}} = RT\ln x_1 \tag{2.19}$$

where $\Delta\bar{S}_{\text{mix}}$ is the partial molar entropy of mixing in an actual solution and $\Delta\bar{S}_{\text{ideal}}$ is the ideal entropy of solution. Combining eqns (2.17)–(2.19), we obtain

$$\ln x_1 = \frac{\Delta H_{\text{melt.Me}_1} - T\Delta S_{\text{melt.Me}_1} - \Delta\bar{H}_{\text{mix}} + T\Delta\bar{S}_{\text{mix}} - T\Delta\bar{S}_{\text{ideal}}}{RT} \tag{2.20}$$

For easier analysis, eqn (2.20) may be converted into

$$\ln x_1 = \frac{\Delta H_{\text{melt.Me}_1}}{R}\left(\frac{1}{T_{\text{melt.Me}_1}} - \frac{1}{T}\right) - \left(\frac{\Delta\bar{H}_{\text{mix}} - T\Delta\bar{S}_{\text{mix}} + T\Delta\bar{S}_{\text{ideal}}}{RT}\right) \tag{2.21}$$

For the case where $\Delta\bar{H}_{\text{mix}} = 0$ and $\Delta\bar{S}_{\text{mix}} = \Delta\bar{S}_{\text{ideal}}$, the solubility of a metal in mercury is described by the Schroder equation [the first member on the right-hand side of eqn (2.21)]. Therefore, the first term on the right-hand side of eqn (2.21) corresponds to the solubility of a metal in an ideal mercury solution ($\ln x_1^{\text{ideal}}$), whereas the second term characterizes the deviation from the ideal metal behavior in a mercury solution ($\ln\gamma_1$). Therefore, eqn (2.21) may be represented as[7]

$$\ln x_1 = \ln x_1^{\text{ideal}} - \ln\gamma_1 \tag{2.22}$$

or

$$\ln x_1 = \ln\left(\frac{x_1^{\text{ideal}}}{\gamma_1}\right) \tag{2.23}$$

$$x_1 = \frac{x_1^{\text{ideal}}}{\gamma_1} \tag{2.24}$$

Using eqn (2.24), one may calculate the solubility of a metal in any given system once the activity coefficients of the components are known, and *vice versa*, if x_1 and $x_1{}^{\text{ideal}}$ are known, one may calculate the activity coefficients of the metals present in a system. Equations (2.21)–(2.24) describe the solubility of a metal in mercury irrespective of its liquid phase state, such as the formation of associates. Analysis of eqns (2.22) and (2.24) indicates that for $\gamma_1>1$ the solubility is less than the ideal solution and for $\gamma_1<1$ the solubility exceeds the ideal solution. At $\gamma_1=1$ the solubility obeys an ideal solution. Consequently, to calculate the solubility of metals in mercury from eqns (2.20)–(2.24), we need data describing the thermodynamic properties of metals in amalgams: $\Delta H_{\text{fusion,Me}_1}$, ΔH_{mix}, $\Delta S_{\text{fusion,Me}_1}$, ΔS_{mix} or γ_1.

2.2 Amalgams with Compounds Formed in the Solid Phase

The temperature dependence of the solubility in mercury of In, Tl, Cd, Zn, Na, Cs, Li, K, Bi, Sn, Pb, Au, Ag, Sm, Pu, Cu, Mn, U, Th, Sb, Ni, Ti, Be, Zr, Cr, Co and Fe has been reported.[6–8,11–26] Many metals that are poorly soluble in mercury, except Th, Sb and Be, feature the same slopes of their curves of $\ln x_1$ *versus* $1/T$. This indicates that these metals should have close heat of solution effects in mercury. Low melting point metals demonstrate different slopes of the curves in question (In, Tl, Cd, Zn, Na, Cs, Li, K, Ga, Bi, Sn, Pb). The solubility of metals in mercury depends on their nature and may vary by 10 orders of magnitude (compare x_1 of indium and iron, indium and cobalt, *etc.*). The actual solubility curves of compound-forming metals such as nickel, manganese, silver and lead at the decomposition temperatures of the peritectic reactions feature distinct breaks, which, according to Jangg and Palman,[30] are due to variations of the activity coefficients of the metals present in the solution. Jangg and Palman[30] suggested the following empirical equation for the formation of compounds in a mercury solution:

$$\ln x_1 = \frac{\Delta H_{\text{fusion}} - T\Delta S_{\text{fusion}} + \alpha(\Delta H_{\text{form}} - T\Delta S_{\text{form}}) - \Delta H_{\text{mix}}}{RT} \tag{2.25}$$

where $\Delta H_{\rm form}$ and $\Delta S_{\rm form}$ are enthalpy and entropy, respectively, of the formation of a mercury compound or associate, $\Delta H_{\rm mix}$ is the enthalpy of mixing between compound and mercury and α is the degree of dissociation of compound in the mercury phase [note that in eqn (2.25) we have replaced the thermochemical system of notations used in Ref. 26 with a thermodynamic system]. The third term on the right-hand side of eqn (2.25) corresponds to change of Gibbs free energy in the course of compound formation:

$$\Delta G_{\rm form} = \alpha(\Delta H_{\rm form} - T\Delta S_{\rm form}) \tag{2.26}$$

Table 2.1 gives values for the melting point, boiling point, enthalpy of melting and enthalpy of vaporization for about 50 different metals.

Nickel forms several compounds with mercury. One compound, $NiHg_4$, is stable at temperatures up to ~224 °C (497 K) and is present as a solid residue and as a dissolved non-dissociated compound:

$$(NiHg_4)_x \rightleftharpoons (NiHg_4)_{x-m} + mNiHg_4 \tag{2.27}$$

According to Jangg and Palman,[30] at temperatures below the peritectic point, the degree of dissociation of $NiHg_4$ makes no major changes and makes the $\ln x_1$ *versus* $1/T$ curve appear as a straight line. Near the peritectic point, due to increasing dissociation, the negative member $\alpha(\Delta H_{\rm form} - T\Delta S_{\rm form})$ becomes progressively smaller and causes the solubility of the metal in mercury in eqn (2.25) to grow faster than it would as a linear function. At temperatures high above the peritectic point, α becomes independent of temperature and equal to zero. Therefore, the curves of $\ln x_1$ *versus* $1/T$ return to linear. In this case, eqn (2.25) reduces to

$$\ln x_1 = \frac{\Delta H_{\rm fusion,Me_1} - T\Delta S_{\rm fusion,Me_1} - \Delta H_{\rm mix}}{RT} \tag{2.28}$$

However, this equation fails to take into account the excess entropy of mixing $\Delta S^{\rm excess}$, which, as seen from eqn (2.18), equals the difference between $\Delta S_{\rm mix}$ and $\Delta S_{\rm ideal}$. Therefore, eqn (2.28) only applies when the dissolution of a metal in mercury comes with entropy of mixing $\Delta S_{\rm mix}$ equal to $\Delta S_{\rm ideal}$, *i.e.*, when the resulting mercury solution of metal Me_1 obeys the laws of regular solutions. In this case, the calculated solubilities of metals in mercury obtained from eqn (2.28) should agree with the experimental data.

According to Barański and Galus,[46] the Ni–Hg system produces three compounds: $NiHg_4$, $NiHg_3$ and $NiHg_2$. $NiHg_2$ is stable up to about 458 K, $NiHg_3$ up to 483 K and $NiHg_4$ up to 493 K. Saturated nickel amalgam is in equilibrium with pure nickel above 493 K. Data on the solubility of nickel in mercury are given in Table 2.2.

By examining the $\ln X_{Ni}$ *versus* $1/T$ curve, according to Barański and Galus,[46] with data for nickel and other metals, we found that the solubility values suggested[46] are seriously understated (by around two orders of

Table 2.1 Physicochemical properties of metals.

Element	T^{melt} *(K)*	T^{boil} *(K)*	ΔH^{fusion} *(J mol^{-1})*	ΔH^{vap} *(kJ mol^{-1})*	*Ref.*
Ag	1234	2436	11 297	283 658	31
Al	933.3	2793	10 795	326 921	31
Au	1336.2	3130	12 552	368 012	31
Ba	1002	2171	7749	182 719	31
Be	1560	2745	11 715	320 013	31,32
Bi	544.5	1837	11 297	209 849	31
Ca	1112	1757	8535	177 749	31
Cd	594.2	1040	6192	111 851	31
Ce	1071	3716	5460	423 195	31,33
Co	1768	3201	16 192	426 847	31
Cr	2130	2945		395 342	31,34
Cs	301.6	944	2092	77 580	31
Cu	1356.6	2836	13 054	335 620	31
Fe	1809	3135	13 807	413 111	31
Ga	302	2478	5590	270 981	31
Gd	1586	3546	10 054	398 932	31,33
Ge	1210.4	3107	36 945	371 707	31
Hf	2500	4876	15 100	618 881	31,35
In	429.8	2346	3264	243078	31
Ir	2716	4898	22 490	660 143	31,36
K	336.4	1032	2335	90 132	31
La	1191	3737	6197	431 303	31,33
Li	453.7	1615	3000	157 800	31
Mg	922	1363	8954	145 243	31
Mn	1517	2335	11 004	282 056	31,37
Mo	2890	4912	32 539	656 553	31
Na	371	1156	2598	107 345	31
Nb	2740	5017	31 100	718 217	31,38
Nd	1294	3347	7142	328 473	31,33
Ni	1726	3187	17 472	428 078	31
Pb	600.6	2023	4799	195 736	31
Pd	1825	3237	16 985	375 832	31,39
Pr	1204	3793	6887	356 837	31,33
Pt	2042	4149	21 330	565 000	31,40
Pu	913	3503	2845	352 167	31
Rb	312.6	961	2192	82 170	31
Rh	2233	4114	27 300	551 840	31,41
Sb	904	1860	19 874	264 228	31
Sc	1812	3109	14 096	376 112	31
Si	1685	3540	50 551	455 638	31
Sn	505.1	2876	7029	301 374	31
Th	2023	4795	13 817	575 614	31,42
Ti	1943	3562	13 000	467 131	31,43
Tl	577	1746	4142	181 594	31
Tm	1818	2223	16 841	233 413	31,33
U	1405	4407	8519	522 866	31
V	2183	3682	21 500	510 946	31,44
Y	1799	3618	11 397	423 785	31,33
Zn	692.7	1180	7322	129 867	31
Zr	2125	4682	30 500	607 488	31,45

Table 2.2 Solubility of nickel in mercury.

T *(K)*	T *(°C)*	*Mole fraction Pt*	*Ref.*
293	20	4.8×10^{-7}	30
293	20	1.5×10^{-9}	46
323	50	1.2×10^{-6}	30
323	50	1.8×10^{-8}	46
373	100	3.7×10^{-6}	30
373	100	4.1×10^{-7}	46
423	150	0.0000085	30
423	150	0.0000047	46
473	200	0.000017	30
473	200	0.000030	46
498	225	0.000021	30
503	230	0.000029	30
505	232	0.000032	30
507	234	0.000034	30
509	236	0.000038	30
516	243	0.000041	30
523	250	0.000044	30
573	300	0.000075	30
623	350	0.00011	30
673	400	0.00015	30
723	450	0.00019	30
773	500	0.00022	30
773	500	0.00030	47
773	500	0.0013	47
823	550	0.00035	47
885	612	0.00043	47
898	625	0.00024	47
923	650	0.00066	47
938	665	0.0041	47
973	700	0.00055	47
973	700	0.00099	47
998	725	0.0012	47
1023	750	0.00071	47
1023	750	0.0027	47

magnitude). At higher temperatures, different authors have suggested figures that are in close agreement between themselves.[11]

Let us compare the actual solubility curves with an ideal curve plotted using eqn (2.7) in coordinates' of β–$\ln x_1$, where

$$\beta = \frac{\Delta H_{\text{melt.Me}_1}}{R}\left(\frac{1}{T_{\text{melt}}} - \frac{1}{T}\right)$$

The curves for the solubility of metals in mercury in coordinates of β–$\ln x_1$ have basically the same slope in the dissolved amalgams section and, with the exception of Th, Pt and Mn, run parallel to each other. Of all metals, the solubility of platinum in mercury is the closest to ideal. We believe that the data for the Pt–Hg system, agreeing with eqn (2.7), are wrong, as the system produces compounds PtHg, $PtHg_2$ and $PtHg_3$.[48,49] Our calculations[7,50] have shown the solubility of platinum in mercury at 298 K to be 3.1×10^{-7} at.%. Jangg and Dortbudak[51] studied the solubility of platinum in mercury from 374 to 593 K and proved it to increase from 3.14×10^{-7} to 9×10^{-6} mole fraction, respectively (Table 2.3). By extrapolating these data in coordinates of $\ln x_{Pt}$–$1/T$ to 298 K, we obtained a solubility of platinum in mercury of 1.8×10^{-6} at.%. This agrees well with the calculated value quoted in the literature.[7,50]

The actual solubility curves are not parallel to the ideal solubility curve. The difference $\ln x_1^{ideal} - \ln x_1$ becomes less pronounced as the temperature increases. In this case, the activity coefficients of metals in mercury should decrease [see eqn (2.22)]. The curves of β *versus* $\ln x_1$ vary substantially with temperature near the melting point of the metal dissolving in mercury. The metal–mercury systems approach complete mutual solubility of the components at the melting point of the solute metal. The straightness of β *versus* $\ln x_1$ curves over a broad interval of temperatures, especially for metals demonstrating poor solubility in mercury, was used to determine accurately the solubility of metals in mercury at different temperatures *via* our graphical method.

In our analysis of the temperature dependence of the solubility of platinum in mercury in coordinates of β–$\ln x_1$, we estimated the solubility of platinum in mercury to be between 3.1×10^{-7} and 1.8×10^{-6} at.%. These were obtained through extrapolation of data of Guminski and Galus.[11] In our view, the often quoted figures for the solubility of platinum in mercury (0.102 at.% at 297.15 K[9,49,53] and 2.8×10^{-2} at.% at 293 K[54]) fail to reflect the actual solubility of platinum in mercury. Clearly, the value $x_{Pt} = 5\times10^{-4}$ at.% quoted in the literature[11,55,56] also overstates the actual solubility of platinum in mercury. It

Table 2.3 Solubility of platinum in mercury.

T *(K)*	T *(°C)*	*Mole fraction Pt*	*Ref.*
298	25	5×10^{-6}	52
374	101	3.14×10^{-7}	51
374	101	3.64×10^{-7}	51
397	124	7.08×10^{-7}	51
397	124	6.96×10^{-7}	51
424	151	6.66×10^{-7}	51
424	151	8.73×10^{-7}	51
445	172	1.13×10^{-6}	51
445	172	1.46×10^{-6}	51
465	192	1.73×10^{-6}	51
465	192	1.10×10^{-6}	51
473	200	2.86×10^{-6}	51
473	200	3.19×10^{-6}	51
523	250	5.51×10^{-6}	51
523	250	5.28×10^{-6}	51
555	282	7.87×10^{-6}	51
555	282	7.71×10^{-6}	51
581	308	8.50×10^{-6}	51
581	308	8.36×10^{-6}	51
593	320	9.06×10^{-6}	51

will be observed that the solubilities of chromium, cobalt and iron according to Kozin[7] are much smaller those reported by various other sources,[9,17–21] and correlates well with other data.[55,56]

The solubilities of rare earth metals in mercury seem to be understated compared with the experimental figures.[16] The alkali metals sodium, cesium, lithium and potassium have solubilities of 3.0, 1.51, 0.56 and 0.45 at.%, respectively.[6,7]

It should be mentioned that the graphical method of finding the solubility of metals in mercury through extrapolation in coordinates of β–lnx_1, which was developed earlier,[7] has proven very reliable. The solubility values yielded by our graphical method were benchmarked against high-precision experimental solubility data for copper, gold, manganese and silver. The results demonstrated a very high degree of correlation.

The x_{Me_1} (solubility) values provided for many metals and elements cover much of the Periodic Table. However, many of those figures are ill-founded, which adds appeal to the calculation methods and allows us to estimate, *a priori*, the probable solubility in mercury for all known elements. Table 2.4 gives a comparison of experimental solubility values and those predicted by eqns (2.29) and (2.30).

Kozin[50,57] suggested two equations for finding the probable solubility of metals in mercury: one based on the difference between the entropies of melting:

$$\ln N_1 = -\frac{\left(\Delta H_{\mathrm{melt.Me}_1}/T - \Delta H_{\mathrm{melt.Me}_1}/T\right)^{1.39}}{1.896} = \frac{\Delta S^{1.39}}{1.896} \tag{2.29}$$

Table 2.4 Measured and calculated solubilities of metals in mercury at 298.15 K (x, mole fraction).

Element	*Solubility at 25 °C*	*Ref. 10*	*Eqn (2.29)*	*Eqn (2.30)*
Ag	0.00071	7.6×10^{-4}	–	3.6×10^{-4}
Al	<0.001	1.6×10^{-4}	1.7×10^{-3}	1.0×10^{-4}
Au	0.00467 (80 °C)	1.4×10^{-3}	–	1.2×10^{-3}
Bi	0.0112 (22.5 °C)	1.3×10^{-2}	–	1.4×10^{-2}
Cd	0.105 (28.5 °C)	9.53×10^{-2}	–	$(3–4.5)\times10^{-2}$
Co	2.0×10^{-8} (160 °C)	1×10^{-9}	1.3×10^{-10}	4.5×10^{-11}
Cu	0.00006–0.0001	1.0×10^{-4}	6.3×10^{-5}	5.5×10^{-5}
Fe	5.4×10^{-6}	$<10^{-9}$	5.1×10^{-8}	1.6×10^{-8}
In	0.70	0.70	–	–
Mn	4.4×10^{-5}	4.5×10^{-5}	6.7×10^{-5}	$(0.07–3.1)\times10^{-4}$
Ni	5×10^{-7}	2×10^{-9}	8.4×10^{-8}	$(0.26–3.5)\times10^{-7}$
Pb	0.0165	1.63×10^{-2}	–	6.5×10^{-3}
Pd	5×10^{-5}	5.1×10^{-5}	–	3.2×10^{-5}
Pt	5×10^{-6}	5×10^{-6}	6.4×10^{-8}	$(0.4–6.5)\times10^{-7}$
Pu	0.000161	1.5×10^{-4}	2.0×10^{-4}	4.5×10^{-4}
Sn	0.0127	1.26×10^{-2}	–	1.10^{-4}
Tl		0.427	–	–
Tm		4×10^{-6}	–	3.3×10^{-6}
Zn	0.0696 (30 °C)	6.32×10^{-2}	–	1.8×10^{-2}
Zr		6×10^{-8}	–	3.6×10^{-8}

and the other based on the bond energies of the crystalline lattice of the dissolved metal:

$$\ln N_1 = -\chi \frac{E_{MeMe}}{2.3RT} \tag{2.30}$$

where χ is a constant equal to 0.414 and E_{MeMe} is the metal–metal bond energy, found from the equation

$$E_{MeMe} = \frac{2\Delta H^{subl}}{J} \tag{2.31}$$

where ΔH^{subl} is the heat of sublimation and J is the coordination number of the metal's crystalline lattice.

For the calculation of $\ln N_1$ using eqn (2.29), heats of melting $\Delta H_{melt,Me_1}$ and melting temperatures T_{melt} were taken from the literature;[31–45,58–63] for the calculation of interatomic bond energies, published heats of sublimation of the metals were used;[58,66] and coordination numbers for solid and liquid states were taken from other sources.[59,60,64–71]

Using melting temperatures and enthalpies of melting and sublimation, one can use eqns (2.29) and (2.30) to calculate likely solubility of metals in mercury at 293.15 K. The X_{Me} values obtained using these equations are given in Table 2.4. For comparison purposes, the table also offers known experimental data on the solubility of metals in mercury (columns two and three). Compared with data in the literature,[6–8,72] many calculated X_{Me1} values were updated using more credible published initial parameters ($\Delta H_{melt.Me1}$, ΔH^{subl}, $T_{melt.Me1}$, J).

As can be seen from Table 2.4, for many metals the theoretical solubility values correlate well with the experimental data. For some metals, owing to inaccurate coordination numbers (different values according to different authors), the calculated solubility values are scattered. In this case, the calculated solubilities obtained using mean coordination numbers demonstrated good correlations with the experimental data. For some low-melting metals, such as gallium, indium, thallium, tin and lead, the calculated solubilities obtained from eqn (2.30) were extremely low. Clearly, the coordination numbers used for these metals were not accurate. According to Lamoreaux,[55] elements with predominantly covalent or ionic bonds have coordination numbers <8, whereas metals generally have coordination numbers of 8–12.

References

1. L. F. Kozin, *Electrodeposition and Dissolution of Polyvalent Metals*, Naukova Dumka, Kiev, 1986.
2. L. F. Kozin and A. G. Morachevsky, *Physical Chemistry and Metallurgy of High-Purity Lead*, Metallurgiya, Moscow, 1986.
3. B. F. Bilenkoy and A. K. Filatova, *Mercury Sulfide: Production and Application*, Lvov University, Lvov, 1988.

4. L. L. Chang and K. Ploog (eds), *Molecular-Beam Epitaxy and Heterostructures*, Nijhoff, Moscow, 1985.
5. K. Kutzsche, *Neue Hütte*, 1989, **34**, 189–190.
6. L. F. Kozin, *Physicochemical Basics of Amalgam Metallurgy*, Nauka, Alma-Ata, 1964.
7. L. F. Kozin, *Amalgam Pyrometallurgy*, Nauka, Alma-Ata, 1973.
8. L. F. Kozin, R. Sh. Nigmetova and M. B. Dergacheva, *Thermodynamics of Binary Amalgam Systems*, Nauka, Alma-Ata, 1979.
9. M. T. Kozlovsky, A. I. Zebreva and V. P. Gladyshev, *Amalgams and Their Application*, Nauka, Alma-Ata, 1971.
10. C. Guminski, *J. Mater. Sci.*, 1989, **24**, 2661–2676.
11. C. Guminski and Z. Galus, *Metals in Mercury, Solubility Data Series*, vol. 25, C. Hirayama, Pergamon Press, Oxford, 1986.
12. L. F. Kozin, *Proc. Inst. Org. Catal. Electrochem. Acad. Sci. Kazakh. SSR*, 1972, **3**, 3–30.
13. K. K. Lepesov and L. F. Kozin, *Proc. Inst. Org. Catal. Electrochem. Acad. Sci. Kazakh. SSR*, 1984, **24**, 28–54.
14. L. F. Kozin, *Physicochemical Research of Amalgam Methods Used to Obtain High-purity Metals: Synopsis of Thesis*, Doctorate of Chemical Sciences, Alma-Ata, 1964, p. 35.
15. L. F. Kozin, *Amalgam Metallurgy*, Tekhnika, Kiev, 1970, p. 272.
16. V. P. Shvedov, A. Z. Frolkov and G. D. Nikitin, *Radiokhimii*, 1971, **13**, 252–255.
17. M. Hansen and K. Anderko, *Constitution of Binary Alloys*, McGraw-Hill, New York, 1958.
18. R. P. Elliott, *Constitution of Binary Alloys*, First Supplement, McGraw-Hill, New York, 1965.
19. F. A. Shunk, *Constitution of Binary Alloys*, Second Supplement, McGraw-Hill, New York, 1969.
20. G. Petot-Ervas, C. Petot and E. Bonnier, *Bull. Soc. Chim. Fr.*, 1968, **10**, 3972–3975.
21. O. A. Barabash and Y. N. Kovalev, *Structure and Properties of Metals and Alloys*, Naukova Dumka, Kiev, 1986.
22. M. E. Drits and L. L. Zusman, *Alloys of Alkali and Alkaline-Earth Metals in Mercury*, Metallurgia, Moscow, 1986.
23. V. S. Fomenko, *Emission Properties of Materials*, Naukova Dumka, Kiev, 1981.
24. W. H. Young, *Can. J. Phys.*, 1987, **65**, 241–265.
25. K. Hain and E. Burig (eds), *Solidification from Melts: a Handbook*, Metallurgia, Moscow, 1987, p. 320.
26. S. Tamaki, *Can. J. Phys.*, 1987, **65**, 286–308.
27. M. B. Dergacheva and L. F. Kozin, *Zh. Fiz. Khim.*, 1977, **51**, 417–420.
28. O. F. Devero, *Problems of Metallurgical Thermodynamics*, Metallurgia, Moscow, 1986.
29. G. Jangg and H. Palman, *Z. Metallkd.*, 1963, **54**, 364–369.

30. R. Hultgren, P. D. Desai, D. T. Hawkins, M. Gleiser and K. K. Kelley, *Selected Values of the Thermodynamic Properties of the Elements*, American Society for Metals, Metals Park, OH, 1973.
31. A. J. Stonehouse, *J. Vac. Sci. Technol. A*, 1986, **4**, 1163.
32. K. A. Gschneidner Jr and L. Eyring (eds), *Handbook on the Physics and Chemistry of Rare Earths*, Elsevier, Amsterdam, 1993, Cumulative Index, vol. 1–15, pp. 509–521.
33. S. V. Lebedev, A. I. Savvatimskii and Yu.B. Smirnov, *High Temp.*, 1971, **9**, 578–581.
34. P.-F. Paradis, T. Ishikawa and S. Yoda, *Int. J. Thermophys.*, 2003, **24**, 239.
35. L. B. Hunt, *Platinum Group Met.*, 1987, **31**, 32–41.
36. S. Sato and O. J. Kleppa, *J. Chem. Thermodyn.*, 1979, **11**, 521.
37. A. Cezairliyan and J. L. McClure, *Int. J. Thermophys.*, 1987, **8**, 577.
38. C. Cagran and G. Pottlacher, *Platinum Met. Rev.*, 2006, **50**, 144.
39. A. K. Chaudhuri, D. W. Bonnell, L. A. Ford and J. L. Margrave, *High Temp. Sci.*, 1970, **2**, 203.
40. J. W. Arblaster, *Platinum Met. Rev.*, 1996, **40**, 62.
41. W. Stoll, *Thorium and Thorium Compounds. Ullmann's Encyclopedia of Industrial Chemistry*, Wiley-VCH, Weinheim, 2000.
42. J. L. McClure and A. Cezairliyan, *Int. J. Thermophys.*, 1992, **13**, 75.
43. J. F. Smith, *Bull. Alloy Phase Diagr.*, 1981, **2**, 40.
44. O. J. Kleppa and S. Watanabe, *Met. Trans.*, 1982, **13B**, 391.
45. A. Barański and Z. Galus, *J. Electroanal. Chem.*, 1973, **48**, 289–305.
46. J. R. Weeks, *Corrosion*, 1967, **23**, 98–106.
47. A. R. Miedema, P. F. De Chatel and F. R. De Boer, *Physica*, 1980, **101 B + C**, 1–28.
48. I. N. Plaksin and N. A. Suvorovskaya, *Zh. Fiz. Khim.*, 1941, **15**, 978–980.
49. L. F. Kozin, *Izv. Akad. Nauk Kazakh. SSR, Ser. Khim.*, 1972, (3), 34–49.
50. G. Jangg and T. Dortbudak, *Z. Metallkd.*, 1973, **64**, 715–719.
51. C. Guminski, H. Roslonek and Z. Galus, *J. Electroanal. Chem.*, 1983, **158**, 357.
52. E. Bauer, H. Nowotny and A. Stempfl, *Monatsh. Chem.*, 1953, **84**, 211–212.
53. J. H. Butler and A. C. Makrides, *Trans. Faraday Soc.*, 1964, **60**, 938–946.
54. R. H. Lamoreaux, *Melting Point, Gram-Atomic Volumes and Enthalpies of Atomization for Liquid Elements*, US Energy Research Development Agency Report LBL-4995, Lawrence-Berkeley Laboratory, Berkeley, CA, 1976. Available online at http://www.osti.gov/energycitations/servlets/purl/7273452-6b48Ir/7273452.pdf. Accessed June 13, 2013.
55. A. K. Niessen, F. R. De Boer and R. Boom, *et al.*, *CALPHAD: Comput. Coupling Phase Diagrams Thermochem.*, 1981, **7**, 51–70.
56. L. F. Kozin, *Physical Chemistry and Metallurgy of High-Purity Mercury and Its Alloys*, Naukova Dumka, Kiev, 1992.
57. G. V. Samsonov, *Properties of the Elements (Handbook)*, Metallurgiya, Moscow, (1976), part 1, p. 810.
58. K. J. Smittles, *Metals Handbook*, Metallurgia, Moscow, 1980.

59. O. Kubaschewski and C. B. Alcock, *Metallurgical Thermochemistry*, Metallurgia, Moscow, 1982.
60. K. Saito, S. Hayakawa, F. Takei and H. Yamadera, *Chemistry and the Periodic Table*, Iwanami Shoten, Tokyo, 1979.
61. D. R. Stull and G. C. Sinke, Thermodynamic Properties of the Elements, *Advances in Chemistry, American Chem. Soc.*, 1956, **18**, 800.
62. V. P. Glushko (ed.), *Thermal Constants of Substances*, Moscow, VINITI, 1968–1981, No. 3–10.
63. G. Schulze, *Metallophysics*, Mir, Moscow, 1971.
64. E. M. Sokolovskaya and L. S. Guzei, *Metallochemistry*, Moscow University Press, Moscow, 1986.
65. L. S. Darken and R. W. Gurry, *Physical Chemistry of Metals*, McGraw-Hill, New York, 1953.
66. D. R. Wilson, *Structure of Liquid Metals and Alloys*, Metallurgia, Moscow, 1972.
67. D. K. Belashchenko, *Transfer Phenomena in Liquid Metals and Semiconductors*, Atomizdat, Moscow, 1970.
68. P. P. Arsentyev and L. A. Koledov, *Metallic Melts and Their Properties*, Metallurgia, Moscow, 1976.
69. I. Vaseda, *Liquid Metals*, Metallurgia, Moscow, 1980, pp. 182–193.
70. A. R. Ubbelohde, *The Molten State of Matter*, Metallurgia, Moscow, 1982.
71. L. F. Kozin, *Electrochem. Solutions Met. Syst.: Proc. Inst.Chem. Sci. Acad. Sci. Kazakh. SSR*, 1962, **9**, 101 121.
72. J. H. Vernik, *Phys. Met. Sci.*, 1967, **1**, 220–277.
73. D. R. Hudson, *Metallurgia*, 1943, **28**, 203.
74. H. A. Leibhafsky, *J. Am. Chem. Soc.*, 1949, **27**, 1468.
75. C. J. De Gruyer, *Recl. Trav. Chim. Pays-Bas*, 1925, **44**, 937.
76. A. A. Sunier and E. B. Gramkee, *J. Am. Chem. Soc.*, 1929, **51**, 1703.
77. C. Rolfe and W. Hume-Rothery, *J. Less-Common Met.*, 1967, **13**, 1.
78. H. C. Bijl, *Z. Phys. Chem.*, 1902, **41**, 641.
79. R. E. Mehl and C. S. Barrett, *Trans. AIME*, 1930, **89**, 575.
80. L. F. Kozin and N. N. Tananaeva, *Russ. J. Inorg. Chem.*, 1961, **6**, 463.
81. A. S. Moshkevich and A. A. Ravdel, *Zh. Prikl. Khim.*, 1970, **43**, 71.
82. H. E. Thompson, *J. Phys. Chem.*, 1935, **39**, 655.
83. P. Dumas, L. Bougarfa and J. Bensaid, *J. Phys.*, 1984, **45**, 1543.
84. M. M. Haring, M. R. Hatfield and P. T. Zapponi, *Trans. Electrochem. Soc.*, 1939, **75**, 473.
85. G. V. Yan-Sho-Syan, M. V. Nosek, N. M. Semibratova and A. E. Shalamov, *Tr. Inst. Khim. Nauk Akad. Nauk Kazakh. SSR*, 1967, **15**, 139.
86. G. Petot-Ervas, M. Gaillet and P. Desre, *C. R. Acad. Sci.*, 1967, **264**, 490.
87. N. A. Puschin, *Z. Anorg. Allg. Chem.*, 1903, **36**, 210.
88. N. A. Pushin, *Z. Anorg. Chem.*, 1903, **34**, 201.
89. E. Cohen and K. Inouye, *Z. Phys. Chem.*, 1910, **71**, 625.

CHAPTER 3

Diffusion of Metals in Mercury

3.1 Effect of Atomic Size on Diffusion

The most commonly used equation to calculate the diffusion coefficients of metals in mercury is

$$D_{\mathrm{Me}_i} = \frac{kT}{A\pi\eta r_{\mathrm{Me}_i}} \tag{3.1}$$

where k is the Boltzmann constant ($1.38\times10^{-23}\,\mathrm{J\,K^{-1}}$), η is the dynamic viscosity of pure mercury and r_{Me_i} is the radius of the diffusing particle. Equation (3.1) is the so-called Stokes–Einstein (if $A=6$) or Einstein–Sutherland (if $A=4$) relation. Along with an A value, the size of the diffusing particle is the subject of discussion when experimental data are analyzed.[1–18] According to Stromberg and Zakharova[1] and Gladyshev,[2] the theory best agrees with the experiment when $A=6$ and when r_{Me_i} is equal to the ionic radius of the diffusing metal. According to Zakharov,[3] also the metallic one. Galus[7] and Ma *et al.*[4] preferred the use of $A=4$.

If $A=4$, the experimental values obtained for some metals (Pb, Bi, In, Sb, Hg, Sn, Cd, Ga, Zn[4–7]) match well with those predicted by the Einstein–Sutherland relation. Experimental data concerning the diffusion coefficients of metals in mercury (system Me_i–Hg) have been reported.[1–10,19,20] Table 3.1 lists values of D_{Me_i} obtained mostly from Guminski,[19] amended only for Zn,[4] Cd, Hg[8] and Ca.[10] As can be seen, the values of D_{Me_i} depend on the nature of the metal and are at a minimum for La ($5.0\times10^{-6}\,\mathrm{cm^2\,s^{-1}}$) and at a maximum for Zn (1.67–$1.89\times10^{-5}\,\mathrm{cm^2\,s^{-1}}$).

To demonstrate how the values are affected by the size factor, Figure 3.1 shows the experimental data (Table 3.1) in coordinates of $1/D_{\mathrm{Me}_i} - r^0_{\mathrm{Me}_i}$, where $r^0_{\mathrm{Me}_i}$ is the ionic radius of a metal.[11]

Mercury Handbook: Chemistry, Applications and Environmental Impact
By Leonid F Kozin and Steve Hansen

Published by the Royal Society of Chemistry, www.rsc.org

Table 3.1 Diffusion coefficients $r^0_{\mathrm{Me}_i}$ of metals in mercury at 298 K, effective atomic/ionic radii and radii of particles diffusing in mercury phase compared with those existing in liquid and solid phases of the constitution diagram.

Element	$D_{\mathrm{Me}_i} \times 10^5$ $(cm^2\ s^{-1})$[19]	$r^{\mathrm{eff}}_{\mathrm{Me}_i}$ (nm)[19]	$r^0_{\mathrm{Me}_i}$ (nm)[8]	$r^{\mathrm{diff}}_{\mathrm{Me}_i}$ (nm)	Δr_{Me_i} (nm)
Ag	1.05 ± 0.03	0.127	0.16	0.212	0.052
Al	1.6 ± 0.2	0.126	0.125	0.138	0.013
Au	0.85 ± 0.04	0.126	0.135	0.256	0.121
Bi	1.35 ± 0.1	0.162	0.16	0.16	0.00
Ca	0.69 (283 K)	0.172	0.18	0.314	0.134
Cd	1.64–1.53 ± 0.03	0.137	0.155	0.155	0.00
Ce	0.60 ± 0.06	0.161	0.185	0.41	0.225
Co	0.84 ± 0.04	0.11	0.135	0.42	0.285
Cs	0.65 ± 0.1	0.24	0.26	0.338	0.078
Cu	1.00 ± 0.08	0.112	0.155	0.218	0.063
Fe	1.84 ± 0.13	0.112	0.14	0.12	–0.020
Ga	1.64 ± 0.08	0.133	0.13	0.136	0.006
Hg	1.59–1.60 ± 0.05	0.143	0.15	0.15	0.00
In	1.38 ± 0.1	0.146	0.155	0.164	0.009
K	0.79 ± 0.08 (293 K)	0.188	0.22	0.278	0.058
La	0.50 ± 0.05	0.165	0.195	0.442	0.247
Li	0.92 ± 0.1	0.137	0.145	0.21	0.065
Mn	0.90 ± 0.08	0.114	0.14	0.244	0.104
Na	0.84 ± 0.15	0.168	0.18	0.259	0.079
Nd	0.78 ± 0.08	0.16	0.185	0.284	0.099
Ni	0.65 ± 0.03	0.109	0.135	0.342	0.207
Pb	1.25 ± 0.04	0.154	0.18	0.18	0.00
Pr	0.60 ± 0.06	0.16	0.185	0.41	0.225
Rb	0.75 ± 0.08	0.223	0.235	0.291	0.056
Sb	1.40 ± 0.1	0.154	0.145	0.155	0.010
Sm	0.52 ± 0.06	0.158	0.185	0.424	0.239
Sn	1.48 ± 0.04	0.148	0.145	0.164	0.019
Sr	0.96 ± 0.1 (293 K)	0.188	0.200	0.310	0.110
Tb	0.82 ± 0.08	0.156	0.175	0.266	0.091
Tl	1.05 ± 0.05	0.151	0.19	0.206	0.016
U	0.6 ± 0.1	0.135	0.175	0.366	0.191
Zn	1.67–1.89	0.123	0.135	0.134	–0.001

3.1.1 Effect of Atomic Radius

From Figure 3.1, it follows that only a limited number of 'elementary' metals, *i.e.* Fe, Zn, Al, Ga, Ge, Sn, Sb and Bi, fit the theoretical curve of D_{Me_i} *versus* ionic radius within experimental error. Silver, lead, indium and thallium, even though they form mercury compounds with only small changes in free energy, demonstrate slight deviations from the theoretical curve.

Alkali metals (Li, Na, K, Rb, Cs), lanthanides (La, Sm, Nd, Tb, Pr, Ce), alkaline earth metals (Ca, Mg, Sr, Ba) and metals marginally soluble in mercury (Ni, Co, Mn, Cu, Ag, Sb) have D_{Me_i} values that are smaller than those found from eqn (3.1) with $A = 4$ and correlate with the atomic radius of the metal.

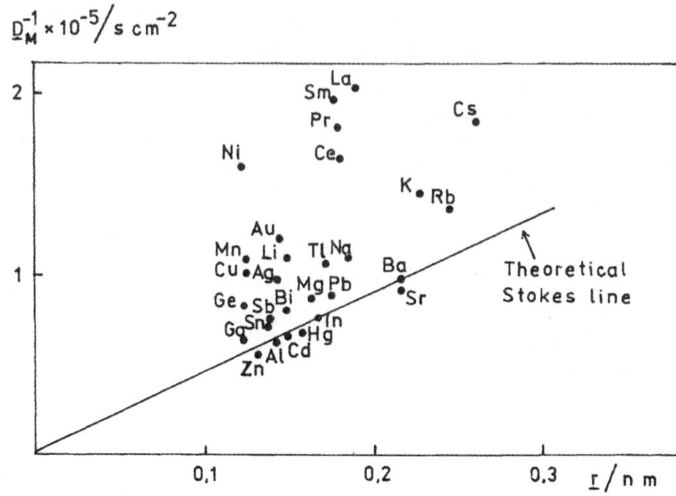

Figure 3.1 Graph of $1/D_{Me_i}$ *versus* ionic radius of diffusing metals, $r^0_{Me_i}$. Line, theoretical curve with $A=4$; ●, experimental values of $1/D_{Me_i}$. Reproduced with kind permission from © IUPAC, Z. Galus, Diffusion coefficients of metals in mercury, *Pure Appl. Chem.*, 1984, **56**, 635 (Ref. 7).

Figure 3.1 helps to visualize the real radii of particles diffusing in mercury ($r^{\text{diff}}_{\text{Me}_i}$). The values obtained are given in column five of Table 3.1. Column six of Table 3.1 gives the difference $r^{\text{diff}}_{\text{Me}_i} - r^0_{\text{Me}_i}$ of atomic and ionic potentials, according to Slater.[11] As can be seen, for simple metals, $\Delta r_{\text{Me}_i} = r^{\text{diff}}_{\text{Me}_i} - r^0_{\text{Me}_i}$ approaches or equals zero. For alkali metals, the value of Δr_{Me_i} is at its maximum for Li (0.096 nm) and at its minimum for Rb (0.056 nm). Δr_{Me_i} is somewhat greater for the alkaline earth metals (0.097 nm for Ba, 0.162 nm for Ca) and the maximum values overall are demonstrated by the rare earth metals (0.247 nm for La, 0.225 nm for Ce, Pr, with the exception of Nd, 0.099 nm, and Tb, 0.091 nm). For mercury, $r^0_{\text{Hg}} = r^{\text{diff}}_{\text{Hg}} = 0.150$ nm, *i.e.* $\Delta r_{\text{Hg}} = 0$. Using the values for Δr_{Me_i} and Na and Hg as an example, one can examine the composition of the diffusing particle based on the sum of the radii $r^0_{\text{Na}} = 0.180$ nm and $r^0_{\text{Hg}} = 0.150$ nm. Thus, the total radius of a diatomic NaHg particle equals 0.330 nm, while the calculated radius $r^{\text{diff}} = 0.259$ nm. However, the NaHg system demonstrates a volumetric compression of 18% as its components react with each other.[12] In the case where n = 6, the compression amounts to

$$\Delta r_{\text{NaHg}} = \frac{0.330 - 0.259}{0.330} \times 100 = 21.5\%$$

The value of $\Delta r_{\text{Na-Hg}}$ agrees well with experiment. For alkaline earth metals in the course of intermetallic compound formation, *e.g.* CaHg, SrHg and BaHg, the compression[26] amounts to $\Delta r_{\text{CaHg}} = 4.85\%$, $\Delta r_{\text{SrHg}} = 11.42\%$ and

$\Delta r_{BaHg} = 14.5\%$ again when n = 6 in Equation 3.1. Therefore, volumetric compression in these systems increases with transition from Ca to Ba.

For the early lanthanides, La, Ce, Pr and Sm, the values of $r^{diff}_{Me_i}$ exceed the sum $r^0_{Me} + r^0_{Hg}$ – a characteristic of diatomic particles Me_iHg (La $r^{diff}_{La} = 0.442 > 0.340$ nm). Clearly, such systems produce stable triatomic particles (associates) $MeHg_2$ in liquid mercury. Calculated compression values $\Delta r_{Me_iHg_2}$ calculated for diffusing species $LaHg_2$, $CeHg_2$, $PrHg_2$ and $SmHg_2$ are as follows, for n = 6 in Equation 3.1:

$LaHg_2$	10.7%
$CeHg_2$	15.5%
$PrHg_2$	15.5%
$SmHg_2$	12.6%

The compression values obtained are comparable to those observed for metal systems with high affinity for each other.[26] Values of Δr_{Me_i} for Nd and Tb are comparable to those for alkali metals. Apparently, Nd and Tb in liquid mercury form diatomic associates NdHg and TbHg. In this case, the compression calculation produced the results $\Delta r_{NdHg} = 15.2\%$ and $\Delta r_{TbHg} = 18.2\%$. These data indicate that diffusion in dilute solutions of these metals in mercury is limited to relatively simple diatomic associates of the type Me_iHg for Li, Na, K, Rb, Cs, Ca, Sr, Mg, Nd and Tb and triatomic associates for La, Ce, Pr and Sm.

Concepts for the diffusion of associates, Me_jHg_m,[10,13,19] are based on the idea of solvation of the metal diluted in mercury as put forth by Hildebrand.[14] Compression values are small[15] for metals whose D_{Me_i} values show basically no deviation from the theoretical curve in the coordinates $1/D_{Me_i} - r^0_{Me_i}$. Nigmetova *et al.*[15] demonstrated that alloy formation between mercury and Cd, In, Tl, Sn, Pb and Bi results in the following compression values for n = 6 in Equation 3.1:

Cd	0.31%
In	0.46%
Tl	0.77%
Sn	0.15%
Pb	1.07%
Bi	1.67%

Hence in the case of 'elementary' metals with a weak affinity for mercury, there is only a slight compression. This explains why these components diffuse in a mercury solution as atoms. The experimental values for D_{Me_i} may disagree with the theoretical values, $D_{Me_i analysis}$, owing to an inaccurate account of the viscosity η, which takes place in the course of the electrolysis of the system $Me_{i(j)}$–Hg, because the surface concentration of $Me_{i(j)}$ atoms may differ from the volume concentration. Indeed, Regel' and Patyanin[22] discovered the phenomenon of surface viscosity: surface layers of Hg, Ca, Na and K, as thick as 10^{-6} cm, demonstrate viscosity that is one or more orders of magnitude greater than the viscosity in the mass (volume). The latter is due to different molecular order at the surface and in the volume and lower fluidity in the surface layer. The above analysis of compression of the atoms of metals and mercury in the course of alloy formation, and also literature data,[12,15–17] indicate that associates of metals formed with mercury contain only a limited number of atoms. Experimentally determined diffusion coefficients obtained for 36 metals have been consolidated.[1–7,9,20,31]

3.2 Temperature Dependence of Diffusion in Amalgams

Chhabra[9] suggested the relation

$$D_{\mathrm{Me}_i} = \frac{(0.63 d_{\mathrm{Hg}})^2 B_{\mathrm{Hg}} RT}{2V_{\mathrm{Hg}}} \left(\frac{V - V_0}{V_0}\right)_{\mathrm{Hg}} \left(\frac{d_{\mathrm{Hg}}}{d_{\mathrm{Me}_i}}\right) \tag{3.2}$$

for calculating the diffusion coefficients of metals in mercury at different temperatures, where B_{Hg} is the intrinsic constant of mercury introduced by Hildebrand into the fluidity equation (the inverse of viscosity),[18] V is the atomic volume of mercury at zero fluidity, approximately equal to the atomic volume at its melting temperature, and d_{Hg} and d_{Me_i} are atomic diameters of mercury as solvent and the diffusing metal, respectively. Chhabra[9] consolidated many previously known experimental data, especially those related to temperature dependence, into D_{Me_i} and compared these with data obtained *via* eqn (3.2). The calculations were made using the values of B and V_0 for mercury, according to Hildebrand;[18] the atomic volumes of mercury at different temperatures were calculated from the temperature dependence of its density. Theoretical values obtained for D_{Me_i} normally exceed the experimental data by 17–99% (for Hg, Ag, Ba, Cd, *etc.*). Calculated D_{Me_i} values for cadmium and potassium in mercury, are compared with the experimental D_{Me_i} in ref. 7–9, 32. As can be seen, the most credible experimental values for D_{Me_i} fit the lines predicted *via* the equation

$$\log D_{\mathrm{Me}_i} = \log D_{0\mathrm{Me}_i} - \frac{E_{\mathrm{Me}_i}}{2.303RT} \tag{3.3}$$

where D_0 is a pre-exponential factor and E_{Me_i} is the activation energy for diffusion of Me_i particles. The diffusion coefficient for cadmium is $D_{0,\mathrm{Cd}} = 1.42 \times 10^{-4}\,\mathrm{cm^2\,s^{-1}}$ and for potassium $D_{0,\mathrm{K}} = 3.64 \times 10^{-4}\,\mathrm{cm^2\,s^{-1}}$. The activation energies of diffusion for particles of cadmium and potassium are 5.38 and 22.21 kJ mol^{-1}, respectively.

The temperature dependence of D_{Me_i} for Ag, Al, Au, Sn, Pb, Zn, Ni and Sb[7,9,31] is not completely clear. For these metals, the function $\log D_{\mathrm{Me}_i} - 1/T$ may be only nominally obeyed. The explanation of the discrepancy should lie in the dependence of D_{Me_i} values on the experimental conditions. The key factors affecting D_{Me_i} are:

1. Purity of the original reagents (no other metallic impurities are allowed).
2. Formation of insoluble compounds (since they cause uncontrolled changes to the analytical signal). The formation of compounds of the

analyzed metal may take place as the metal in contact with mercury dissolves in it (*e.g.* Pt and Ag).

3. The use of capillaries of suboptimal sizes for electrochemical measurements. Ma and Kao[4] demonstrated that values of D_{Me_i} increase with increasing radius of the active electrode and they reach their limit at a certain value R_i. The error in calculating D_{Me_i} may reach 56% due to this factor alone.[4]
4. Convection currents in the capillaries used to determine D_{Me_i}. Convection currents overstate values of D_{Me_i} owing to the input from the convective diffusion coefficient $D_{convection}$. In this case, an experiment is used to determine the apparent diffusion coefficient ($D_{apparent}$), which equals the sum of D_{Me_i} and $D_{convection}$:

$$D_{apparent} = D_{Me_i} + D_{convection} \tag{3.4}$$

5. The analyzed Me_i–Hg systems should be homogeneous and free from heterogeneous Me_i particles.

3.3 Concentration Effects on Diffusion

In theory, the diffusion coefficient should not depend on the concentration of metals in the mercury solution (amalgam). However, this condition is true only if the concentration of the analyzed metals is very low (Table 3.2). At higher Me_i concentrations, D_{Me_i} has been observed to depend on C_{Me_i}. Indeed, Ravdel and Moshkevits[25] have shown that D_{Pb} and D_{Zn} decrease as Me_i concentration increases.

At the same time, according to Ignatova,[6] the dependence of D_{Me_i} (at a significance level of 0.05) on factors such as electrode radius, r, initial concentration of metal in amalgam, $C^0_{Me_i}$, time, τ, and amalgam viscosity, η, is complex and depends on the nature of the amalgamative metal. Diffusion coefficients D_{Me_i} (where Me_i = Cd, Zn, Sb) do not depend on r, $C^0_{Me_i}$ or τ, whereas D_{Cu} depends on C^0_{Cu} and η.[6]

Ignatova[6] offered a detailed analysis of diffusion coefficients of five metals, Cu, Zn, Cd, Sn and Sb, in mercury at 298 K (Table 3.3). These values of D_{Me_i} are the most credible of all known values. The dependence of D_{Me_i} on x_{Me_i} has also been measured in the Tl–Hg system.[34] This system is characterized by a high solubility of thallium in mercury (43.7 at.%) at 298 K (25 °C). The dependence of D_{Tl} on x_{Tl} was studied using the capillary method with

Table 3.2 Upper and lower limits for concentration-independent diffusion.[4]

Metal	*Me_i (lower) (mol L^{-1})*	*Me_i (upper) (mol L^{-1})*
Zn	1.6×10^{-4}	2.5×10^{-2}
K, Na	5.0×10^{-5}	4.0×10^{-4}
Cd	5.0×10^{-5}	1.2×10^{-2}
Mn	2.0×10^{-4}	1.3×10^{-2}
Cu, Tl	4.5×10^{-4}	2.8×10^{-3}

Table 3.3 Diffusion coefficients of Cu, Zn, Cd, Sn and Sb at 298 K.[6]

Metal	$D_{Me_i} \times 10^5$ *(cm² s⁻¹)*
Cu	0.98 ± 0.05
Zn	1.80 ± 0.06
Cd	1.60 ± 0.007
Sn	1.45 ± 0.026
Sb	1.63 ± 0.007

radioactive thallium tracer (^{204}Tl). The diffusion coefficient of thallium in thallium amalgam was calculated[34] using the equation

$$\frac{C_{\text{average}} - C_\infty}{C_1 - C_\infty} = \frac{8}{\pi^2} \sum_{n=0}^{n=\infty} \frac{1}{(2n+1)^2} \exp\left(\frac{2n+1}{2l}\right)^2 \pi^2 Dt \tag{3.5}$$

where C_{average} is the average composition of thallium amalgam in a capillary with filled end and length l ($d = 1$ mm, $l = 3$ cm), formed within time t. C_1 and C_∞ are the initial concentration of radioactive thallium in the capillary and in the volume, respectively. Under the experimental conditions described,[34] $C_\infty = 0$, so by measuring C_1, C_{average}, l and t, it is possible to calculate D_{Tl}. The diffusion coefficient of thallium decreases in a linear fashion from 0.98×10^{-5} cm^2 s^{-1} at 0.75 at.% to 0.47×10^{-5} cm^2 s^{-1} at 28.57 at.% thallium. A thallium concentration of 28.57 at.% corresponds to the compound Tl_2Hg_5 detected in liquid amalgam. At $x_{\text{Tl}} > 35$ at.%, the change of D_{Tl} is moderate, yet the curve D_{Tl}–x_{Tl} passes through a minimum. The same curve behavior has also been observed in the system K–Hg, in which with increasing concentration of potassium in the amalgam, D_K decreases and reaches a minimum at $x_K = 33.3$ at.%.[32] Such behavior of D_{Me_i} demonstrated by thallium and potassium, according to Galus[7] and Guminski,[19] indisputably proves the existence of the compounds Tl_2Hg_5 and KHg_2 in the liquid phase. Earlier we offered other evidence confirming this point.[26–28,35]

We should also consider the steady increase of diffusion coefficients D_{Me_i} with increasing concentration of diffusing metal Me_i. Thus, a study of diffusion coefficients of indium in the system In–Hg at $x_{\text{In}} = 3.5$–20.0 at.% revealed a rise in D_{In}[36] with increasing amalgam concentration, according to the equation

$$D_{\text{In}} = 2.71 \times 10^{-3}(1 + 2.66x_{\text{In}})\exp\left(-\frac{326a}{RT}\right) \tag{3.6}$$

Foley and Reid[30] suggested that diffusion involves particles that are more complex than atoms of indium or ions. Indeed, from the literature,[37,38] it follows that in the In–Hg system the compound $InHg_6$ ($T_{\text{melt}} = 257.5$ K) has, at 258 K, already dissociated according to the equation

$$2InHg_6 \rightarrow InHg_3 + InHg + 8Hg \tag{3.7}$$

Table 3.4 Diffusion coefficients of metals in mercury (compiled by Galus[7]).

Metal	$D_M \times 10^5$ *(cm² s⁻¹)*	T (°C)	*Ref.*
Ag	1.05	25	36
Au	0.85	25	37
Bi	1.35	20	38
Ce	0.62	25	39
Cu	1.00	25	40
Ge	1.32	25	41
K	0.85	20	5
Mg	1.20	25	42
Mn	0.94	20	5
La	0.50	25	43
Pb	1.25	20	44
Sb	1.40	25	45
Sn	1.48	20	5
Zn	1.81	25	40

The compound $InHg_6$ exists only in the liquid state. Measurements of the electrical conductivity of indium amalgams[26] have shown that $InHg_3$ and InHg are the stable compounds. It has been reported[32] that the equilibrium constant of the reaction

$$InHg_3 \rightleftharpoons In + 3Hg \tag{3.8}$$

at 298 K is equal to 5.6×10^{-2} (mole fraction)[3]. Predel and co-workers[33,34] believed that in the In–Hg system there exist structural clusters of the two compounds $InHg_2$ and InHg. According to others,[34] the dissociation constant, k_d, of InHg:

$$InHg \rightleftharpoons In + Hg \tag{3.9}$$

is 0.26.

The system Al–Hg forms a degenerate eutectic phase diagram. Aluminum does not form compounds with mercury. However, it has been stated[41] that the system demonstrates an increase in the diffusion coefficient of aluminum with increasing amalgam concentration.

Table 3.4 gives recommended values of the diffusion coefficient of metals in mercury at ambient temperatures, compiled by Galus.[7]

3.4 Diffusion of Mercury in Solid Metals

Warburton and Turnbull[46] discussed the nature of the diffusion of noble and late transition metals in mercury. Diffusion often involves and interstitial mechanism. Table 3.5 gives the experimental results for mercury diffusion in Ag, Au, Cu, Pb, Sn and Zn. A diffusion mechanism by interstitial–vacancy pairs was developed to explain the results of diffusion of Hg in Pb.[47]

Table 3.5 Diffusion of mercury in solid pure metals.

Metal	T *(°C)*	T *(K)*	D_0 *($cm^2 s^{-1}$)*	Q *($kJ mol^{-1}$)*	*Ref.*
Ag			0.08	159.4	49
Au	600–1000	873–1273	0.116 + 0.13–0.06	156.4 ± 6.7	50
Cu			0.35	184.1	49, 54
Pb	200–300	473–673	1.05 ± 0.24	95.0 ± 0.84	51
Sn[110]	175–232	448–505	30 + 20–12	112.1 ± 2.1	52
Sn[001]	175–232	448–505	7.5 + 6.4–3.5	105.9 ± 2.5	52
Zn‖	260–413	533–686	0.056 ± 0.002	82.4 ± 0.2	53
Zn⊦	260–423	533–686	0.073 ± 0.006	84.4 ± 0.4	53

Shires *et al.*[48] reported Hg diffusion in the solid amalgam phase Ag_2Hg_3. They determined the activation energy for volume diffusion to be 30.3 kJ mol^{-1} and the diffusion prefactor to be 1.2×10^{-4} cm^2 s^{-1} at temperatures between 50 and 102 °C. In the same study, the grain boundary diffusion rate constants were determined as $Q_D = 19.1$ kJ mol^{-1} and $D_0 = 22.8$.

References

1. A. G. Stromberg and E. A. Zakharova, *Zh. Fiz. Khim.*, 1966, **40**, 81.
2. V. P. Gladyshev, *Elektrokhimiya*, 1971, **7**, 1423.
3. M. S. Zakharov, *Zh. Fiz. Khim.*, 1965, **39**, 509.
4. X. S. Ma and H. Kao, *J. Electroanal. Chem.*, 1983, **151**, 179.
5. A. Baranski, S. Fitek and Z. Galus, *J. Electroanal. Chem.*, 1975, **60**, 175.
6. L. A. Ignatova, Issled. mech. kin. processes anodnogo rastvoreniya amalgam organ. ob'ema metodom amalgam. Chronoamperometrii, dissertation Kand. Khim. Nauk, Tomsk, 1978.
7. Z. Galus, Diffusion coefficients of metals in mercury, *Pure Appl. Chem.*, 1984, **56**, 635.
8. V. M. M. Lobo and R. Mills, *Electrochim. Acta*, 1982, **27**, 969.
9. R. P. Chhabra, *Metall. Trans.*, 1986, **17A**, 355.
10. V. N. Korshunov, *Amalgam Systems: Equilibrium Potentials of Simple Amalgams*, Izdatelstvo Moskovskogo Univesiteta, Moscow, 1990.
11. J. C. Slater, *Quantum Theory of Molecules and Solids. Symmetry and Energy Bonds in Crystals*, McGraw-Hill, New York, 1965, vol. 2.
12. W. Van der Lugt and W. Geerstma, *Can. J. Phys.*, 1987, **65**, 326.
13. V. N. Korchunov, *Elektrokhimiya*, 1981, **17**, 295.
14. J. Hildebrand, *J. Am. Chem. Soc.*, 1913, **35**, 501.
15. R. S. Nigmetova, V.F. Kiselev and L.F. Kozin, *Zh. Fiz. Khim.*, 1982, **56**, 2080.
16. R. S. Nigmetova, A. K. Kulikovskii, V. F. Kiselev and L. F. Kozin, *Zh. Fiz. Khim.*, 1986, **60**, 1782.

17. O. J. Kleppa, M. Kaplan and C. E. Thalmayer, *J. Phys. Chem.*, 1961, **65**, 843.
18. J. H. Hildebrand, *Viscosity and Diffusivity: a Predictive Treatment*, Wiley-Interscience, New York, 1977.
19. C. Guminski, *J. Mater. Sci.*, 1989, **24**, 2661.
20. D. K. Belachenko, *Yavlenia Perenosa v Zhidkikh Metal i Poluprovodihakh*, Atomizdat, Moscow, 1970.
21. L. F. Kozin, *Physico-Chemishkie Osnovi Amalgamnoi*, Metallurgia, Alma-Ata, 1964.
22. A. R. Regel' and S. I. Patyanin, *Atomic Energy*, 1963, **14**, 114.
23. C. Guminski and Z. Galus, *Metals in Mercury, Solubility Data Series*, ed. C. Hirayama, Pergamon Press, Oxford, 1986.
24. J. B. Edwards, E. E. Hucke and J. J. Martin, *J. Electrochem. Soc.*, 1968, **115**, 488.
25. A. A. Ravdel and A. S. Moshkevits, *Zh. Prikl. Khim.*, 1971, **44**, 178.
26. W. T. Foley and M. T. H. Lui, *Can. J. Chem.*, 1964, **42**, 2607.
27. L. F. Kozin, *Amalgam Pyrometallurgy*, Nauka, Alma-Ata, 1973.
28. L. F. Kozin, *Physicochemical Research of Amalgam Methods Used to Obtain High-Purity Metals: Synopsis of Thesis*, Doctoral of Chemical Sciences, Alma-Ata, 1964.
29. L. F. Kozin, R. S. Nigmetova and M. B. Dergacheva, *Thermodynamics of Binary Amalgam Systems*, Nauka, Alma-Ata, 1979.
30. W. T. Foley and L. E. Reid, *Can. J. Chem.*, 1963, **41**, 1782.
31. V. P. Shvedov, A. Z. Frolkov and G. D. Nikitin, *Radiokhimii*, 1971, **13**, 252.
32. L. F. Kozin and M. B. Dergacheva, *Zh. Fiz. Khim.*, 1969, **43**, 249.
33. B. Predel and D. Rothacker, *Acta Met.*, 1967, **15**, 135.
34. B. Predel and G. Oehme, *Z. Metallkd.*, 1974, **65**, 525.
35. S. Ziegel, E. Peled and E. Gileadi, *Electrochim. Acta*, 1979, **24**, 513.
36. A. W. Castleman Jr and J. J. Conti, *Phys. Rev. A*, 1970, **2**, 1975.
37. C. Guminski and Z. Galus, *J. Electroanal. Chem.*, 1977, **83**, 139.
38. H. Kao and C. G. Chang, *Acta Sci. Nat. Univ. Nanking*, 1965, **9**, 326.
39. K. Zh. Sagadeva, A. I. Zebreva and T. L. Badamova, *Elektrokhimiya*, 1979, **15**, 210.
40. E. A. Zakharova, G. A. Kataev, L. A. Ignateva and V. E. Morozova, *Tr. Tomsk. Gos. Univ.*, 1973, **249**, 103.
41. Z. J. Karpinski and Z. Kublik, *J. Electroanal. Chem.*, 1977, **81**, 53.
42. P. C. Mangelsdorf Jr, *Proc. Met. Soc. Conf.*, 1961, **7**, 429.
43. K. Zh. Sagadeva, A. I. Zebreva, R. M. Dzholdasova and B. M. Isaberlina, *Elektrokhimiya*, 1977, **13**, 1378.
44. W. Cohen and H. R. Bruins, *Z. Phys. Chem.*, 1924, **109**, 397.
45. A. G. Stromberg and E. A. Zakharova, *Elektrokhimiya*, 1965, **1**, 1036.
46. W. K. Warburton and D. Turnbull, *Thin Solid Films*, 1975, **25**, 71.

47. W. K. Warburton, *Phys. Rev. B*, 1973, **7**, 1341.
48. P. J. Shires, A. L. Hines and T. Okabe, *J. Appl. Phys.*, 1977, **48**, 1734.
49. D. Lazarus, *Solid State Phys.*, 1960, **10**, 71.
50. A. J. Mortlock and A. H. Rowe, *Philos. Mag.*, 1965, **11**, 1157.
51. W. K. Warburton, *Phys. Rev. B*, 1973, **7**, 1330.
52. W. K. Warburton, *Phys. Rev. B*, 1972, **6**, 2161.
53. A. P. Batra and H. B. Huntington, *Phys. Rev.*, 1967, **154**, 569.
54. A. Sawatzky and F. E. Jaumot Jr, *J. Met.*, 1957, **9**, 1207.

CHAPTER 4

Purification of Mercury Using Chemical and Electrochemical Methods

4.1 Technical Requirements for Mercury

Mercury is obtained by burning mercury concentrates that contain cinnabar at 773 K. Elementary mercury formed in the process evaporates and then condenses into dedicated chambers.[1] The crude mercury is rich with impurities that come to it in the course of cinnabar smelting, since the cinnabar typically contains iron, chromium, silver, gold, copper, zinc, lead, cobalt, nickel, antimony, arsenic, manganese and other substances.[2] The amount and nature of the impurities depend on the ore where the mercury concentrate was mined. Cadmium, platinum and tin impurities may also be present. Crude organic substances and various gases may be introduced into mercury during the pyrometallurgical processing of cinnabar. The solubility of these substances and gases depends on the temperature.

According to a Russian national standard, the quality of mercury is determined by measurement of the amount of non-volatile residues remaining when subliming mercury after filtration through chamois. The sublimation is carried out in a porcelain crucible, then the residue is baked at 500 °C until its mass is stable. The amount of primary substance and the non-volatile residue should meet the requirements indicated in Table 4.1.

No national standards (GOSTs) have been established for high-purity mercury. The Nokitosk Mercury Plant produced high-purity mercury in compliance with the specification of P10-6 grade. In this case, the content of impurity metals (bismuth, zinc, silver, copper, nickel, iron, gallium, titanium, manganese) is 1×10^{-7}–5×10^{-8}% (by mass). The Experimental Chemical–Metallurgical Plant Giredmeta also manufactured high-purity

Mercury Handbook: Chemistry, Applications and Environmental Impact
By Leonid F Kozin and Steve Hansen

Published by the Royal Society of Chemistry, www.rsc.org

Table 4.1 Requirements for different grades of mercury.

Mercury grade	*Mercury content, not less than (%)*	*Non-volatile residue content, not less than (%)*
P0	99.9992	0.0008
P1	99.999	0.001
P2	99.990	0.010
P3	99.900	0.100

mercury of RVCh-1 grade. Impurities in the high-purity mercury (%) were as follows:

Iron	1×10^{-6}
Tin	8×10^{-7}
Lead	3×10^{-6}
Manganese	3×10^{-7}
Nickel	1×10^{-6}
Chromium	1×10^{-6}
Silver	3×10^{-7}
Cobalt	2×10^{-6}

Many methods of treatment for obtaining high-purity mercury are known. These methods combine various technologies based on the use of different physical and chemical properties of mercury. The most common are complex technologies that include chemical and pyrometallurgical processes, distillation and fractionation in a carrier gas flow or in vacuum, electrolysis and zone melting. Here we review some of these processes and, based on analysis of the results achieved, select the best combination. The remaining methods are discussed in detail elsewhere.[3]

4.2 Chemical Methods for Mercury Treatment

Chemical methods for mercury treatment are widely applied both in laboratories and in industry.[1,4–15] There are a number of chemical methods that can be separated into dry[1,7,9,10,13] and wet[1,4–15] methods. The dry methods are based on oxidation of the impurities and their transformation to oxides, which are then removed through filtration.[7,8,11] Air oxygen, pure oxygen or ozone is used as the oxidant. When using oxygen, the process is run at 423 K (150 °C) by blowing compressed air or treated oxygen[10] through liquid mercury or combining this process with mercury distillation in vacuum.[16] However, oxidation of impurities in mercury by air oxygen is slow. In a method reported by Kuzmenkov,[10] oxygen from which traces of organic substances had been removed was blown at 323 K through four quartz vessels containing mercury and connected in series. This process was run for 24 h and the oxides that were formed were removed by filtration. However, such methods do not allow for removal of electrically negative impurities. Generally, the degree of mercury purification from impurities is low for these methods. To improve the treatment efficiency, a layer of HNO_3 is poured on the top of mercury and air is passed through the mercury for 1 week. Subsequently, traces of the acid are washed out and the mercury is dried[17] and then distilled several times in an air flow at 543 K using the procedure described by Hulett and Minchin.[18] To accelerate the oxidation reaction, small droplets of mercury are passed five times through a column (1.5 m × 5 cm i.d.) filled with $HNO_3 + Hg_2(NO_3)_2$, then the mercury is washed, dried and distilled three times in a stream of oxygen.[19] According to Moore,[20] the column (1.5 m × 5 cm i.d.) was charged with an 8% solution of HNO_3, whereas Spicer and Banick[21] used a 40% solution of HNO_3. In the latter case, air was blown through mercury for several hours to oxidize the impurities then, in order to remove the oxides, the mercury was filtered, passed through a column containing a 40% solution of HNO_3, washed to remove the acid, dried and distilled in vacuum.[21] The treatment process proceeds faster if the air to be used is heated to 423–433 K and contains vapors of acids.[1,9,10,22]

Tin is completely removed by blowing hot air, that has been passed through a vessel containing fuming HCl, through mercury for 12 h. Lead is removed by passing air at 423 K through mercury.[1,10] However, the process efficiency is low in this case; further, this method does not allow for removal of impurities that are electrically positive with respect to mercury. According to Melnikov,[1] a major shortfall of this method is the need for ultrapurification of large amounts of released gases from mercury to reduce mercury losses and protect the environment.

Much more productive is the ozone-based dry process, which also has simple implementation requirements. Ozone has high oxidative activity; it not only oxidizes metal and sulfide impurities, but also decomposes mercury-soluble organic compounds.[21] In this case, treatment is effective for electrically negative impurities (zinc, lead) but not so effective for precious metals. Thus, the silver content could only be reduced from 1×10^{-3} to 4×10^{-4}%.[1] In real situations, when ozone is used as oxidant of mercury impurities, a weak nitric acid solution is introduced into the mercury reactor.[1]

Wet chemical treatment methods include the following sequence of operations: (a) removal of mechanical impurities; (b) removal of organic compounds; (c) removal of metallic impurities; and (d) washing and removal of traces of moisture.

Mechanical impurities, which normally remain on the surface of mercury, are removed by filtering through porous barriers (chamois, cheese cloth, layers of gauze, filter-paper perforated with a thin quartz needle) or a funnel with an in-built capillary tube bent 45–60° off-horizontal.[7] The filtration rate may be increased by applying negative pressure. The implementation of the filtration process was described by Pugachevich.[7] To reduce spattering, the filtering operation may be combined with a process for organic compound removal by filtering mercury into a concentrated solution of sodium hydroxide or potassium hydroxide (20–30%)[7,8] contained in a column assembly.

The column has the following operating principle. Mercury is run many times through the column containing the alkaline solution. Often, to remove organic impurities, mercury lying below a layer of the alkaline solution is rapidly stirred with a mechanical mixer, shaken or exposed to ultrasound. Following the removal (saponification) of organic compounds, mercury is washed with distilled water to remove alkali and then treated with acidic solutions of oxidants, *e.g.* nitric acid, to remove metallic impurities. Then mercury is vacuum distilled or electrochemically purified.

The standard potentials of mercury half-reactions, E^0, as shown in Chapter 5, are as follows:

$$E^0_{Hg_2^{2+}/Hg^0} = 0.7973\ V$$

$$E^0_{Hg^{2+}/Hg^0} = 0.854\ V$$

$$E^0_{Hg^{2+}/Hg_2^{2+}} = 0.920\ V$$

i.e., they are electropositive. Therefore, mercury ions (Hg_2^{2+},Hg^{2+}) are strong complex-forming agents since they create complex compounds with many non-organic and organic ligands (see Chapter 5). Accordingly, E^0 of mercury half-reactions depends on the nature of ligands present in a solution. Table 4.2 gives standard potentials E^0 of mercury for different solutions.

Table 4.2 Standard electrode potentials of half-reactions of mercury at 298 K (*versus* NHE).

Half-reaction	E^0 *(V)*	*Ref.*
$Hg + H^+ + e \rightleftharpoons HgH$	–2.281	23
$HgS + 2e \rightleftharpoons Hg + S^{2-}$	–0.70	24
$HgS + 2e \rightleftharpoons Hg + S^{2-}$	–0.69	28
$Hg(CN)_4^{2-} + 2e \rightleftharpoons Hg + 4CN^-$	–0.370	24,27,28
$Hg_2(CN)_2 + 2e \rightleftharpoons Hg + 2CN^-$	–0.360	24
$Hg(SCN)_4^{2-} + 2e \rightleftharpoons Hg + 4SCN^-$	–0.092	23
$Hg_2(SCN)_2 + 2e \rightleftharpoons Hg + 2SCN^-$	+0.22	24
$Hg_2O + H_2O + 2e \rightleftharpoons Hg + 2OH^-$	+0.123	24–26
$HgO + H_2O + 2e \rightleftharpoons 2Hg + 2OH^-$	+0.098	27
$Hg_2I_2 + 2e \rightleftharpoons 2Hg + 2I^-$	–0.0405	24,28
$HgI_4^{2-} + 2e \rightleftharpoons Hg + 4I^-$	–0.038	24,28
$Hg_2Br_2 + 2e \rightleftharpoons 2Hg + 2Br^-$	0.1397	28,26
$HgBr_4^{2-} + 2e \rightleftharpoons Hg + 4Br^-$	0.223	28
$Hg(OH)_3^- + 2e \rightleftharpoons Hg + 3OH^-$	0.231	25
$Hg_2Cl_2 + 2e \rightleftharpoons 2Hg + 2Cl^-$	0.267	24
$HgCl_4^{2-} + 2e \rightleftharpoons Hg + 4Cl^-$	0.380	24
$Hg_2C_2O_4 + 2e \rightleftharpoons 2Hg + C_2O_4^{2-}$	0.417	24,26
$Hg_2CrO_4 + 2e \rightleftharpoons 2Hg + CrO_4^{2-}$	0.540	26
$Hg_2SO_4 + 2e \rightleftharpoons 2Hg + SO_4^{2-}$	0.6151	24,26
$2HgCl_2 + 2e \rightleftharpoons Hg_2Cl_2 + 2Cl^-$	0.630	24
$Hg_2HPO_4 + H^+ + 2e \rightleftharpoons 2Hg + H_2PO_4^-$	0.638	25,26
$HgF^+ + 2e \rightleftharpoons Hg + F^-$	0.602[a]	23

[a]Our calculation.

As can be seen from Table 4.2, the standard electrode potentials of mercury in the presence of ligands that create strongly complexing compounds with mercury (SCN^-, CN^-, S^{2-}, Na_2S_2, Na_2S_n, Cl^-, Br^-, I^- in excessive amounts), and also in equimolar solutions of halides due to the formation of poorly soluble mercury salts (Hg_2^{2+}), shift the equilibrium toward the electronegative side, with the half-reaction potentials E^0 in strongly complex-forming environments (SCN^-, CN^-, S^{2-}) changing by as much as 1.07–1.55 V. Therefore, in order to remove impurities that are more negative and stand higher in the electrochemical series with respect to mercury (Table 4.3), simple electrolytes must be used (HNO_3, $HClO_4 + HNO_3$, H_2SO_4, CCl_3–COOH, dilute $HCl + HNO_3$, *etc.*), the anions (ligands) of which do not form stable complexes with mercury ions Hg_2^{2+} and Hg^{2+}. Metallic impurities are removed from mercury via exchange reactions:

$$\frac{z}{2}Hg_2^{2+} + Me_i(Hg)_x \rightleftharpoons Me_i^{z+} + (x+2)Hg \tag{4.1}$$

$$\frac{z}{2}Hg^{2+} + Me_i(Hg)_x \leftrightarrow Me_i^{z+} + (x+1)Hg \tag{4.2}$$

and include metals and metalloids more negative than mercury (As, Cu, Bi, Sb, Pb, Sn, Ni, Co, Tl, In, Cd, Fe, Cr, Ga, Zn, Mn, Al) and other more electronegative impurities.

Comparison of the data in Tables 4.2 and 4.3 shows that in complex forming media and also in chloride, bromide and iodide solutions, the difference in potentials

$$\Delta E = E_{Hg_i^{2+}/Hg^0} - E_{Me^{z+}+Me^0}\ (i = 1, 2)$$

decreases or even becomes negative for some electrolytes. The nature of the electrolytes affects the electrode potentials of mercury and metals in different ways (compare Tables 4.2 and 4.3). An example is the way in which the potentials of gold change compared with mercury in cyanide solutions (CN^-) and solutions formed by simple (H_2O) electrolytes (Table 4.4).

It can be seen that $\Delta E_{Me_2^{2+}/Me^0}$ and $\Delta E_{Me^{z+}/Me^0}$ for mercury and gold are 1.14 and 2.31 V, respectively; whereas gold, more positive in the solutions of simple electrolytes compared with mercury, acquires a negative potential in cyanide electrolytes. Chapter 5 demonstrates that the introduction of ligands into the solution of the system $Hg_{liquid}{}^0$–Hg_2X_2–HgX_2–HY–H_2O (where X is a halide and Y is an anion ligand) results in convergence of standard potentials and a shift of half-reaction potentials (see Tables 4.2, 4.3 and 5.4) towards the negative side. At the same time, K_{reprop} values decrease and $K_{disprop}$ values (as $K_{disprop} = 1 / K_{reprop}$) increase and conversion of potentials takes place:

$$E^0_{Hg^{2+}/Hg_2^{2+}} < E^0_{Hg_2^{2+}/Hg} > E^0_{Hg^{2+}/Hg}$$

and lower valence ions of mercury become unstable in the system $Hg_{liquid}{}^0$–Hg_2X_2–HgX_2–HY–H_2O.

Table 4.3 Standard electrode potentials of half-reactions of metals at 298 K (*versus* NHE).[23–25,27–29]

Half-reaction	E^0 *(V)*	*Half-reaction*	E^0 *(V)*
$Ti^{2+}+2e=Ti$	–1.750	$Cu(CN)_2^-+e=Cu+2CN^-$	–0.430
$Be^{2+}+2e=Be$	–1.700	$Cd^{2+}+2e=Cd$	–0.404
$Al^{3+}+3e=Al$	–1.670	$Pd(CN)_4^{2-}+2e=Pd+4Cn^-$	–0.400
$Zr^{4+}+4e=Zr$	–1.530	$PbI_2+2e=Pb+2I^-$	–0.365
$ZnS+2e=Zn+S^{2-}$	–1.440	$Cu_2O+H_2O+2e=2Cu+OH^-$	–0.361
$Zn(CN)_4^{2-}+2e=Zn+4CN^-$	–1.260	$In^{3+}+3e=In$	–0.340
$Zn(OH)_2+2e=Zn+2OH^-$	–1.245	$Tl^++e=Tl$	–0.338
$CdS+2e=Zn+S^{2-}$	–1.230	$PtS+2H^++2e=Pt+H_2S$	–0.300
$V^{2+}+2e=V$	–1.180	$Ag(CN)_2^-+e=Ag+2CN^-$	–0.300
$Nb^{3+}+3e=Nb$	–1.100	$CuCNS+e=Cu+CNS^-$	–0.270
$Mn^{2+}+2e=Mn$	–1.050	$PbCl_2+2e=Pb+2Cl^-$	–0.268
$Tl_2S+2e=2Tl+S^{2-}$	–1.040	$CuS+2H^++2e=Cu+H_2S$	–0.259
$FeS(a)+2e=Fe+S^{2-}$	–1.010	$Sb_2O_3+6H^++6e=2Sb+3H_2O$	–0.255
$In(OH)_3+3e=In+3OH^-$	–1.000	$Ni^{2+}+2e=Ni$	–0.250
$PbS+2e=Pb+S^{2-}$	–0.980	$SnF_6^{2-}+4e=Sn+6F^-$	–0.250
$SnS+2e=Sn+S^{2-}$	–0.940	$Cu(OH)_2+2e=Cu+2OH^-$	–0.224
$Cd(CN)_4^{2-}+2e=Cd+4CN^-$	–0.900	$CuI+e=Cu+I^-$	–0.187
$Rh(CN)_6^{2-}+e=Rh(CN)_3^{2-}+CN^-$	–0.900	$AgI+e=Ag+I^-$	–0.151
$Cr^{2+}+2e=Cr$	–0.900	$Sn^{2+}+2e=Sn$	–0.140
$NiS(a)+2e=Ni+S^{2-}$	–0.860	$Pb^{2+}+2e=Pb$	–0.126
$SbS_2^-+3e=Sb+2S^{2-}$	–0.850	$OsO_2{\cdot}2H_2O+4e=Os+4OH^-$	–0.120
$PtS+2e=Pt+S^{2-}$	–0.830	$WO_3+6H^++6e=W+3H_2O$	–0.090
$Ni(CN)_4^{2-}+2e=Ni(CN)_3^{2-}+CN^-$	–0.820	$O_2+H_2O+2e=HO_2^-+OH^-$	–0.076
$Cd(OH)_2+2e=Cd+2OH^-$	–0.810	$AgCN+e=Ag+CN^-$	–0.040
$Zn^{2+}+2e=Zn$	–0.762	$RuO_2+2H_2O+4e=Ru+4OH^-$	–0.040
$TlI+e=Tl+I^-$	–0.760	$CuI_2+e=Cu+2I^-$	0.0
$CuS+2e=Cu+S^{2-}$	–0.760	$HOsO_5^-+4H_2O+8e=Os+9OH^-$	0.000
$CrCl_2^++3e=Cr+2Cl^-$	–0.740	$CuBr+e=Cu+Br^-$	0.033
$Co(OH)_2+2e=Co+2OH^-$	–0.730	$Rh_2O_3+3H_2O+6e=2Rh+6OH^-$	0.040
$Cr^{3+}+3e=Cr$	–0.710	$CuBr_2^-+e=Cu+2Br^-$	0.050
$HgS+2e=Hg+S^{2-}$	–0.700	$Pd(ON)_2+2e=Pd+2OH^-$	0.070
$TlBr+e=Tl+Br^-$	–0.658	$AgBr+e=Ag+Br^-$	0.073
$Au(CN)_2^-+e=Au+2CN^-$	–0.600	$Pt(CN)_4^{2-}+2e=Pt+4CN^-$	0.090
$ReO_4^-+4H_2O+7e=Re+8OH^-$	–0.584	$HgO+H_2O=Hg+2OH^-$	0.098
$PbO+H_2O+2e=Pb+2OH^-$	–0.578	$Pd(OH)_2+2e=Pd+2OH^-$	0.100
$PbS+H_2O+2e=Pb+OH^-+SH^-$	–0.560	$Ir_2O_3+3H_2O+6e=2Ir+6OH^-$	0.100
$TlCl+e=Tl+Cl^-$	–0.557	$CuCl+e=Cu+Cl^-$	0.124
$Cu_2S+2e=2Cu+S^{2-}$	–0.540	$Hg_2Br_2+2e=2Hg+2Br^-$	0.139
$Ga^{3+}+3e=Ga$	–0.520	$Pd(SCN)_4^{2-}+2e=Pd+4SCN^-$	0.140
$BiOOH+H_2O+3e=Bi+3OH^-$	–0.460	$ReO_4+8H^++7e=Re+4H_2O$	0.150
$Fe^{2+}+2e=Fe$	–0.441	$Sb_2O_3+6H^++e=2Sb+3H_2O$	0.152
$BiCl_4^-+3e=Bi+4Cl^-$	0.160	$RuCl_5^{2-}+3e=Ru+5Cl^-$	0.600
$BiOCl+2H^++3e=Bi+H_2O+Cl^-$	0.160	$RuO_2+4H^+=Ru+2H_2O$	0.680
$Cu^{2+}+e=Cu^+$	0.167	$IrCl_6^{3-}+3e=Ir+6Cl^-$	0.720
$PdI_4^{2-}+2e=Pd+4I^-$	0.180	$PtCl_4^{2-}+2e=Pt+4Cl^-$	0.730
$CuCl_2^-+e=Cu+2Cl^-$	0.190	$Fe^{3+}+e=Fe^{2+}$	0.771
$Pt(OH)_2+2e=Pt+2OH^-$	0.200	$Hg_2^{2+}+2e=2Hg$	0.789

Table 4.3 (*Continued*)

Half-reaction	E^0 *(V)*	*Half-reaction*	E^0 *(V)*
$HgBr_4^{2-}+2e=Hg+4Br^-$	0.210	$Ag^++e=Ag$	0.799
$AgCl+e=Ag+Cl^-$	0.222	$Rh^{3+}+3e=Rh$	0.800
$Hg_2Cl_2+2e=2Hg+2Cl^-$	0.244	$OsO_4+8H^++8e=Os+4H_2O$	0.840
$Ru^{3+}+e=Ru^{2+}$	0.249	$IrCl_6^{3-}+3e=Ir+6Cl^-$	0.860
$Hg_2Cl_2+2e=2Hg+2Cl^-$	0.267	$AuBr_4^-+3e=Au+4Br^-$	0.870
$Cu^{2+}+2e=Cu$	0.345	$2Hg^{2+}+2e=Hg_2^{2+}$	0.920
$HgCl_4^{2-}+2e=Hg+4Cl^-$	0.380	$PdCl_6^{2-}+4e=Pd+6Cl^-$	0.920
$PtI_4^{2-}+2e=Pt+4I^-$	0.400	$AuBr_4^-+e=Au+2Br^-$	0.960
$PtI_6^{2-}+4e=Pt+6I^-$	0.400	$Pd^{2+}+2e=Pd$	0.987
$RhCl_6^{3-}+3e=Rh+6Cl^-$	0.440	$AuCl_4^-+3e=Au+4Cl^-$	1.000
$AuI+e=Au+I^-$	0.500	$RuO_4+8H^++8e=Ru+4H_2O$	1.040
$Cu^++e=Cu$	0.522	$RhO_4^{2-}+8H^++6e=Rh+4H_2O$	1.100
$Hg_2CrO_4+2e=2Hg+CrO_4^{2-}$	0.540	$AuCl_2^-+e=Au+2Cl^-$	1.130
$Te^{4+}+4e=Te$	0.568	$Pt^{2+}+2e=Pt$	1.200
$PtBr_4^{2+}+2e=Pt+4Br^-$	0.580	$RhCl_6^{2-}+e=RhCl_6^{3-}$	1.200
$PdCl_4^{2-}+2e=Pd+4Cl^-$	0.590	$Au^{3+}+2e=Au^+$	1.290
$OsCl_6^{3-}+3e=Os+6Cl^-$	0.600	$Au^{3+}+3e=Au$	1.500
$PdBr_4^-+2e=Pd+4Br^-$	0.600	$Au^++e=Au$	1.700

Table 4.4 Change in potentials (V) of gold and mercury in different solutions.

Potential	H_2O	CN^-	ΔE *(V)*
$E_{Hg_2^{2+}/Hg^0}$	0.789	–0.360	1.14
E_{Au^+/Au^0}	1.70	0.61	2.31
E_{Hg^{2+}/Hg^0}	0.854	–0.370	1.22
E_{Au^{3+}/Au^0}	1.50	–0.810	2.31

This is why chemical treatment methods commonly rely on solutions of nitric acid, nitrates and perchlorates of mercury(I) and -(II). Exchange reactions (4.1) and (4.2) take place upon contact between mercury-containing impurities and a mercury salt solution. Equilibrium exchange reactions (4.1) between ions of mercury Hg_2^{2+} and more electronegative impurities may be described by the equation

$$K_p = \frac{a_{Me^{z+}} a_{Hg}^{x+2}}{a_{Me(Hg)} a_{Hg_2^{2+}}^{z/2}} \tag{4.3}$$

Equilibrium in the system $Me(Hg)/Hg_2^{2+}$ is reached if there are equal potentials $E_{Hg^{2+}/Hg} = E_{Me^{z+}/Me^0} = E$:

$$E_{Hg_2^{2+}/Hg^0} = E^0_{Hg_2^{2+}/Hg^0} + \frac{RT}{2F}\ln\left[\frac{C_{Hg_2^{2+}}\gamma_{Hg_2^{2+}}}{C_{Hg^0}\gamma_{Hg^0}}\right] \tag{4.4}$$

$$E_{Hg_2^{2+}/Hg^0} = E^0_{Hg_2^{2+}/Hg^0} + \frac{RT}{2F} \ln\left[\frac{C_{Me^{z+}}\gamma_{Me^{z+}}}{C_{Hg^0}\gamma_{Hg^0}}\right] \quad (4.5)$$

where γ_i are activity coefficients.

By subtracting eqn (4.5) from eqn (4.4) and considering that $[C_{Hg_2^{2+}}\gamma_{Hg_2^{2+}}C_{Me^0}\gamma_{Me}/C_{Me^{z+}}\gamma_{Me^{z+}}C_{Hg^0}C_{Hg^0}] = K_p$, one finds the equilibrium constant to be[4,30]

$$\ln K_p = \frac{(2-z)FE + zFE^0_{Me^{z+}/Me^0} - 2FE^0_{Hg_2^{2+}/Hg^0}}{RT} + \ln\left(\frac{\gamma_{Hg_2^{2+}}\gamma_{Me^0}}{\gamma_{Me^{z+}}\gamma_{Hg^0}}\right) \quad (4.6)$$

At $\gamma_{Hg_2^{2+}} = \gamma_{Me^{z+}} = 1$, $\gamma_{Me^0} = \gamma_{Hg^0}$ and $z = 2$, eqn (4.6) is reduced to[4,30]

$$\ln K_p = \frac{2F}{RT}\left(E^0_{Me^{z+}/Me^0} - E^0_{Hg_2^{2+}/Hg}\right) \quad (4.7)$$

Equation (4.7) shows that larger potential differences have a greater equilibrium constant, K_p, of the exchange reaction. For exchange reaction (4.2), we obtain equations appearing similar to eqns (4.6) and (4.7) if $E^0_{Hg_2^{2+}/Hg}$ is replaced with $E^0_{Hg^{2+}/Hg}$ and $\gamma_{Hg_2^{2+}}$ is replaced with $\gamma_{Hg^{2+}}$.

The principles of exchange reactions in systems with liquid mercury and amalgam electrodes have been set out.[4,6,31–33] Exchange reactions described by eqns (4.1) and (4.2) occur at a very high rate. In reactions (4.1) and (4.2), the exchange process normally obeys first-order kinetics:

$$\ln\frac{C}{C_0} = kt \quad (4.8)$$

where k is the rate constant and t is time and depends on the nature of the metals and the electrolyte.

Kinetic rates of exchange reactions of Hg_2^{2+} ions in a nitrate solution [0.5 M $Hg_2(NO_3)_2 + 2$ M HNO_3] with amalgams of copper, lead and cadmium containing 0.1 at.% metal are given in ref. 3. The exchange rate increases in the sequence Cu→Pb→Cd ($k_{Cu} = 1.1\times10^{-4}$, $k_{Pb} = 2.1\times10^{-4}$, $k_{Cd} = 4.0\times10^{-4}$ s^{-1}). The same sequence is characterized by increasing solubility of the metals in mercury (in the experiments copper amalgam was heterogeneous) and increasing atomic and ionic radii. According to Kozin,[4] the exchange reaction rate constant depends on the diffusion coefficient (D), Hg(Me_i)/solution phase boundary (S), diffusion layer thickness (δ) and reaction system volume (ν):

$$k = \frac{DS}{\delta\nu} \quad (4.9)$$

Therefore, increases in D and S and decreases in δ and ν yield higher exchange reaction rates. To reduce δ, one needs to keep the solution in active motion near the surface of mercury.[4] The phase exchange rate depends on the hydrodynamic conditions of the solution's motion relative to mercury.

It will be observed that if crude mercury containing a range of metal impurities is used, exchange reactions take place not only between ions of mercury and the impurities, but also between electropositive impurities contained in the solution and electronegative impurities contained in mercury according to the reaction

$$Me_1^{z+} + Me_2(Hg) \rightleftharpoons Me_2^{z+} + Me_1(Hg) \qquad (4.10)$$

In this case, the exchange rate also depends on the experimental conditions. The rate of exchange reaction in $\log(C/C_0)$ *versus* t and η *versus* t coordinates as a function of the stirring rate of cadmium sulfate solution and zinc amalgam has been reported.[34] The exchange rate depends strongly on the stirring rate. The curves depicting the isolation of cadmium from the solution as a function of contact time between zinc amalgam and the solution (η *versus* t) suggest a logarithmic relationship. Indeed, the $\log(C/C_0)$ *versus* t relationship is linear, which points to diffusion control of the phase exchange.

The relationship between the concentration of cadmium ions and the rate of exchange ratios has also been reported.[34] The exchange reaction rate constant does not depend on the concentration of cadmium ions in the electrolyte. The data indicate that higher temperatures cause higher exchange reaction rates. At 283 K, 99.9% of cadmium ions take as long as 385 min to engage in an exchange reaction with zinc amalgam, whereas at 348 K the same process takes only 48 min. The phase exchange rate constants (s^{-1}) are as follows:

283 K	4.6×10^{-5}
298 K	9.1×10^{-51}
323 K	1.6×10^{-41}
348 K	2.9×10^{-4}

The temperature dependence of the effective constants of exchange rates is used to calculate the activation energies of exchange reactions. Experimentally determined exchange reaction rate constants as functions of the reciprocal of temperature can easily fit a straight line. The expected activation energy of the exchange reaction between cadmium ions and zinc amalgam is 23.0 kJ mol^{-1} and suggests that the exchange process is restricted by concentration limitations.

Kozin *et al.*[34] considered the exchange reaction mechanism based on the $Cd^{2+}/Zn(Hg)_x$ system:

$$Cd^{2+} + Zn(Hg) \rightleftharpoons Cd(Hg) + Zn^{2+} \qquad (4.11)$$

The polarization curves show the discharge–ionization of zinc and cadmium at amalgam electrodes (5.0 at.% amalgam) in the sulfuric acid electrolytes analyzed. The same curves also show that ionization of both zinc from amalgam and cadmium takes place with some polarization. The discharge of cadmium ions at an amalgam electrode occurs with considerable polarization, probably because in sulfuric acid solutions cadmium is bound into a negatively charged anionic complex.[35] The cadmium ion concentration in the solution increases as the polarization in the course of discharge decreases. According to Smirnova *et al.*,[36], the zero discharge potential for 5.6 at.% zinc and 5.6 at.% cadmium amalgams is 0.45 and 0.42 V, respectively.[36,37] Therefore, the Cd^{2+}–Zn(Hg) exchange reaction takes place on the negatively charged surface of zinc amalgam. The equilibrium potential of 5.0 at.% zinc amalgam is 0.805 in the presence of 1 g L^{-1} Zn^{2+} and 0.740 V at 100 g L^{-1} Zn^{2+}.

The exchange reaction between ions of cadmium (Cd^{2+}) and atoms of zinc present in the amalgam produces Zn^{2+}, which are adsorbed on the negatively charged surface of zinc amalgam, making it difficult for Cd^{2+} ions to reach the amalgam surface. By this reaction, Zn^{2+} ions produced in the course of the exchange reaction push Cd^{2+} ions back from the near-electrode layer. The exchange reaction then becomes inhibited. Apparently, the rate of the anode reaction:

$$Zn(Hg) \rightleftharpoons Zn^{2+} + Hg + 2e \tag{4.12}$$

may be much greater than the rate of cathode reaction:

$$Cd^{2+} + Hg + 2e \rightleftharpoons Cd(Hg) \tag{4.13}$$

The cathode reaction is the limiting factor of the exchange reaction (4.11). It is expected that a flow of zinc ions generated on the surface of the amalgam and bearing the same charges as cadmium ions will prevent the cadmium ions from reaching the surface of the amalgam due to an emerging electrostatic repulsion. Suppose that if an additional electrode is submerged and positioned in contact with zinc amalgam and the solution, the ionization of zinc atoms from the amalgam and the cathodic discharge of cadmium ions may be separated by some space. The reaction rate should now increase.

The increase of the exchange reaction rate constant for Cd^{2+} ions achieved *via* spatial separation of the processes taking place at the amalgam–solution phase boundary with the help of an additional electrode, on which cadmium settles the most, occurs because the arrangement avoids the limiting phase of Cd^{2+} ion reduction at the amalgam electrode. Experiments have shown that only 10% of Cd^{2+} ions penetrate the layer of Zn^{2+} ions on their way to the amalgam–solution phase boundary.

The effect of electrostatic repulsion of Bi^{3+} ions by the positively charged outer double layer of the amalgam has been described.[38] In this case, if electrons were diverted with the help of an additional electrode, the exchange reaction rate constant increased 4.2-fold. It has been demonstrated that if electrons are diverted during the reduction of the bismuth ions, only 8–10% of bismuth enters the zinc amalgam.

The relatively low exchange rates in the Hg_2^{2+}/Hg system are also apparently due to the electrostatic repulsion effect. The zero charge potential of mercury is 0.21 V (see Chapter 6); consequently, the mercury surface is positively charged, which causes a mutual repulsion of the like positive charges of mercury from the surface and from the Hg_2^{2+} ions and ultimately reduces mass transfer in the system.

Another reaction takes place at high nitric acid concentration along with the exchange reaction, in which HNO_3 reacts with mercury and the impurities:

$$6Hg + 8HNO_3 \rightarrow 3Hg_2(NO_3)_2 + 2NO + 4H_2O \tag{4.14}$$

In concentrated HNO_3 solutions the following reaction takes place:

$$3Hg + 8HNO_3 \rightarrow 3Hg(NO_3)_2 + 2NO + 4H_2O \tag{4.15}$$

The resulting mercury(II) ions react with mercury according to

$$Hg(NO_3)_2 + Hg \rightarrow Hg_2(NO_3)_2 \tag{4.16}$$

Impurities contained in the mercury react as follows:

$$Hg(NO_3)_2 + Me(Hg)_x \rightarrow Me(NO_3)_2 + (x + 1)Hg \tag{4.17}$$

The nitrate solution-based treatment process is lengthy[39,40] and, given a standard implementation setup, requires ceaseless attention and causes sizable material losses.[26]

To upgrade productivity, reduce mercury losses and improve labor conditions, semiautomatic[41] and automatic[42,43] chemical treatment installations were designed. Mercury is purified[42] after many cycles of operation, using two processes: air blasting and chemical treatment. Air is supplied through a porous glass filter to improve treatment quality and avoid unintentional impurities. Stable column operation is achieved by increasing water flow. To initiate the purification process, the column is loaded with 0.1 L of mercury and 0.75 L of process liquid. Mercury is added to the column in 0.075 L portions from time to time during column operation. Mercury circulates at around 0.120 L min^{-1} under normal operating conditions and completes ~100 cycles in 1 h. Based on this design, Artamonov[42] created a three-stage mercury purification column. The first column used a 20% solution of sodium hydroxide, the second a 5% solution of nitric acid and the third was filled with distilled water. Each portion of prefiltered mercury passes through all the three columns in succession. If all columns are activated, the installation produces 0.075 L h^{-1} of high-purity mercury.

Krolikiewicz *et al.*[40] proposed a high-performance chemical treatment installation. The installation consists of (a) an accumulation container for crude mercury equipped with a drain pipe and an overflow outlet in the upper part to drain excess water in case of overfilling, (b) three identical columns connected in series and (c) a pure mercury receptacle. The columns are cylinders with hemispherical bottoms. Each column has a pipe built into it along the centerline; the pipe runs full length and narrows at the top. Inside the shell, a U-shaped siphon drain pipe enters from the bottom, reaching around half of the column height. The space between the pipe and the walls is filled with glass balls. The container is topped with a dish-shaped cover with holes to accept mercury.

Another high-performance mercury treatment method, based on the use of nitric acid solutions with $Hg_2(NO_3)_2$ in a column equipped with a graphite–glass nozzle, has been suggested.[43] The process allows continuous purification of mercury flows in solutions of constant acidity and mercury(I) ion concentration. As mercury becomes free of impurities, it loses its capacity to form a uniform liquid layer and gradually turns into a mass consisting of 1–1.5 mm balls ('fish eggs'). These balls merge only if stirring ceases and mercury is allowed to settle out for 10–15 min. Mercury can then be washed with doubly distilled water to remove electrolyte and dried by filtration through a filter-paper perforated with thin quartz needles.

The method of Klinsky *et al.*[44] may also be used with some modifications. They proposed a mercury treatment method to remove zinc, cadmium, copper, bismuth, lead and tin in which mercury is exposed to mercurous compounds bound by an ion-exchange resin (KU-1, KU-2, KRS, KB-4 and various types of sulfocationites) in the form of monovalent cations of mercury(I) using resin grain sizes of 0.02–0.60 mm. Mercury is pumped through a layer of ion-exchange resin at the rate of 185–2700 kg Hg m^{-2} h^{-1}. The method yields excellent results. With original zinc, cadmium, copper, lead, bismuth and tin impurities at 0.5 mol L^{-1} (in separate tests), after mercury had been pumped through a layer of ion-exchange resin at the rate of $\nu = 185$–810 kg Hg m^{-2} h^{-1}, the impurities had decreased to the following levels (mol L^{-1}):

Zn	5×10^{-7}
Cd	1×10^{-6}
Cu	2×10^{-9}
Bi	5×10^{-6}
Pb	8×10^{-6}
Sn	8×10^{-6}

For mercury containing zinc, cadmium and copper at 0.17, 0.15 and 0.18 mol L^{-1} at v = 2700 kg Hg m^{-2} h^{-1}, the post-treatment impurity contents (mol L^{-1}) were

Zn	5×10^{-7}
Cd	1×10^{-6}
Cu	2×10^{-8}

The treatment efficiency depended on the mercury throughput rate. Thus, for mercury containing 0.33 mol L^{-1} of lead, 0.33 mol L^{-1} of bismuth and 0.34 mol L^{-1} of tin, the following post-treatment concentrations were obtained, depending on throughput rate: v = 46–2700 kg Hg m^{-2} h^{-1} gave a maximum of 10^{-6} mol L^{-1}, v = 3000 a maximum of 10^{-5} mol L^{-1} and v = 3500 a maximum of 10^{-2} mol L^{-1}. This shows that the throughput rate should not exceed 2700 mol L^{-1}. There is also evidence that the treatment efficiency depends on the grain size of the ion-exchange resin. Impurity analyses found a maximum of 10^{-6} mol L^{-1} at a grain size of 0.01–0.60 mm, 10^{-5} mol L^{-1} at 0.650 mm and 10^{-2} mol L^{-1} at 0.750 mm.[44] Data presented by Klinsky *et al.*[44] suggest that the impurity separation coefficient

$$\xi = \frac{\sum \mathrm{Me}_i \nu_{\mathrm{initial}} \mathrm{Hg}}{\sum \mathrm{Me}_i \nu_{\mathrm{treated}} \mathrm{Hg}} \tag{4.18}$$

is between 3.4×10^5 and 1.0×10^7. This method is much more efficient than other known chemical methods for mercury treatment. However, this method, as with all other chemical methods, has the problem that mercury does not lose its impurities of precious metals – silver, gold, rhodium, iridium, platinum – and also electronegative metals bound into intermetallic phases.[4,5,45,46]

Horizontal and vertical magnetohydrodynamic (MHD) pumps[4,5,12,38] are used in industrial applications for the chemical treatment of technical mercury. With standard chemical methods, if crude mercury arrives with an array of impurities, frequent replacement of spent solutions is necessary and considerable losses of mercury are common. Kozin and Ivanchenko[38] created high-performance fully automatic industrial plants capable of producing $0.2–1.0\,t\,h^{-1}$ of mercury containing 99.99–99.9995% of base metal. The impurity concentration was reduced from 0.05% wt% (500 ppm) in the original mercury down to $(1–2)\times10^{-3}$ wt% after treatment. Mercury losses were very small, typically $(1–3)\times10^{-2}$ wt%.

Ordinary chemical treatment methods involve losses between 2 and 5 wt%. Such high loss figures are due to the high oxidation capacity of nitric acid with respect to metallic mercury and to mercury engaging in soluble solutions of $Hg_2(NO_3)_2$ and $Hg(NO_3)_2$. Therefore, sizable amounts of mercury join metal impurities as part of the solution and often mercury still contains impurities that cannot be completely removed.[39,40]

Some of the disadvantages mentioned above can be partially overcome by a method for mercury purification based on $H_2SO_4 + KMnO_4$ solutions,[47], which successfully removes the following amount (wt%):

Cd	3
In	2
Tl	2
Pb	1
Sn	0.5
Zn	1.0
Cu	0.003
Fe	0.003

The treatment process consists of three phases. In Phase 1, equal amounts of a 9 N solution of H_2SO_4 and a saturated aqueous solution of $KMnO_4$ are shaken with mercury until the solution is completely colorless. Then new portions of the solution are added while shaking until balls appear on the mercury surface. Subsequently, mercury is washed with water to remove impurities. In Phase 2, the 'last traces of impurities' are removed. To do that, mercury is treated with a 2.0 N solution of H_2SO_4 + a 0.1 N solution of $KMnO_4$ while shaking until there is a large mass of small mercury balls. Once again, mercury is washed with water and shaken with 2 N H_2SO_4 until there is a complete mass of mercury balls (Phase 3). Finally, mercury is washed with water and 2 N H_2SO_4 yet again until the balls ('fish eggs') disappear.

If mercury containing impurities is shaken with a concentrated solution of iron(III) tetraoxosulfate(IV) in H_2SO_4 or a mixture of this solution with

$KMnO_4$, the impurity removal rates increase considerably. The rapid impurity removal rate is due to catalytic acceleration of $KMnO_4$ reduction as it interacts with $FeSO_4$ – the product of chemical reaction between $Fe_2(SO_4)_3$ and mercury impurities. At the same time, iron(II) sulfate is converted into iron(III) sulfate, which reacts rapidly with the impurities contained in the mercury. With the help of a 1 N solution of $Fe_2(SO_4)_3$ in a 2 N solution of H_2SO_4, it took 10 s of shaking to remove 1 g of lead from 200 g of mercury and 30 s to remove 7 g of zinc.[47] With the help of a concentrated solution of $KMnO_4$ in a 6 N solution of $H_2SO_4 + 20\,cm^3$ of a 0.1 N solution of $Fe_2(SO_4)_3$, Russell and Evans[47] managed to remove 14 g (total) of Zn, Cd, Sn, Pb solder and Bi from 480 g of mercury without losing any material. A dedicated experiment demonstrated that the order in which metals are removed from mercury by oxidants contained in the solutions is different from their position in the standard potential series only for Cr, Mn, Fe, Co and Ni (atomic numbers 24–28).[47,48] The authors explained such behavior by a manifestation of passivity.

We have modified the method of Russell and Evans.[47] The treatment quality was controlled using potentiometry. If the measurable potential of mercury in a 1 M solution was in the range 0.600–0.610 V (*versus* NHE) and was constant with time, we assumed that metal impurities more electronegative with respect to mercury were completely removed. If the mercury potential tended to migrate with time towards the negative side, that would indicate that more electronegative impurities were bound into intermetallic phases. The intermetallic phases dissociate and remove the dissolved portion of electronegative impurities (see Chapter 2).[4,5] The high effectiveness of sulfuric acid solution-based treatment methods is due to the electrochemical properties of the Hg_2SO_4/Hg system. This system demonstrates a positive potential and Hg_2SO_4 is poorly soluble in water and water-based solutions. Therefore, when oxidants are introduced into a sulfuric acid solution, they are used only to oxidize the impurities, since, after the impurities have been removed, the surface of the mercury is passivated with a very thin film. The thin film causes mercury to break into small balls and turn into 'fish eggs' if stirring is applied. This is why the Hg_2SO_4/Hg mercury treatment system is easy to automate.

Krolikiewicz *et al.*[40] suggested their own implementation of the chemical method. Crude (contaminated) mercury is purified by passing it through a cascade of columns. After passing through a layer of deionized water, contaminated mercury enters the first column. Prior to initiation of the treatment system, the columns are filled with pure mercury. The method is noteworthy because mercury is treated with an oxalic acid solution containing 1% tartaric (citric) acid, 0.5% progalite and 1% (30%) H_2O_2.[48] The basis of the method relies on crude mercury being subjected to rapid mixing with an oxalic acid solution containing additives ($\omega = 100$ rpm) at 263–353 K. In the course of treatment, a circulation pump continuously forces the solution through a layer of mercury. The method produces mercury containing 99.9% of base metal by mass.

Kobza and Grudina[22] developed a method to decontaminate mercury of greases, lubricants and mineral oils. A container is loaded with treated mercury and a 2–5% aqueous solution of sodium chloride. An Hg:electrolyte volume ratio of about 1:2 is used. AC current is supplied to mutually isolated electrodes introduced into the electrolyte and mercury. Some time later, the unit is de-energized, electrolyte is replaced and the cycle is repeated. It takes 3–5 replacements of electrolyte or the use of a flowing electrolyte to finally obtain high-purity mercury.[22]

4.3 Single-stage Electrochemical Methods for Obtaining High-purity Mercury

Electrolytic purification of mercury uses the principles of anodic dissolution and single- or multistage reprecipitation on mercury cathodes. The principle of staged anodic dissolution of impurities[5] may also be employed. In the case of anodic dissolution of mercury, the solution acquires more electronegative impurity metals, while mercury retains the more electropositive metals – silver, gold, rhodium – and also the more electronegative metals bound into intermetallic phases.[5,45,46] The anodic dissolution process is normally initiated at a current density of 0.2 A dm^{-2} in electrolyte containing a 2% solution of nitric acid or a mixture of 5% solutions of nitric and sulfuric acids. The electrolytic process removes bismuth, antimony, arsenic, iron, tin and lead, although not completely enough. Anodic dissolution methods yield the same degree of purification as the chemical methods.

Technologically, the single-stage electrolysis-based deep treatment method consists of dissolution of the original crude mercury, which contains more electronegative (Me_j) and electropositive (Me_i) metallic impurities ($HgMe_iMe_j$), in the electrolyzer's anode space:

$$HgMe_iMe_j \rightleftharpoons Hg_{x-1}Me_i + Hg^{z+} + Me_j^{m+} + (z+m)e \tag{4.19}$$

and cathodic deposition of mercury ions Hg^{z+} at the mercury cathode in the electrolyzer's cathode space:

$$Hg_x + Hg^{z+} + ze \rightleftharpoons Hg_{x+1} \tag{4.20}$$

where $z = 1$ or 2 depending on the nature of the electrolyte. For complexes with composition $HgX_z^{(z-n)}$ ($X = Cl^-$, Br^-, CN^-, SCN^-, *etc.*), $z = 2$. In simple electrolyte solutions ($X = ClO_4^-$, NO_3^-, ClO_3^-), $z = 1$.

A basic outline of the processes occurring during single-stage electrolytic purification of mercury was given by Kozin.[3] When crude mercury (1) is added to the anode space, the anode receives more electropositive and more electronegative impurity metals. The single-stage mercury purification process is based on mass transfer of mercury by the action of electric current with its ionization at the mercury anode together with electronegative impurities, according equation (4.19), and electrodeposition of mercury ions on the cathode, according to equation (4.20). In theory, more positive impurities should remain in the mercury anode and stay away from the electrode process. However, the behavior of electropositive impurities depends on the experimental conditions.

Kozin[3] also showed concurrent (including 'parasitic') processes 3–9, explained as follows. Processes 3 and 4 are related to the reduction of the ions of the treated metal–mercury (Hg_2^{2+}, Hg^{2+}, HgX_2, HgX_4^{2-}, *etc.*) by electronegative metal impurities (Me_i^0) contained in crude mercury. At the same time, more electronegative impurities migrate into the electrolyte solution according to eqn (4.1) or (4.2). Processes 5 and 6 are due to migration of cathodic mercury, Hg_x, into the solution and electropositive impurities, Me_j, into cathodic mercury as a consequence of phase exchange reactions:

$$Hg_x + Me^{z+}_{jaq} \rightleftharpoons Hg^{z+} + (Me_j)Hg_x \quad (4.21)$$

The principles of exchange reactions are covered in the literature.[4,5,7,49] Process 7 consists of electrochemical dissolution of electropositive metals contained in the mercury anode (crude mercury):

$$(Me_j)Hg_x \rightleftharpoons Me_j^{z+} + Hg_x + ze \quad (4.22)$$

Process 8 is due to the interaction between more electrically positive trace metals in the electrolyte with the mercury being purified in the course of exchange reaction:

$$Me_j^{z+} + Hg_x \rightleftharpoons (Me_j)Hg_{x-1} + Hg^{z+} \quad (4.23)$$

The results of a single-stage purification of mercury have been presented.[50,51] Usually, purification is performed in electrolyzers with separate anode and cathode spaces in an electrolyte composed of 235 g L^{-1} perchloric acid and 200 g L^{-1} mercury oxide.[50] During electrolysis, the mercury(II) perchlorate is reduced to univalent mercury. Studies of the behavior of trace metals such as zinc, cadmium, tin, bismuth, copper, silver and gold introduced in the mercury to be purified in amounts of 0.02–0.1% during electrolysis have shown that in a number of experiments copper and gold were found in some batches of purified mercury.

The perchloric acid electrolyte was prepared[1] by dissolving 20 cm^3 of a 75% solution of $HClO_4$ in 80 cm^3 of water and then dissolving 20 g of mercury oxide in the solution obtained. Mercury(II) perchlorate is formed during electrolysis and by contact of metallic mercury is reduced to $Hg_2(ClO_4)_2$. Along with reduction of mercury(II) ions to mercury(I) ions in the course of electrolysis, reduction of mercury(I) to metallic mercury takes place.

Feryanchich[51] carried out the electrolytic purification of mercury in a solution of nitric acid (1:2) at an initial current density of 0.2 A dm^{-2}. As the electrolyte was being saturated with mercury ions, the anode current density increased to 1.5 A dm^{-2}. During the electrolysis, the content of impurities in the refined mercury decreased from 0.12 to 0.007 wt%. According to Lorenz,[51], the anode residue mainly contained iron.

For mercury treatment under laboratory conditions,[38] a thick-walled crystallizer of diameter 20–25 cm was suggested. A 5% solution of HNO_3 serves as the electrolyte. Between 3 and 4 kg of raw mercury is loaded into the crystallizer and covered with the solution. Current is fed to a mercury cathode and anode through platinum wires soldered into tubes. The electrolysis takes place with agitation of anode mercury and the electrolyte with an agitator. The voltage applied to the electrodes is 5–6 V and the electrolysis current is 5–8 A. Metallic impurities pass from the purified mercury to the solution and, in accordance with the separation factors, are formed on the cathode.

Depending on the degree of mercury contamination, the impurities turn the electrolyte dark in 5–10 min. The electrolyte is then drained using a siphon and a fresh portion of a 5% solution of HNO_3 is loaded. The electrolyte will be changed 3–8 times until coloration stops.[38] This suggests that electrolytic treatment of raw mercury is ineffective and that chemical treatment should be performed first. The use of a stationary mercury cathode also reduces the efficiency of single-stage electrolytic purification.

Lorenz[52] reported a design of an electrolyzer where efficient agitation of the anode and cathode is ensured. A current of 10 A and voltage of 12 V is applied. Impurities (Zn, Co, Pb, Sn, Sb, Cu, Ag) are removed through anodic dissolution of mercury at a current density of 50–60 A dm^{-2}. Heat is dissipated using air or water cooling. Noble metals – Au and Pt – remain in the mercury. To obtain high-purity mercury (99.9999%) after anodic treatment, mercury is subjected to vacuum distillation.[52] Electrolyzers of this design can also be used for a single-stage electrolytic treatment.

Analysis of the literature on mercury purification[5,7–10,50,52] shows that electrolytes are generally composed of nitric acid solutions and have the disadvantage of being unstable over time. From experience, the single-stage method cannot achieve ultrapurification of mercury from associated impurities. We believe that a more promising method to obtain ultrapure mercury is multistage electrolysis that can be performed in electrolyzers with bipolar mercury electrodes.

Further methods of purification and ultrapurification of mercury, including vacuum distillation and multi-stage electrolysis, were described in considerable detail by Kozin.[3]

References

1. S. M. Melnikov, *Metallurgy of Mercury*, Metallurgiya, Moscow, 1971.
2. A. A. Saukov, *Geochemistry*, Gosgeolizdat, Moscow, 1950.
3. L. F. Kozin, *Physico-Chemistry and Metallurgy of High-Purity Mercury and Its Alloys*, Naukova Dumka, Kiev, 1992.
4. L. F. Kozin, *Amalgam Metallurgy*, Tekhnika, Kiev, 1970.
5. L. F. Kozin, *Physicochemical Basics of Amalgam Metallurgy*, Nauka, Alma-Ata, 1964.
6. L. A. Niselson and A. G. Yaroshevsky, *High-Purity Substances*, 1988, **6**, 8–20.
7. P. P. Pugachevich, *Handling Mercury in Laboratory and Industrial Conditions*, Khimiya, Moscow, 1972.
8. N. N. Rozanov, *Collected Scientific Papers*, Giredmet, Metallurgizdat, Moscow, 1959, vol. 1, pp. 779–788.

9. J. Soucek, *Chem Listy*, 1964, **58**, 1203.
10. L. N. Kuzmenkov, *Studies in the Field of Physico-Chemical Measurements*, Standards Publishing, Moscow, 1968, Issue 96/156, pp. 79–91.
11. A. G. Stromberg and I. Bykov, *Zavod. Lab.*, 1953, **19**, 171.
12. L. F. Kozin and A. V. Abrosimov, *Electrode Processes in Solid and Liquid Electrodes*, Nauka, Alma-Ata, 1964, vol. 12, p. 194.
13. S. V. Ptitsin, *J. Tech. Phys.*, 1935, **5**, 309.
14. M. C. Wilkinson, *Chem. Rev.*, 1972, **72**, 575.
15. C. L. Gordon and E. Wichers, *Ann. N. Y. Acad. Sci.*, 1957, **65**, 369.
16. L. F. Kozin, *Bulletin of Inventions*, 1960, (5), 38–39; Author's Certificate 126 613, USSR.
17. A. A. Sunier and C. M. White, *J. Am. Chem. Soc.*, 1930, **52**, 1842.
18. G. A. Hulett and H. D. Minchin, *Phys. Rev.*, 1905, **21**, 388.
19. A. A. Sunier and C. B. Hess, *J. Am. Chem. Soc.*, 1928, **50**, 662.
20. C. J. Moore, *J. Am. Chem. Soc.*, 1910, **32**, 971.
21. W. M. Spicer and C. J. Banick, *J. Am. Chem. Soc.*, 1953, **75**, 2268.
22. F. Kobza and K. Grudina, *Method for Mercury Treatment from Organic Substances*, 15 September 1987; Author's Certificate 244 495 Czechoslovakia.
23. G. Milazzo and S. Caroli, *Tables of Standard Electrode Potentials*, Wiley, New York, 1978.
24. D. Dobosh, *Electrochemical Constants*, Mir, Moscow, 1980.
25. S. G. Bratsch, *J. Phys. Chem. Ref. Data*, 1989, **18**, 1.
26. A. M. Sukhotin (ed.), *Handbook of Electrochemistry*, Khimiya, Leningrad, 1981.
27. W. M. Latimer, *Oxidation States of Elements and Their Potentials in Aqueous Solutions*, Foreign Literature Publishing, Moscow, 1954.
28. N. E. Khomutov, *Elektrokhimiya*, 1964, **1**, 7–58.
29. R. N. Goldberg and L. G. Hepler, *Chem. Rev.*, 1968, **68**, 229.
30. V. A. Lebedev, V. I. Kober and L. F. Yamschikov, *Thermochemistry of Alloys of Rare-Earth Metals and Actinide Elements, Handbook*, Metallurgiya, Chelyabinsk, 1989.
31. L. F. Kozin, S. P. Bukhman and Z. N. Nysakbaeva, Electrodes and Electrolytes, *Izv. Akad. Nauk Kazakh., Ser. Khim.*, 1967, (15), 112.
32. S. P. Bukhman, *Cementation by Metal Amalgams*, Nauka, Alma-Ata, 1986.
33. L. F. Kozin, *Kinetics and Electrode Processes in Aqueous Solutions*, Naukova Dumka, Kiev, 1983, pp. 37–63.
34. L. F. Kozin, T. I. Saprykina and A. T. Zamulyukin, *Izv. Akad. Nauk Kazakh., Ser. Khim.*, 1972, (2), 72–77.
35. I. Leden, *Acta Chem. Scand.*, 1952, **6**, 971.
36. M. G. Smirnova, V. A. Smirnov and L. I. Antropov, *Research Papers Novocherkas Polytech. Inst.*, 1956, **34/48**, 69–75.
37. A. N. Frumkin and F. J. Service, *Zh. Fiz. Khim.*, 1930, **1**, 52–64.
38. L. F. Kozin and N. I. Ivanchenko, *Izv. Akad. Nauk Kazakh., Ser. Khim.*, 1975, **3**, 84–90.
39. V. M. Foliforov and V. S. Gorovits, *Application of Magnetohydrodynamic and Electrohydrodynamic Devices in Industrial Power and Metallurgy*,

Publishing House of the Institute for High Temperatures of the USSR Academy of Sciences, Moscow, 1989, pp. 27–39.

40. R. Krolikiewicz, K. Karwotski and A. Szmidkowski, *Pol. Pat.*, 133 028, *Polish Patent Bulletin*, 1986, No.10.
41. V. Grmela, *Chem. Zvesti*, 1966, **20**, 615.
42. V. G. Artamonov, *Zavod. Lab.*, 1965, **31**, 254.
43. V. G. Prihodchenko and O. V. Bogmat, *Bulletin of Inventions*, 1964, (4), 38; Author's Certificate 160 588, USSR.
44. G. D. Klinsky, A. F. Mazenko, D. A. Knyazev, *et al.*, *Bulletin of Inventions*, 1985, (32), 120; Author's Certificate 1 175 971, USSR.
45. L. F. Kozin, *Amalgam Pyrometallurgy*, Nauka, Alma-Ata, 1973.
46. L. F. Kozin, R. Sh. Nigmetova and M. B. Dergacheva, *Thermodynamics of Binary Amalgam Systems*, A Nauka, lma-Ata, 1977.
47. A. S. Russell and D. C. Evans, *J. Chem. Soc.*, 1925, **Pt. 2**, 2221.
48. D. Knokhe, *East Ger. Pat.*, 207 311, *Patent Bulletin*, 1984, (8).
49. L. F. Kozin, A. G. Morachevsky and I. A. Shcheka, *Ukr. Khim. Zh.*, 1989, **55**, 495.
50. E. Newberg and S. M. Naube, *Trans. Electrochem. Soc.*, 1933, **64**, 189.
51. F. A. Feryanchich, *Tr. Komis. Anal. Khim. Akad. Nauk SSSR*, 1947, **2**, 87.
52. J. R. Lorenz, *KFKI Közlemények*, 1964, **12**, 329.

CHAPTER 5

Chemical Properties of Mercury

5.1 Inorganic Mercury Compounds

Chemical compounds of mercury and metallic mercury have been widely studied in theoretical and applied inorganic and organic chemistry, in electrochemistry and in the instrumentation industry. Many compounds of mercury possess extraordinary and valuable properties (physical and chemical) on the one hand, but on the other hand, they are extremely toxic and, thus, hazardous for the environment. The chemical and physical properties of mercury compounds have been extensively studied and a considerable amount of quantitative data has been obtained and will be discussed in this chapter. A number of reviews[1–10] have covered the analysis of mercury and its compounds.

5.1.1 Disproportionation in Hg_2^{2+} and Hg^{2+}

A peculiar feature of the chemistry of Hg_2X_2 compounds is the ability of Hg_2^{2+} ions to exhibit disproportionation (disprop) in the presence of excess Cl^- and Br^- ions, and also some ligands in solution. The Hg_2^{2+} ion forms stable complexes with OH^-, S^{2-}, CN^- and I^- ions and molecules of amines and alkyl sulfides. Under these conditions, the Hg_2^{2+} ions disproportionate according to

$$Hg^{2+}_{2(aq)} \rightleftharpoons Hg^{2+}_{(aq)} + Hg^{0}_{(aq)} \tag{5.1}$$

with an equilibrium constant

$$K_{\text{disprop}} = \frac{[Hg^{2+}][Hg^{0}_{(aq)}]}{[Hg_2^{2+}]} = 5.5 \times 10^{-9}\,\text{M} \tag{5.2}$$

The ratio between the concentrations of Hg^{2+} and Hg_2^{2+} ions in the presence of metallic mercury in solution, *i.e.*, $[Hg^{2+}]/[Hg_2^{2+}]$ in the $Hg^0 - Hg_2^{2+} - Hg^{2+}$

Mercury Handbook: Chemistry, Applications and Environmental Impact
By Leonid F Kozin and Steve Hansen

Published by the Royal Society of Chemistry, www.rsc.org

system, is equal to 0.0183 according to some citations[1-4] and 0.0060–0.0120 according to Sidgwick.[5]

If the solubility of free metallic mercury in water is taken into account [which according to Sidgwick[5] is 3.0×10^{-7} M and $(3.01\pm0.12)\times10^{-7}$ mol kg^{-1}], then $K_{\text{disprop}}=(1.8\text{–}5.5)\times10^{-9}$ M. Moser and Voigt[6] described in detail the equilibrium in the $Hg-Hg_2^{2+}-Hg^{2+}$ system at an ionic strength of 0.05–0.1 M and different concentrations of Hg_2^{2+} $(0.38\text{–}1.52)\times10^{-5}$ M and Hg^0 $(1.1\text{–}3.2)\times10^{-7}$ M and flow rates of clean nitrogen (1.5–2.5 L min^{-1}). The disproportionation reaction is rapid. Elementary mercury that is generated in this reaction is easily carried off by the air flow from the water solution to the gaseous phase according to

$$Hg^0_{(aq)} \xrightarrow{k} Hg_{(gas)} \tag{5.3}$$

where k is a velocity constant that characterizes the generation of Hg° in the velocity-driving stage of the disproportionation reaction. The role of the disproportionation reaction of Hg(I) ions in the process of carrying mercury to the atmosphere was discussed by Toribara *et al.*[8] The velocity of Hg(I) ion disproportionation follows the equation

$$V_{Hg_2^{2+}} = -\frac{d[Hg_2^{2+}]}{dt} = \frac{kK_{\text{disprop}}[Hg_2^{2+}]}{[Hg^{2+}]} \tag{5.4}$$

The values of k were determined in a separate experiment. The results of studies of the disproportionation reaction of the Hg(I) ion in the temperature range 238–308 K are given in Table 5.1. K_{disprop} was calculated using solutions with an ionic strength of $\mu=0.1$ and $[Hg_2^{2+}]=7.6\times10^6$ M saturated with metallic mercury at each temperature.

5.1.2 Solubility of Metallic Mercury in Water

Values of the solubility of metallic mercury have been studied over a broad range of temperature. Experimental data in the temperature range 278.15–407.95 K are given in Table 5.2.

According to Sanemasa,[10] the solubility of metallic mercury in water at 298 K is 3.2×10^{-7} M, whereas Glew and Hames[11] suggested a value of

Table 5.1 Reaction rate constants and disproportionation equilibrium constants of Hg_2^{2+} ions at various temperatures.[7]

T (K)	k *(min^{-1})*	kK$_{disprop}\times10^9$ *(min^{-1})*	K$_{disprop}\times10^9$ *(M)*
238	1.7 ± 0.1	8.2 ± 0.5	4.8 ± 0.1
293	2.0 ± 0.1	16 ± 1	7.9 ± 0.1
298	2.2 ± 0.1	23 ± 2	11 ± 1
303	2.3 ± 0.1	36 ± 1	16 ± 1
308	2.5 ± 0.1	55 ± 1	22 ± 1

Table 5.2 Solubility of metallic mercury in water.

		Mercury solubility					
T (K)	*T (°C)*	*g-atom L^{-1} (mol L^{-1})*	*μg L^{-1} (ng g^{-1}, ppb)*	*μg g^{-1} (ppm)*	*$x_{Hg}\times10^9$ (mole fraction)*	*mg per 100 mL*	*Ref.*
278.15	5	9.57E–08	19.2	0.0192	1.72	0.00192	10
283.15	10	1.38E–07	27.7	0.0277	2.49	0.00277	10
293.15	20	2.24E–07	45	0.045	4.04	0.0045	10
303.15	30	4.05E–07	81.3	0.0813	7.30	0.00813	10
313.15	40	6.83E–07	137	0.137	12.3	0.0137	10
323.15	50	1.09E–06	218	0.218	19.6	0.0218	10
333.15	60	1.83E–06	368	0.368	33.1	0.0368	10
343.15	70	2.79E–06	560	0.56	50.3	0.056	10
353.15	80	4.24E–06	850	0.85	76.3	0.085	10
373.15	100	5.98E–06	1200	1.2	107.8	0.12	10
393.15	120	8.97E–06	1800	1.8	161.7	0.18	10
298.15	25	3.15E–07	63	0.063	5.7	0.01	13[a]
308.15	35	5.43E–07	109	0.109	9.8	0.01	13[a]
323.15	50	8.82E–07	177	0.177	15.9	0.02	13[a]
338.15	65	1.08E–06	217	0.217	19.5	0.02	13[a]
353.15	80	1.30E–06	261	0.261	23.4	0.03	13[a]
363.15	90	1.67E–06	334	0.334	30.0	0.03	13[a]
298.15	25	2.99E–07	60	0.060	5.4	0.01	6
298.15	25	3.05E–07	61.2	0.061	5.5	0.0061	12
313.15	40	5.12E–07	103	0.103	9.2	0.0104	12
323.15	50	7.43E–07	149	0.149	13.4	0.0150	12
333.15	60	1.078E–06	216	0.216	19.4	0.0216	12
343.15	70	1.333E–06	267	0.267	24.0	0.0267	12
353.15	80	1.637E–06	328	0.328	29.5	0.0328	12
307.15	34.0	3.61E–07	72	0.072	6.5	0.0072	16
307.15	34.0	3.55E–07	71	0.071	6.4	0.0071	16
307.65	34.5	4.44E–07	89	0.089	8	0.0089	16
312.45	39.3	5.55E–07	111	0.111	10	0.0111	16
312.95	39.8	5.55E–07	111	0.111	10	0.0111	16
327.15	54.0	6.66E–07	134	0.134	12	0.0134	16
330.65	57.5	8.33E–07	167	0.167	15	0.0167	16
331.65	58.5	1.11E–06	223	0.223	20	0.0223	16
333.45	60.3	9.99E–07	200	0.200	18	0.0200	16
333.45	60.3	6.66E–07	134	0.134	12	0.0134	16
333.45	60.3	9.44E–07	189	0.189	17	0.0189	16
351.95	78.8	1.22E–06	245	0.245	22	0.0245	16
362.75	89.6	2.44E–06	490	0.490	44	0.0490	16
363.65	90.5	2.16E–06	434	0.434	39	0.0434	16
364.45	91.3	2.00E–06	401	0.401	36	0.0401	16
364.45	91.3	1.78E–06	356	0.356	32	0.0356	16
375.15	102.0	2.11E–06	423	0.423	38	0.0423	16
375.15	102.0	2.11E–06	423	0.423	38	0.0423	16
375.15	102.0	2.22E–06	445	0.445	40	0.0445	16
377.35	104.2	2.72E–06	546	0.546	49	0.0546	16
380.65	107.5	3.05E–06	612	0.612	55	0.0612	16
404.65	131.5	5.77E–06	1158	1.158	104	0.1158	16
407.65	134.5	5.05E–06	1013	1.013	91	0.1013	16
407.95	134.8	5.11E–06	1024	1.024	92	0.1024	16

[a]Original data were corrected by a factor of 1000.

3.05×10^{-7} M. As stated by Hepler and Olofsson,[1] the solubility of metallic mercury in water in the temperature range 273.15–393.15 K is described by the equation

$$\log X_{Hg} = -43.3343 - 20.9053 / (T / 100\ K) + 15.7778 \ln(T / 100\ K) \quad (5.5)$$

with a standard deviation of 3.9×10^{-9} mole fraction of mercury (x_{Hg}). Values for the solubility of mercury in water over the temperature range 277–343 K given by Choi and Tuck[12] are close to those in Table 5.2. According to Choi and Tuck,[12] the solubility of mercury in water follows the equation

$$\log m_{Hg} = -126.345 + \frac{4715.2}{T} + 42.0288 \log T \quad (5.6)$$

where m = moles Hg–1000 g H_2O^{-1}, over the temperature range 273.15–393.15 K (0–120 °C).[13] Figure 5.1 summarizes the data on the solubility of mercury in water.[10]

The solubility of mercury in water at various temperatures and pressures in given in Table 5.3.[17] Inversion of the solubility in water and in NaCl solution takes place at 54 °C. Systematic studies of the solubility of metallic mercury in solutions of electrolytes (NaCl, NaOH) were conducted by Chviruk and co-workers.[18–21] Analysis of the data on mercury solubility in a log$S$–1/T coordinate system must correlate the experimental points with a straight line having a slope corresponding to the Henry constant.

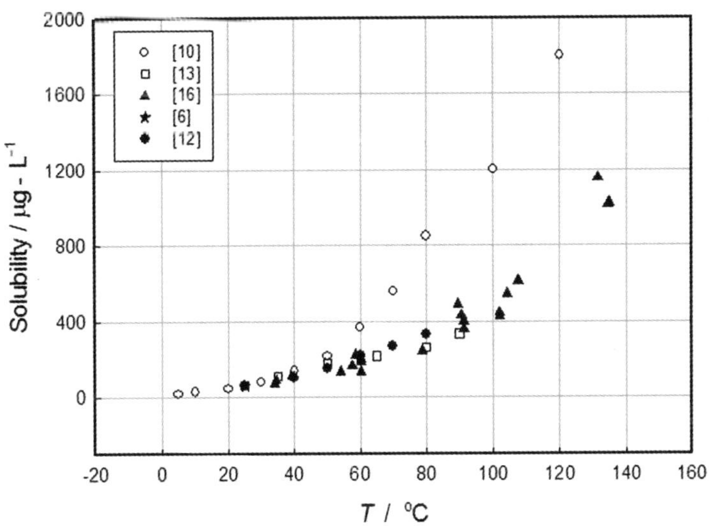

Figure 5.1 Comparison of the solubilities of the mercury in pure water as a function of temperature.
Adapted from Refs. 6, 10, 12, 13, 16.

Table 5.3 Solubility of mercury in water at elevated temperatures and pressures.[17]

T (K)	T (°C)	P (atm)	P (bar)	Mercury solubility		
				$g\ kg^{-1}$	$mol\ kg^{-1}$	$x_{Hg}\times 10^9$ (mole fraction)
573	300	500	507	0.29	0.0014	0.0260
573	300	640	648	0.24	0.0012	0.0216
571	298	900	912	0.19	0.0009	0.0171
673	400	400	405	3.37	0.0168	0.302
673	400	500	507	2.76	0.0138	0.248
673	400	495	502	3.22	0.0161	0.289
673	400	700	709	2.47	0.0123	0.222
673	400	700	709	2.80	0.0140	0.251
671	398	920	932	2.23	0.0111	0.200
674	401	910	932	2.13	0.0106	0.191
773	500	500	507	24.12	0.1202	2.16
775	502	510	517	23.71	0.1182	2.12
773	500	520	527	20.21	0.1008	1.81
768	495	755	765	18.45	0.0920	1.65
780	507	700	709	19.90	0.0992	1.78
771	498	990	1003	16.36	0.0816	1.47
776	503	960	972	13.41	0.0667	1.20

At 298 K (25 °C), the solubility of mercury in water is $(2.93\text{–}3.15)\times 10^{-7}$ M. Therefore, the difference in values for $K_{disprop}$ of Hg_2^{2+} ions in water must be small and should depend mainly on the accuracy of determination of mercury(I) and -(II) ion concentrations and their proportion. Moser and Voigt[6] found that $K_{disprop}$ at 298 K is equal to $(1.1\pm 0.1)\times 10^{-8}$ M, which is twice the value of $K_{disprop} = 5.5\times 10^{-9}$ M reported by Sidgwick.[5] This difference of $K_{disprop}$ values may be due to specifics of the experiments related to the dispersion of mercury even in the presence of traces of surfactants.

5.1.3 Solubility of Mercury in Ionic Solutions

The solubility of Hg_2Cl_2 in water depends on the concentration of chloride ions:

$$Hg_2Cl_{2(solid)} \rightarrow Hg_2^{2+} + 2Cl^- \tag{5.7}$$

and, therefore, $K_{disprop}$ of reaction (5.1) will also change. The thermodynamic equilibrium constant of reaction (5.7) is $K_s = (1.42\pm 0.02)\times 10^{-18}$.[10] Thompson *et al.*[9] demonstrated that relatively high concentrations of mercury(I) ions in the presence of liquid mercury cause the Hg_{liquid}–Hg_2^{2+}–Hg^{2+} system to produce colloidal mercury. The high concentrations of mercury(I) ions in the disproportionation reaction shift the equilibrium of reaction (5.1) to the left and cause $K_{disprop}$ to decrease. Atoms of mercury should be in equilibrium with Hg_2^{2+} and Hg^{2+} ions to obtain the true $K_{disprop}$ value in the system $Hg_{(aq)}$–Hg_2^{2+}–Hg^{2+}.

Equilibrium in the Hg_{liquid}–Hg_2^{2+}–Hg^{2+} system is established very quickly. Ions of Hg_2^{2+} are easily created during the reduction of mercury(II) salts and are easily oxidized to Hg^{2+} and reduced to the metallic state. The reduction of Hg_2^{2+} to Hg^0 is determined by positive half-reaction potentials towards many metals and reagents. Table 5.4 gives standard half-reaction potentials and ratios between Hg_2^{2+} and Hg^{2+} ions in the system Hg^0–Hg_2X_2–HgX_4^{2-}. Standard electrode potentials may be used to calculate the ratios $a_{Hg_2^{2+}/Hg^{2+}}$ and $a_{Hg^{2+}/Hg_2^{2+}}$, using the equation

$$\ln K^{\circ} = \ln\left(\frac{a_{Hg^{2+}}}{a_{Hg_2^{2+}}}\right) = \frac{2F\left(E^0_{Hg_2^{2+}} - E^0_{Hg_2^{2+}/Hg^{2+}}\right)}{RT} \tag{5.8}$$

Table 5.4 Standard half-reaction potentials, temperature coefficients (dE°/dT) and ratios of ions Hg_2^{2+}/Hg^{2+} and Hg^{2+}/Hg_2^{2+}.

Half-reaction	E° *(V)*	*Ref.*	dE°/dT *(mV K⁻¹)*	$\frac{Hg_2^{2+}}{Hg^{2+}}$	$\frac{Hg^{2+}}{Hg_2^{2+}}$
Acid solutions					
$Hg_2^{2+}+2e^- \rightarrow 2Hg$	0.789	22		1.66×10^2	6.02×10^{-3}
$2Hg^{2+}+2e^- \rightarrow Hg_2^{2+}$	0.920	22	–	1.66×10^2	6.02×10^{-3}
$Hg^{2+}+2e^- \rightarrow 2Hg$	0.854	22		1.66×10^2	6.02×10^{-3}
$Hg_2^{2+}+2e^- \rightarrow 2Hg_{liq}$	0.796	22	−0.327	6.13×10^3	1.63×10^{-4}
$2Hg^{2+}+2e^- \rightarrow Hg_2^{2+}$	0.908	22	0.095	6.13×10^3	1.63×10^{-4}
$Hg^{2+}+2e^- \rightarrow Hg_{liq}$	0.852	22	−0.116	6.13×10^3	1.63×10^{-4}
$Hg_2Br_2+2e^- \rightarrow 2Hg + 2Br^-$	0.1390	22	−0.142	2.55×10^2	3.92×10^{-3}
$HgBr_4^{2-}+2e^- \rightarrow Hg + 4Br^-$	0.210	23	−0.42	2.55×10^2	3.92×10^{-3}
$HgBr_4^{2-}+e^- \rightarrow HgBr_2^- + 2Br^-$	0.281	24	–	2.55×10^2	3.92×10^{-3}
$Hg_2I_2+2e^- \rightarrow 2Hg + 2I^-$	−0.0405	22	0.019	1.48	0.67
$HgI_4^{2-}+2e^- \rightarrow Hg + 4I^-$	−0.040	24	0.04	1.48	0.67
$HgI_4^{2-}+e^- \rightarrow HgI_2^- + 2I^-$	−0.0395	24	–	1.48	0.67
$HgSO_4+2e^- \rightarrow 2Hg_{liq}+SO_4^{2-}$	0.61257	25	−0.826	–	–
$Hg_2(SCN)_2+2e^- \rightarrow 2Hg_{liq}+2SCN^-$	0.22	23	–	–	–
$Hg(CN)_4^{2-}+2e^- \rightarrow Hg_{liq}+4CN^-$	−0.37	23	0.78	–	–
$Hg(CN)_2+2e^- \rightarrow Hg + 2CN^-$	−0.304	23	–	7.25×10^8	1.38×10^{-9}
$Hg(CN)_2+2e^- \rightarrow 2Hg + 2CN^-$	−0.435	23	–	7.25×10^8	1.38×10^{-9}
$HgS_{(Solid\_red)}+2H^+ + 2e^- \rightarrow Hg_{liq}+H_2S$	−0.096	1	–	–	–
$HgS_{(solid\_black)}+2H^+_{gas}+2e^- \rightarrow Hg_{liq}+H_2S$	−0.085	1	–	–	–
$Hg_2(N_3)_{2(solid)}+2e^- \rightarrow 2Hg_{liq}+2N_3^-$	0.260	1	–	–	–
$Hg_2CO_{3(solid)}+2e^- \rightarrow 2Hg_{liq}+CO_3^{2-}$	0.309	1	–	–	–
$Hg_2Ac_{2(solid)}+2e^- \rightarrow 2Hg_{liq}+2Ac^-$	0.51163	1	0.8995	–	–
$Hg(CN)_2+e^- \rightarrow HgCN_2^-$	−0.173[a]		–	–	–

[a]Obtained by the analytical method developed by Chviruk and Koneva.[20]

Table 5.5 Thermodynamic parameters for the formation of $Hg^{2+}_{(gas)}$ and $Hg^{2+}_{2(gas)}$ ions at 298.15 K.[26]

Ion	ΔG°_{form} *(kJ mol⁻¹)*	ΔH°_{form} *(kJ mol⁻¹)*	S° *(J mol⁻¹ K⁻¹)*
Hg_2^{2+}	153.607 ± 0.105	166.816 ± 0.210	–65.816 ± 0.837
Hg^{2+}	164.703 ± 0.105	170.163 ± 0.210	–36.233 ± 0.837

Contact of Hg^{2+} ions with metallic mercury creates Hg_2^{2+} ions as a result of the reproportionation reaction

$$Hg_{liquid} + Hg^{2+} \rightleftharpoons Hg_2^{2+} \tag{5.9}$$

Equilibrium in the Hg–Hg_2X_2–HgX_2 system also depends on both the nature of the ligands ($X = Cl^-$, Br^-, I^-, ClO_4^-, SO_4^{2-}, Ac^-, *etc.*) and their concentration. If no ligands are present in the solution, the equilibrium of reaction (5.9) at $a_{Hg_{liquid}} = 1$ will shift to the right to achieve the ratio

$$K = \frac{a_{Hg_2^{2+}}}{a_{Hg^{2+}}} = 1.66 \times 10^{-2} \tag{5.10}$$

Introduction of Hg^0_{liquid}–Hg_2X_2–HgX_2 into a solution of halides and other ligands leads to changes of the standard electrode potentials of mercury half-reactions and shifts their values towards the negative side as seen in Table 5.4. At the same time, $K_{eqn\,(5.9)}$ values decrease and $K_{eqn\,(5.8)}$ values increase. Notably, the standard electrode potentials of mercury in the presence of ligands that form stable mercury complexes (CN^-, I^-) acquire negative values. According to Hepler and Olofsson,[1] the dissociation constant of the mercury(I) ion:

$$Hg^{2+}_{2(aq)} \rightleftharpoons Hg^{+}_{(aq)} \tag{5.11}$$

is $K < 10^{-7}$. This K value allows $Hg^+_{(aq)}$ ions to register directly, with the help of electron spin resonance spectra, in mercury(I) perchlorate solutions.

Thermodynamic parameters for the formation of $Hg^{2+}_{(gas)}$ and $Hg^{2+}_{2(gas)}$ ions at 298.15 K, according to Vanderzee and Swanson,[26] are given in Table 5.5.

5.2 Mercury(I) and Mercury(II) Halides and Pseudohalides

The synthesis of mercury halides is well known and was summarized by Simon *et al.*[27] Calomel can be produced by sealing a mixture of mercury and mercury(II) chloride in iron or fused-silica tubes and heating at 525 °C. A cooled condenser is attached that receives calomel vapor condensate.[28,29] Synthesis from the elements is also possible.[30] Very finely divided mercury(I) chloride can also be obtained by precipitation from a dilute nitric acid solution of mercury(I) nitrate and sodium chloride.[28]

Mercury(II) chloride can be synthesized from the elements in sealed and heated vessels, such as quartz. Mercury is oxidized with chlorine; the reaction occurs with flames at temperatures >300 °C. The reaction product is condensed in cooled collectors as fine crystals. Calomel is not formed when excess chlorine is present.[30]

Mercury and chlorine also react in the presence of water; in this case, intensive stirring is necessary. The chloride formed precipitates as crystals after the solubility limit has been exceeded. If an alkali metal chloride solution is used in place of water, solutions of chloro complex salts are formed, which are used mainly for the production of other compounds of divalent mercury.[31] Mercury(II) chloride can also be prepared from other mercury compounds. Mercury(II) sulfate, for example, is heated in the dry state with sodium chloride and the evolved mercury(II) chloride vapor is condensed to a solid in receivers. In another synthesis method, a warm sublimate solution is obtained from the reaction of mercury(II) oxide and a stoichiometric amount of hydrochloric acid; the chloride separates as crystals on cooling.[27]

$HgBr_2$ is produced from mercury and bromine in the presence of water by dissolution of mercury(II) oxide in hydrobromic acid or by precipitation from a nitric acid solution of mercury(II) nitrate with addition of sodium bromide.[32] Methods for the synthesis of HgI_2 are similar to those for mercury(II) bromide.[33-35]

5.2.1 Mercury(I) Fluoride – Hg_2F_2

Physical properties of Hg_2F_2 are given in Table 5.6. In the solid state, Hg_2F_2 forms yellow tetragonal crystals, which quickly darken when exposed to light. Mercury(I) fluoride hydrolyzes in water into HF and unstable mercury(I) hydroxide and disproportionates:

$$2HgOH \rightarrow Hg + Hg(OH)_2 \quad (5.12)$$

$$Hg(OH)_2 \rightleftharpoons HgO + H_2O \quad (5.13)$$

Hg_2F_2 is more soluble than Hg_2Cl_2 in water; however, the literature provides no reliable information concerning its solubility.[3,4]

Figure 5.2 illustrates the coordination of Hg_2X_2 (X = F, Cl, Br, I).

5.2.2 Mercury(II) Fluoride – HgF_2

HgF_2 forms colorless crystals with a cubic structure and lattice parameter $a = 0.555$ nm.[44] Each Hg atom in the lattice is surrounded by eight closely set atoms of fluorine ($z = 8$). HgF_2 is synthesized by exposing Hg_2F_2 to chlorine at 543 K or NOF·3HF at 473 K. Mercury(II) fluoride, HgF_2, can also be obtained *via* reaction of gaseous fluorine with $HgCl_2$ or $HgBr_2$. Mercury(II) fluoride

Table 5.6 Physical properties of Hg_2F_2.

Property	*Value*	*Ref.*
Molecular weight ($g\,mol^{-1}$)	439.18	
T^{melt} (K)	decomposes > 843 K	36
$H^{\circ}_{f,298.15}$ ($kJ\,mol^{-1}$)	–492 at 25 °C	37
$G^{\circ}_{f,298.15}$ ($kJ\,mol^{-1}$)	–437.5 at 10 °C	38
$S^{\circ}_{298.15}$ ($J\,mol^{-1}\,K^{-1}$)	40 (estimated)	1
Crystal structure	Tetragonal	39
Color	Yellow	36
Distances:		
d(Hg–Hg) (nm)	0.243	40
	0.251	41
	0.243	42
	0.253	36
d(Hg–X) (nm)	0.241	40
	0.214	41
	0.231	42
d(X–X) (nm)	0.385	40

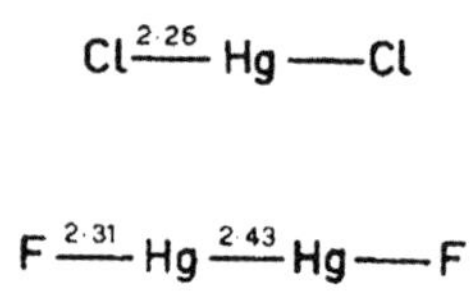

Figure 5.2 Structures of halides Hg_2X_2 (X=F, Cl, Br, I). Bond angles are not exactly 180 °.
Reproduced with permission from Ref. 43.

easily forms a dihydrate, $HgF_2{\cdot}2H_2O$. The dihydrate has a density of 5.72 g cm^{-3},[24] and crystallizes in an orthorhombic unit cell.

In dilute HF solutions (0–4.3%), the solubility in the HgO + HF + H_2O system at 298 K has shown that the solid phase is HgO.[45] Equilibrium of the same system in concentrated HF (5.9–100%) has shown that at 5.9–18.4% HF the solid-state phase is HgOHF and at 23.6–76.7% HF it is $HgF_2{\cdot}H_2O$.[46] The solubility of HgO decreases from 15.0 to 2.8% as the HF concentration increases from 23.6 to 76.7%. Of all of the mercury halides, only mercury(II) fluoride is dissociated in aqueous media. It forms unstable complexes in the system Hg(II)–F^-. The ion formation constant in the reaction

$$Hg^{2+} + F^- \rightleftharpoons HgF^+ \qquad (5.14)$$

according to Aylett[3] equals 10 and according to Hepler and Olofsson[1] equals 38. The thermodynamic properties of the hydrated ion $HgF^+_{(aq)}$ have been published.[1,47] Table 5.7 gives physical and thermodynamic properties of HgF_2.

Values of ΔG°_f and ΔH°_f for Hg_2X_2 decrease as the transition from fluorides to iodides occurs. However, $\Delta S^{\circ}_{298.15}$ increases in the same sequence. The molar heat capacity of mercury(I) halides depends only slightly on the nature of the

Table 5.7 Physical properties of HgF_2.

Property	*Value*	*Ref.*
Molecular weight ($g\,mol^{-1}$)	238.39	
Melting point, (K)	decomposes > 918	36
ΔH_{melt} ($kJ\,mol^{-1}$)	23.0 ± 4.2	48
Density ($g\,cm^{-3}$)	8.95 (15 °C)	39
Crystal structure	Cubic	36

halide. Overviews of the physical properties of mercury fluorides have appeared.[46,49] Although mercury(I) fluorides are poisonous, they are widely used for the fluorination of organic compounds.

5.2.3 Mercury(I) Chloride (Calomel) – Hg_2Cl_2

The thermodynamic and physical properties of Hg_2Cl_2 are given in Table 5.8. Hg_2Cl_2 forms colorless tetragonal crystals: $z = 2$, $a = 0.445$ nm, $c = 1.089$ nm.[40] X-ray diffraction analysis of solid mercury(I) halides demonstrates that they all have a similar structure, in which each atom of mercury creates an interatomic bond with two closest located atoms of halide. The melting point (under pressure) is 798 K (525 °C). As a result of disproportionation, mercury(I) chloride sublimes by decomposing into Hg and $HgCl_2$ at 656.25 K (383.7 °C):

$$Hg_2Cl_{2(gas)} \rightarrow HgCl_{2(gas)} + Hg_{(gas)} \tag{5.15}$$

The enthalpy of this reaction is $\Delta H_{disprop} = -506.26\,kJ\,mol^{-1}$.[3] Therefore, only an extremely strong Hg–Hg bond is able to stabilize the gaseous dimer Hg_2^{2+} (or Hg_2Cl_2), whereas the monomer HgCl will be subject to disproportionation. Given that, the phase transformation (melting, boiling) temperatures determined using standard methods will not be correct. There is no doubt that intensive mercury(I) halide disproportionation reactions occur at much lower temperatures than the phase transformation temperatures, which introduces an error into transformation temperature measurements. In addition, Hg_2Cl_2, just as other mercury(I) halides, especially mercury(I) iodide, decomposes through a disproportionation reaction when exposed to light. Mercury(I) chloride oxidizes to $HgCl_2$ under the effect of Cl_2 or $FeCl_3$. SO_2, $SnCl_2$ and Fe(II) restore $HgCl_2$ to Hg_2Cl_2. The solubility product of Hg_2Cl_2 between 278.15 and 318.15 K may be calculated[4,50] using the equation

$$\begin{aligned}\log K_s = &-(17.884 \pm 0.017) + (0.0622 \pm 0.0002)(T - 298.15)\\ &-(3.0 \pm 0.2) \times 10^{-4}(T - 298.15)^2\,mol\,kg^{-1}\end{aligned} \tag{5.16}$$

Values obtained through this equation agree well with published experimental data for K_s.[4] The observed solubility of Hg_2Cl_2 in saturated solutions at

Table 5.8 Physical properties of Hg_2Cl_2.

Property	*α-Hg_2Cl_2*	*β-Hg_2Cl_2*	*Ref.*
Molecular weight (g mol^{-1})	472.086		
Ferroelastic transition (K)	185		51
T^{melt} (K)		798[a]	51
Density (g cm^{-3})		7.18	52
$H^{\circ}_{f,298.15}$ (kJ mol^{-1})		–265.6	1
$G^{\circ}_{f,298.15}$ (kJ mol^{-1})		–210.8	1
$S^{\circ}_{298.15}$ (J mol^{-1} K^{-1})		–183.8	1
Refractive index:			
n_0		1.96	53
n_e		2.62	53
Crystal structure		Tetragonal	51
Distances (nm)			
d(Hg–Hg)		0.260	40
		0.245	36
d(Hg–Cl)		0.236	40
		0.252	54
		0.241	55
d(Cl–Cl)		0.333	54
		0.370	55

[a]Constrained pressure.

Table 5.9 Solubility product of Hg_2Cl_2 (adapted from Ref. 4).

T *(K)*	T *(°C)*	K_s *(mol^3 kg^{-3})*
278.15	5.00	$(5.65 \pm 0.22) \times 10^{-20}$
298.15	25.00	$(1.433 \pm 0.056) \times 10^{-18}$
318.15	45.00	$(1.738 \pm 0.068) \times 10^{-17}$

298.15 K is $(7.5 \pm 0.3) \times 10^{-6}$ M.[30] According to Marcus,[50] the solubility of Hg_2Cl_2 is calculated to be $(8.4 \pm 1.0) \times 10^{-6}$ M.

The solubility product of Hg_2Cl_2[4] at several temperatures is given in Table 5.9.

The relationships between $\log K_{sp}$ and $1/T$ for Hg_2Cl_2 and other mercury(I) halide solubility products are linear over a wide temperature range. Experimental data are in good agreement with the calculated solubility product.

5.2.4 Mercury(II) Chloride (Corrosive Sublimate) – $HgCl_2$

Mercury(II) chloride forms colorless orthorhombic crystals with lattice parameters $z = 4$, $a = 0.5963$ nm, $b = 1.2735$ nm and $c = 0.4325$ nm.[6] The vapor pressure of mercury(II) chloride vapor pressure[57] is found through the equation

$$\ln\left[P_{HgCl_2}\ (\mathrm{Pa})\right] = 28.17 - \frac{9531}{T} \tag{5.17}$$

Dry $HgCl_2$ is stable when exposed to air. When solutions boil, mercury(II) chloride fumes escape with water vapor. Aqueous solutions of mercury(II) chloride have been the subject of many studies. The results of experiments performed to test the solubility of mercury(II) chloride in water and different electrolytes have been summarized.[4] Solubility values for $HgCl_2$ in water are listed in Table 5.10.

The recommended value for the solubility in water at 298.15 K (25 °C) is $0.269 \pm 0.003\ \text{mol kg}^{-1}$. The solubility of $HgCl_2$ between in the temperature range 273.15–378.15 K (0–105 °C) is expressed by the equation

$$\log[HgCl_2] = -56.7732 + \frac{64.4481}{\left(\frac{T}{100}\right)} + 29.7574\ln\left(\frac{T}{100}\right)\text{mol kg}^{-1} \quad (5.18)$$

and in the range 383–508 K (110–235 °C) by the equation

$$\log[HgCl_2] = -14.7003 - \frac{51.8426}{\left(\frac{T}{100}\right)}\text{mol kg}^{-1} \quad (5.19)$$

The calculated solubility is smaller than the experimental values listed given by Skinner *et al.*[34] at 298.15 K, greater at 303.15–368.15 K and again smaller at 373–378 K. The deviation is ± 3.5%. Experimental and calculated solubilities of mercury(II) chloride, bromide and iodide in water *versus* the inverse of temperature feature a break in the ln*m*–1/T plot.[91] Plots for mercury(I) chloride, bromide and iodide *versus* inverse of temperature are straight lines.[91]

Mercury(II) chloride is practically undissociated in aqueous solutions. Equilibrium reaction constants are as follows:[4]

$$HgCl_{2(form)} \rightleftharpoons Hg^{+}_{(aq)} + 2Cl^{-}_{(aq)} \quad K_{form\ eqn(5.20)} = 7.1 \times 10^{-15} M \quad (5.20)$$

$$HgCl_{2(form)} \rightleftharpoons HgCl_{2(aq)} \quad K_{form\ eqn(5.21)} = 0.11\ M \quad (5.21)$$

The ions $HgCl^+$, $HgCl_2$, $HgCl_3^-$ and $HgCl_4^{2-}$ are generated in aqueous solutions with free chloride ion concentrations between 1×10^{-3} and 1.0 M.[92–96] Equilibrium reaction constants for these ions are as follows:

$$Hg^{2+} + X^- \rightleftharpoons HgX^+ \quad (K_1, \Delta H_1) \quad (5.22)$$

$$HgX^+ + X^- \rightleftharpoons HgX_2 \quad (K_2, \Delta H_2) \quad (5.23)$$

$$HgX_2 + X^- \rightleftharpoons HgX_3^- \quad (K_3, \Delta H_3) \quad (5.24)$$

$$HgX_3^- + X^- \rightleftharpoons HgX_4^{2-} \quad (K_4, \Delta H_4) \quad (5.25)$$

$$Hg^{2+} + 2X^- \rightleftharpoons HgX_2 \quad (\beta_2, \Delta H_\beta) \quad (5.26)$$

$$Hg^{2+} + 2X^- \rightleftharpoons HgX_4^{2-} \quad (\beta_4 / \beta_3, \Delta H_3 + \Delta H_4) \quad (5.27)$$

$$Hg^{2+} + 4X^- \rightleftharpoons HgX_4^{2-} \quad (\beta_4, \Delta H_\beta 4) \quad (5.28)$$

Table 5.10 Solubility of mercury(II) chloride in water (adapted from Ref. 4).

T *(K)*	T *(°C)*	*Molality*, m_{HgCl_2} *(mol kg^{-1} H$_2$O)*	*Ref.*
273.15	0.00	0.21	58
273.15	0.00	0.16	59
273.25	0.10	0.149	60
274.05	0.90	0.173	61
277.65	4.50	0.185	60
278.15	5.00	0.1707	62
280.65	7.50	0.197	60
283.15	10.00	0.1920	62
283.15	10.00	0.193	63
286.95	13.80	0.201	60
288.15	15.00	0.2164	62
288.15	15.00	0.211	64
288.71	15.56	0.206	65
289.15	16.00	0.26	58
291.15	18.00	0.229	66
293.15	20.00	0.272	67
293.15	20.00	0.242	64
293.15	20.00	0.320	63
293.15	20.00	0.2407	62
294.05	20.90	0.244	61
298.15	25.00	0.266	68
298.15	25.00	0.272	69
298.15	25.00	0.273	70
298.15	25.00	0.267	71
298.15	25.00	0.271	72
298.15	25.00	0.269	73
298.15	25.00	0.267	74
298.15	25.00	0.2596	75
298.15	25.00	0.268	64
298.15	25.00	0.2658	76
298.15	25.00	0.272	77
298.15	25.00	0.2658	78
298.15	25.00	0.269	61
298.15	25.00	0.27	79
298.15	25.00	0.273	80
298.15	25.00	0.265	81
298.15	25.00	0.2711	62
298.15	25.00	0.263	82
298.15	25.00	0.257	83
298.25	25.10	0.281	60
302.65	29.50	0.302	60
303.15	30.00	0.305	84,85
303.15	30.00	0.305	86
307.15	34.00	0.344	87
308.15	35.00	0.342	88
308.15	35.00	0.345	89
308.15	35.00	0.331	81
311.15	38.00	0.401	60
312.35	39.20	0.380	61
317.15	44.00	0.45	58
318.15	45.00	0.424	81

Table 5.10 (*Continued*)

T *(K)*	T *(°C)*	*Molality*, m_{HgCl_2} *(mol kg^{-1} H$_2$O)*	*Ref.*
322.15	49.00	0.467	60
328.95	55.80	0.566	61
329.15	56.00	0.581	87
334.15	61.00	0.651	60
335.85	62.70	0.687	61
348.55	75.40	0.995	61
353.15	80.00	1.13	60
353.15	80.00	1.12	87
360.15	87.00	1.44	60
364.75	91.60	1.612	61
372.85	99.70	2.110	61
373.15	100.00	2.38	60
373.15	100.00	2.07	87
378.2	105.05	2.35	90
378.8	105.65	2.773	61
389	115.85	3.54	90
394	120.85	5.45	60
396	122.85	4.56	90
400	126.85	8.47	60
402	128.85	5.88	90
406	132.85	6.87	90
413	139.85	12.3	60
414	140.85	8.84	90
418	144.85	10.06	90
423	149.85	13.3	60
430	156.85	14.7	90
432	158.85	14.9	60
433	159.85	16.4	60
437	163.85	17.5	60
438	164.85	16.5	60
448	174.85	23.6	90
455	181.85	29.2	90
468	194.85	39.1	90
479	205.85	48.9	90
496	222.85	67.1	90
508	234.85	88.4	90

where $X^- = Cl^-$, Br^-, I^-. MeX_m^{2-m}/X^- can be found from the following equations:

$$K_1 = \frac{[HgX^+]}{[Hg^{2+}][X^-]} \tag{5.22a}$$

$$K_2 = \frac{[HgX_2]}{[HgX^+][X^-]} \tag{5.23a}$$

$$K_3 = \frac{[HgX_3^-]}{[HgX_2][X^-]} \tag{5.24a}$$

$$K_4 = \frac{[HgX_4^{2-}]}{[HgX_3^-][X^-]} \quad (5.25a)$$

$$\beta_2 = K_1 K_2 = \frac{[HgX_2]}{[Hg^{2+}][X^-]^2} \quad (5.26a)$$

$$\beta_4 / \beta_2 = K_3 K_4 = \frac{[HgX_4^{2-}]}{[HgX_2][X^-]^2} \quad (5.27a)$$

$$\beta_4 = K_1 K_2 K_3 K_4 = \frac{[HgX_4^{2-}]}{[Hg^{2+}][X^-]^4} \quad (5.28a)$$

The interaction of mercury(II) ions in the course of reactions (5.25)–(5.31) is attended by a change in enthalpy, which, in a 3 M solution of $NaClO_4$, is as given in Table 5.11.[97]

The standard thermodynamic functions of chloride complexes of mercury(II), according to different authors, differ considerably, as illustrated in Table 5.12 for two sets of results.[1,97]

Table 5.13 compares the known equilibrium constants ($\log K_i$) for different reactions taking place in Hg(II)–Cl^- systems.[92–95] The consecutive constants of reactions (5.22)–(5.26) differ widely among themselves, which is due to the specifics and accuracy of the different methods used, and also different concentrations of the solutions. The complex formation constants K_1, K_2,. . ., K_i

Table 5.11 Change in enthalpy of mercury(II) ions in a 3 M solution of $NaClO_4$.[97]

Parameter	ΔH *(kJ mol⁻¹)*
ΔH_1	–24.23 ± 1.00
ΔH_2	–27.15 ± 1.51
ΔH_3	–4.31 ± 0.88
ΔH_4	–6.19 ± 1.00
$\Delta H_{\beta 2}$	–51.38 ± 1.13
$\Delta H_3 + \Delta H_4$	–10.50 ± 0.50
$\Delta H_{\beta 4}$	–61.88 ± 1.26

Table 5.12 Standard thermodynamic functions of chloride complexes of mercury(II).[1,97]

Species	*Ref.*	$\Delta G^\circ_{form,\ 298.15}$ *(kJ mol⁻¹)*	$\Delta H^\circ_{form,\ 298.15}$ *(kJ mol⁻¹)*	$S^\circ_{form,\ 298.15}$ *(J mol⁻¹ K⁻¹)*
$[HgCl_3]^-$	1	308.8	389.5	205.0
	97	314.5 ± 1.1	381.1 ± 0.7	252.4 ± 4.4
$[HgCl_4]^{2-}$	1	446.4	554.8	289.0
	97	449.0 ± 1.2	558.4 ± 0.9	288.4 ± 5.0

Table 5.13 Equilibrium constants for the system Hg(II)–Cl^-.[43,61,93,94]

Equation	*LogK$_i$*	*298 K*[43]	*298 K*[61]	*298 K*[93]	*295 ± 1 K*[94]
(5.22)	LogK_1	6.73 ± 0.98	–	6.72 ± 0.02	–
(5.23)	LogK_2		6.30 ± 0.02	–	–
(5.24)	LogK_3	0.85 ± 0.15	0.95 ± 0.03	1.00 ± 0.01	1.25 ± 0.25
(5.25)	LogK_4	–	–	–	1.36 ± 0.25
(5.26)	LogK_3K_4	0.85 ± 0.05	2.00 ± 0.05	1.97 ± 0.05	2.65 ± 0.50

for $HgX_m^{(z-m)-}$ ($z = 2$; $m = 1$–4) given by Hepler and Olofsson[1] and Clever *et al.*[4] are in good agreement with the data presented in the literature.[92–96]

Consecutive constants are of great importance for reactions producing $HgCl^+$ and $HgCl_2$ ($m = 1$ and 2). The equilibrium constant, K_2, involved in the formation of $HgCl_2$ *via* the reaction

$$HgCl^+ + Cl^- \rightleftharpoons HgCl_2 \tag{5.29}$$

is more than 3.7×10^5 times greater than K_3 for the reaction

$$HgCl_2 + Cl^- \rightleftharpoons HgCl_3^- \tag{5.30}$$

Notably, $K_1 > K_2 \gg K_3 > K_4$, whereas K_1 is comparable to K_2 and K_3 to K_4. Such a relationship is also characteristic of other halides (Br^-, I^-) and ions.

Table 5.17 gives the formation constants for the mercury(II) complexes formed with F^-, Cl^-, Br^-, CN^- and SCN^- ions, according to Hepler and Olofsson.[1] It should be mentioned that the sequence F^-, Cl^-, Br^-, I^-, CN^- is characterized by increasing complex ion formation constants for $HgX_m^{(z-m)-}$, which is due to increasing deformability of X^- ions from F^- through CN^-.

Mercury(II) halides, HgX_2, are practically undissociated in aqueous solutions, as mentioned above, yet if exposed to excess alkali metal halides, MeX (Me = Na, K, Rb, Li), or saturated solutions of ammonium halides they form highly soluble and dissociated complexes $Me_2[HgX_4]$ and $(NH_4)_2{\cdot}[HgX_4]$:

$$2MeX + HgX_2 \rightleftharpoons Me_2[HgX_4] \tag{5.31}$$

$$Me_2[HgX_4] \rightleftharpoons Me^{2+} + HgX_4^{2-} \tag{5.32}$$

$$2NH_4Cl + HgX_2 \rightleftharpoons (NH_4)_2[HgX_4] \tag{5.33}$$

The effect of chloride ions on the formation of dissociated forms of complexes is illustrated in Figure 5.3, depicting the production of mono-, di-, tri- and tetrachlorides of mercury(II) as a function of concentration of free chloride ions.[98] The aqueous solubility of dihalides increases at the same time. Thus, the poorly soluble mercury(II) iodide does dissolve if exposed to excess mercury salts, producing particles $[Hg–I–Hg]^{3+}$. This property of mercury dihalides is exploited when designing electrolytes needed to obtain high-purity mercury.

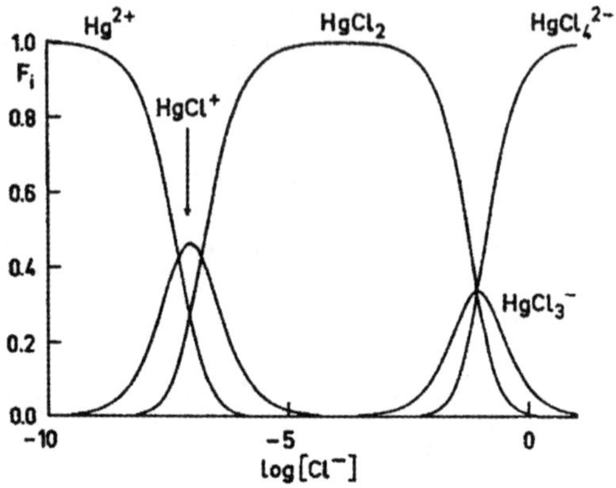

Figure 5.3 Yield of complex ions of mercury(II) as a function of free chloride ion concentration.
Reproduced with permission from Ref. 99.

Table 5.14 Molar solubility of Hg(II) halides in organic solvents at 25 °C.

Solvent	*$HgCl_2$*	*$HgBr_2$*	*HgI_2*	*Ref.*
Benzene	16×10^{-3}	16×10^{-3}	7.5×10^{-3}	100
Benzene	15.6×10^{-3}	16.6×10^{-3}		101
Diethyl ether	0.170	67×10^{-3}	8.3×10^{-3}	100
Acetone	3.22	0.96	31×10^{-3}	100
Acetonitrile	1.83	0.26	7.6×10^{-3}	100
Methanol	2.0	1.5	68×10^{-3}	100
Dimethyl sulfoxide	2.00	3.25	4.25	100
Pyridine	0.90	0.80	0.70	100
Piperidine	44×10^{-3}	0.97	1.20	100
Chloroform	2.2×10^{-3}	3.2×10^{-3}		101
Toluene	22.1×10^{-3}	22.7×10^{-3}		101
o-Xylene	34.9×10^{-3}	38.2×10^{-3}		101

The solubility of mercury(II) halides in organic solvents is given in Table 5.14. Physical properties of $HgCl_2$ are summarized in Table 5.15.

5.2.5 Mercury(I) Bromide – Hg_2Br_2

Physical properties of Hg_2Br_2 are given in Table 5.16. Solid Hg_2Br_2 forms colorless tetragonal crystals with $z=2$, $a=0.465$ nm and $c=1.110$ nm.[109] It does not form hydrates.[54] Hg_2Br_2 dissolves when heated in concentrated nitric acid, hot concentrated sulfuric acid or hot ammonium carbonate solution. Hg_2Br_2 is less soluble than Hg_2Cl_2 in water. The recommended solubility product K_s at 298.25 is 6.40×10^{-23} mol kg^{-3} H_2O.[4] The temperature dependence of the mercury(I) bromide solubility product at zero ionic force can be calculated using the equation

$$\lg K_s = 55.306 - 235.22(T/100) - 25.195\ln(T/100)\ \mathrm{mol^3\,kg^{-3}} \qquad (5.34)$$

Table 5.15 Physical properties of $HgCl_2$.

Property	*Value*	*Ref.*
Molecular weight (g mol^{-1})	271.495	
Melting point (K)	553	36
$\Delta H^{sublimation}$ (kJ mol^{-1})	80.8	102
	77.4	103
	83.1	104
Boiling point (K)	591	36
Critical temperature (K)	972	105
Density (g cm^{-3})	5.44	56
$H^{\circ}_{f,298.15}$ (kJ mol^{-1})	–225.9	1
	–229.2	106
	–226.0	107
$G^{\circ}_{f,298.15}$ (kJ mol^{-1})	–180.3	1
$S^{\circ}_{298.15}$ (J mol^{-1} K^{-1})	–152.9	1
Color	White	36
Crystal structure	Orthorhombic	108
Distances (nm):		
d(Hg–Cl)	0.229	56
d(Cl–Cl)	0.333	56
Cl–Hg–Cl bond angle (°)	178.9	56

*Calculated.

obtained *via* regression analysis of known experimental data[97,110,111] and summarized by Clever *et al.*[4]

The Hg–Br phase diagram (Figure 5.4)[112] shows the phase relations between Hg and Br. Mercury(II) bromide forms by a congruent reaction with liquid whereas Hg_2Br_2 forms by a syntectic reaction between L_1 and L_2. Liquidus points are from Dworsky and Komarek.[115] The phase diagrams of Hg–Cl and Hg–I are similar.[113,114]

Figure 5.5 illustrates the environment of mercury in the structure of mercury(II) chloride, bromide and iodide.

Table 5.18 provides experimental data values for the solubility of mercury(I) bromide in water. These data are internally consistent. According to Table 5.18, as the ionic force increases, the Hg_2Br_2 solubility product increases from 6.40×10^{-23} at $\mu=0$ to 670×10^{-23} at $\mu=3.1$, where μ is the ionic force. Mercury(I) bromide is obtained *via* anodic dissolution of mercury in HBr, by exposing $Hg_2(NO_3)_2$ solution to KBr in nitric acid. Hg_2Br_2 is used in electrolytes used to refine mercury, electrochemical experiments, optoacoustic bulk devices, organomercuric compound synthesis, organic catalysis, *etc.*

5.2.6 Mercury(II) Bromide – $HgBr_2$

Solid $HgBr_2$ forms colorless orthorhombic crystals with lattice parameters $z=4$, $a=0.679$ nm, $b=1.2445$ nm and $c=0.4624$ nm.[109] Its vapor pressure was determined.[121] The vapor pressure of liquid $HgBr_2$[121,122] is also well known.

Table 5.16 Physical properties of Hg_2Br_2.

Property	*α-Hg_2Br_2*	*β-Hg_2Br_2*	*Ref.*
Molecular weight (g mol^{-1})		560.988	
$T^{\alpha\rightarrow\beta}$ (ferroelastic transition) (K)	143		51
T^{melt} (K)		727.35 (decomp.)	115
$\Delta H^{sublimation}$ (kJ mol^{-1})		84.1	112
T^{boil} (K)		618 (sublimes)	112
Density (g cm^{-3})		7.307	116
$H^0_{f,298.15}$ (kJ mol^{-1})		–206.94	1
$G^0_{f,298.15}$ (kJ mol^{-1})		–181.08	117
			118
$S^0_{298.15}$ (J mol^{-1} K^{-1})		86.7	1
Color			
Refractive index:			
n_0		2.12	119
n_e		2.98	119
Crystal structure		Tetragonal	54
Distances (nm)			
d(Hg–Hg)		0.258	54
		0.250	55
d(Hg–Br)		0.249	120
		0.253	54
		0.245	55
d(Br–Br)		0.271	120
		0.340	54
		0.355	55

In aqueous solution, $HgBr_2$ forms complex solutions in which the following forms are in equilibrium: Hg^{2+}, Br^-, $HgBr^+$, $HgBr_2{}^0$, $HgBr_3{}^-$, $HgBr_4{}^{2-}$ and $HgOH^+$. $HgBr_2{}^0$ molecules prevail in aqueous solution. $HgBr_2$ exhibits a high solubility in water[4] if samples contain traces of HBr. This is due to complex formation in the system Hg(II)–Br^-. The aqueous solubility of mercury(II) bromide is lower than that of mercury(II) chloride, being 1.70×10^{-2} mol kg^{-1} at 298.15 K. Solubility can be found from the following equation in the temperature interval 273–353 K:

$$\ln m_{HgBr_2} = 5.3570 - 28.096\left(\frac{T}{100}\right) \text{mol kg}^{-1} \quad (5.35)$$

and at higher temperatures using

$$\ln m_{HgBr_2} = 85.918 - 380.791\left(\frac{T}{100}\right) \text{mol kg}^{-1} \quad (5.36)$$

Table 5.19 gives experimental solubility values for $HgBr_2$ in water at temperatures between 273.15 and 474 K. From analysis of the experimental data, it was found that within the temperature ranges 273.15–437 K (0–164 °C) and 437–509 K (164–236 °C) the solubility curves of log[$HgBr_2$] *versus* $1/T$ have different temperature coefficients. The solubility curves intersect at 437 K. A sharp increase in the solubility of $HgBr_2$ in water occurs at temperatures

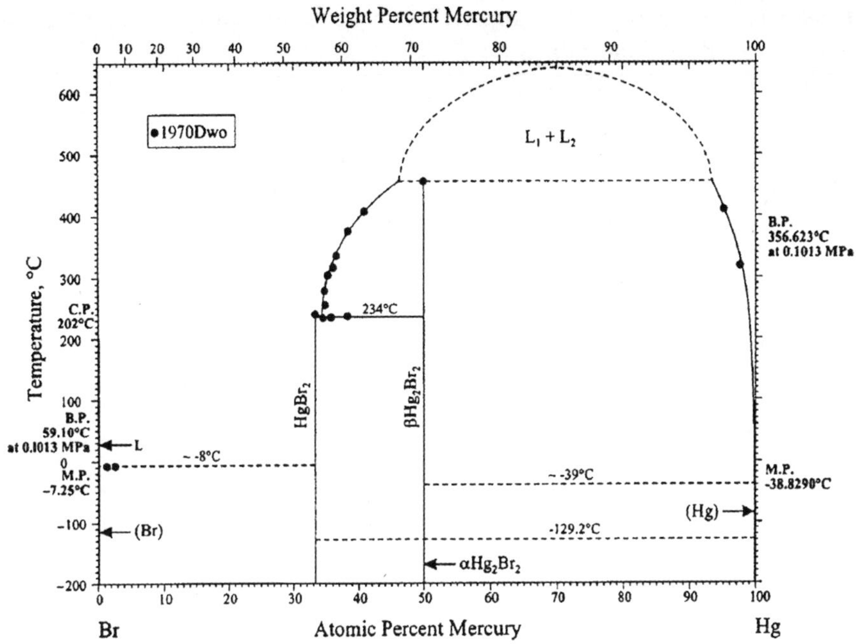

Figure 5.4 The Hg–Br phase diagram. Hg_2Br_2 forms by a syntectic reaction ($L_1 + L_2 \rightarrow Hg_2Br_2$). Hg_2Cl_2 and Hg_2I_2 also form in the same type of reaction.[112–114] Liquidus data points are from Ref. 115.

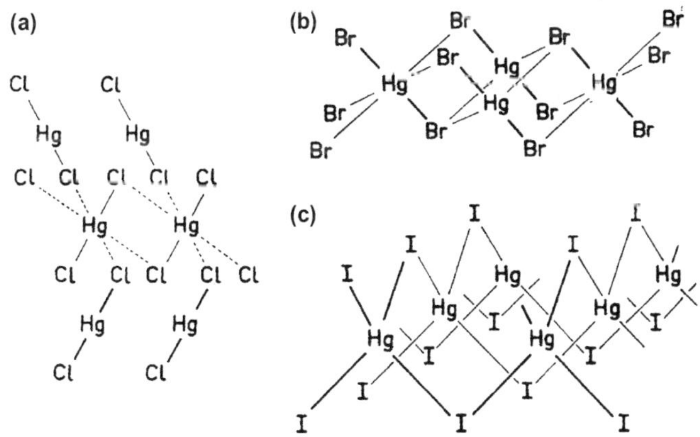

Figure 5.5 Environment of mercury in the crystal structure of (a) $HgCl_2$, (b) $HgBr_2$ and (c) HgI_2.
Reproduced with permission from Ref. 43.

above 437 K (164 °C). A temperature rise of only 28 °C (from 446 to 474 K) increases the solubility 458-fold. This effect has the following explanation. Clearly, the sharp increase in the solubility of $HgBr_2$ in water is due to the acid–base interaction that occurs in the course of dissociation of mercury(II) bromide and formation of Hg^{2+} and Br^-. Br^- and $HgBr_2$ exhibit acidic

Table 5.17 Formation of mercury(II) complexes at 298.15 K according to Refs 1 and 110.

		Formation constant					
Reaction	K_i	F^-	Cl^-	Br^-	I^-	CN^-	SCN^-
$Hg^{2+}+X^- \rightarrow HgX^+$	K_1	38	5.8×10^6	1.1×10^9	6.4×10^{12}	2.0×10^{17}	1×10^9
$HgX^++X^- \rightarrow HgX_2$	K_2	–	2.5×10^6	2.5×10^8	1.3×10^{11}	1.7×10^{17}	1×10^8
$HgX_2+X^- \rightarrow HgX_3^-$	K_3	–	6.7	1.5×10^2	6.2×10^3	5.5×10^3	7×10^2
$HgX_3^-+X^- \rightarrow HgX_4^{2-}$	K_4	–	13	23	1.1×10^2	1.0×10^3	7×10^1
$Hg^{2+}+4X^- \rightarrow HgX_4^{2-}$	β_4	–	1.3×10^{15}	9.2×10^{20}	5.6×10^{29}	1.9×10^{41}	5×10^{21}
$Hg^{2+}+2X^- \rightarrow HgX_2$	β_2	–	1.45×10^{13}	2.75×10^{17}	–	–	–
$HgX_2+2X^- \rightarrow HgX_4^{2-}$	$\beta_4\beta_2$	–	8.96×10^1	3.35×10^3	–	–	–

Table 5.18 Experimental values of the mercury(I) bromide solubility product in aqueous solution as a function of temperature.

T *(K)*	T *(°C)*	K_s *(mol kg^{-1})*	*Ref.*
283.95	10.80	0.545×10^{-23}	123
288.05	14.90	1.00×10^{-23}	123
288.15	15.00	0.968×10^{-23}	124
292.35	19.20	3.89×10^{-23}	123
293.15	20.00	2.56×10^{-23}	124
298.15	25.00	5.50×10^{-23}	123
298.15	25.00	6.43×10^{-23}	124
299.65	26.50	6.95×10^{-23}	123
303.15	30.00	15.21×10^{-23}	124
308.15	35.00	35.85×10^{-23}	124
313.15	40.00	81.33×10^{-23}	124
318.15	45.00	177.6×10^{-23}	124

properties, whereas Hg^{2+}, with basic properties, creates complexes $Hg[HgBr_3]^+$ and $Hg[HgBr_4]$ with greater water solubility.

Table 5.19 demonstrates the large change in the solubility of $HgBr_2$ with temperature. Clever *et al.*[4] calculated the $HgBr_2$ solubility product at 298.15 K to be $K_s = 6.2\times10^{-20}\,mol^3\,kg^{-3}$, which is in good agreement with the value reported by Iwamoto *et al.*[129]

A study of the reaction equilibrium:

$$Hg^{2+} + HgBr_2 \rightleftharpoons 2HgBr^+ \tag{5.37}$$

demonstrated that the equilibrium constant:

$$K_{12} = \frac{[HgBr^+]^2}{[Hg^{2+}][HgBr_2]} \tag{5.38}$$

equals 6.6 ± 0.2.[97] Mercury(II) bromide is also practically undissociated. The reaction equilibrium:

$$HgBr_{2(solid)} \rightleftharpoons Hg^{2+} + 2Br^- \tag{5.39}$$

Table 5.19 Mercury(II) bromide solubility in water as a function of temperature (adapted from Ref. 4).

T (K)	T (°C)	Molarity (mol dm^{-3})	Molality (mol kg^{-1})	Ref.
273.15	0.00		0.008	125
277.65	4.50		0.0075	126
283.55	10.40		0.0119	126
293.15 ± 1.0	20.00	0.00223		127
298.15	25.00	0.017		70
298.15	25.00		~0.011	68
298.15	25.00	0.0167		72
298.15	25.00	0.017		73
298.15	25.00	0.017		74
298.15	25.00		0.0170	128
298.15	25.00		0.0170	126
298.15	25.00	0.017		83
307.15	34.00		0.02	125
353.15	80.00		0.08	125
415	141.85		0.378	90
437	163.85		0.801	90
446	172.85		1.40	90
458	184.85		4.01	90
460	186.85		8.72	90
461	187.85		16.4	90
462	188.85		54.8	90
466	192.85		166	90
474	200.85		641	90

is shifted to the left. The equilibrium of the reverse reaction in aqueous solution containing 3.0 and 0.5 M ClO_4^-:

$$Hg^{2+} + Br^- + H_2O \rightarrow HgBr(OH) + H^+ \quad (5.40)$$

is complicated by hydrolysis.[1] In aqueous solutions with a free bromide ion concentration of 1×10^{-3}–1.0 M, the following forms of $HgBr_2$ are created: $HgBr_2$, $HgBr_3^-$ and $HgBr_4^{2-}$. The reaction equilibrium constants for eqns (5.22)–(5.29), where $X^- = Br^-$, are given in Table 5.17. The physical properties of $HgBr_2$ are given in Table 5.20.

Mercury(II) bromide is fairly soluble in organic solvents. Table 5.17 gives solubility data for $HgCl_2$, $HgBr_2$ and HgI_2 in several organic solvents. Mercury(II) bromide is also soluble in acetone, benzene and carbon disulfide but poorly soluble in diethyl ether. It is used as a catalyst in organic synthesis and analytical chemistry and in the production of high-purity mercury.

5.2.7 Mercury(I) Iodide – Hg_2I_2

Physical properties of mercury(I) iodide are given in Table 5.21. Solid Hg_2I_2 forms tetragonal crystals with parameters $z = 2$, $a = 0.492$ nm and $c = 1.161$ nm.[133] Mercury(I) iodide disproportionates into Hg and HgI_2 when exposed to light. It has a weak solubility in water: the most credible value for

Table 5.20 Physical properties of $HgBr_2$.

Property	*Value*	*Ref.*
Molecular weight ($g\,mol^{-1}$)	360.40	
Melting point (K)	511	115
Boiling point (K)	591	36
Density ($g\,cm^{-3}$)	6.08	4
$\Delta G^{\circ}_{f,298.15}$ ($kJ\,mol^{-1}$)	–153.1	130
$\Delta H^{\circ}_{f,298.15}$ ($kJ\,mol^{-1}$)	–170.7	1
	–175.5	131
	–166.2 ± 4	107
$\Delta S^{\circ}_{298.15}$ ($J\,mol^{-1}\,K^{-1}$)	167	1
Color	Colorless as solution Yellow as liquid	36
Crystal structure	Orthorhombic	132
Distances (nm):		
d(Hg–Hg)		40
d(Hg–Br)		40
d(Br–Br)		40

Table 5.21 Physical properties of Hg_2I_2.

Property	*Value*	*Ref.*
Molecular weight ($g\,mol^{-1}$)	654.989	
T^{melt} (K)	570 (decomp.)	113
ΔH^{fusion} ($kJ\,mol^{-1}$)	10.32	133
Density ($g\,cm^{-3}$)	7.78	134
$H^0_{f,298.15}$ ($kJ\,mol^{-1}$)	–121.34	130
	–123.2 ± 8	135
$G^0_{f,298.15}$ ($kJ\,mol^{-1}$)	–111.0	130
$S^0_{298.15}$ ($J\,mol^{-1}\,K^{-1}$)	233.5	130
Crystal structure	Tetragonal	1
Distances (nm):		
d(Hg–Hg)	0.269	54
d(Hg–I)	0.268	54
d(I–I)	0.355	54

the solubility product at 298.15 K is $K_s = 5.2 \times 10^{-29}\,mol^3\,kg^{-3}$.[1,110] From 283 to 298 K the solubility product can be found from the equation

$$\log K_{s(Hg_2I_2)} = -30.72 + 0.094(T - 273.15)\,mol^3 kg^{-3} \tag{5.41}$$

and between 273.15 and 373.15 K the following equation is suggested:

$$\log K_{s(Hg_2I_2)} = -3.5483 - \frac{7347}{T} + 0.0044 \log T + 0.293 \times 10^{-3} T\,mol^3 kg^{-3} \tag{5.42}$$

Table 5.22 Solubility product of mercury(I) iodide in aqueous solution (adapted from Ref. 4).

T (K)	T (°C)	$K_s \times 10^{29}$ ($mol^3\ kg^{-3}$)	Ref.
283.15	10.0	2.01×10^{30}	123
288.05	14.9	5.10×10^{30}	123
292.35	19.2	1.05×10^{29}	123
298.15	25.0	4.95×10^{29}	123

Table 5.22 gives results for $K_{s(Hg_2I_2)}$.

Mercury(I) iodide does not form hydrates. It disproportionates in potassium iodide solutions through the reaction

$$Hg_2I_2 + 2KI \rightleftharpoons K_2[HgI_4] + Hg^0 \qquad (5.43)$$

with $K_{disprop} = 0.67$. Hg_2I_2 may be synthesized by exposing Hg (in excess) to HgI_2 at $T < 563$ K or by reacting $HgCl_2$ with stannous chloride in an alcoholic solution of KI. Hg_2I_2 dissolves in castor oil and aqueous ammonia solution but does not dissolve in ethanol and diethyl ether.

5.2.8 Mercury(II) Iodide – HgI_2

The physical properties of mercury(II) iodide are given in Table 5.25. Table 5.23 lists solid-state transformations. The vapor pressures of solid and liquid HgI_2 have been published.[122,123,136] The different crystal structures of mercury(II) iodide are quite complex. The current Hg–I phase diagram has been drawn correctly,[46] but the true complexity of the various metastable forms of HgI_2 is not apparent. The metastable yellow (yellow^M) and orange (orange^M) forms produced from solution growth are mechanically unstable and transform into the red form (stable at room temperature and pressure). Orange crystals, when heated to 127 °C, will transform into the stable high-temperature yellow (yellow^{HT}) form of HgI_2.

The red to yellow transformation occurs with a heat of transition of 2.68–2.85 kJ mol^{-1}.[142,143] Mercury(II) iodide does not form hydrates. Table 5.24 gives the aqueous solubility of HgI_2 at various temperatures. Figure 5.6 shows the HgI_2–H_2O phase diagram.[90]

5.2.8.1 *Yellow HgI_2*

There are two yellow polymorphs of HgI_2: one is metastable at room temperature, designated yellow^M, and the other is the stable high-temperature form, yellow^{HT}. The metastable yellow form is formed at room temperature by sublimation or by crystallization from solution. The high-temperature yellow form results from a phase transition at 127–130 °C. The metastable yellow

Table 5.23 Solid-state phase transformations in mercury(II) iodide.[137]

Transformation	*Crystal structure*	*Color*	*Transition temperature (°C)*	*Ref.*
α-HgI_2 → β-HgI_2	Tetragonal → monoclinic	Red → yellow^{HT}	127–130	138,139
Yellow^{M}-HgI_2 → α-HgI_2	Orthorhombic → tetragonal	Yellow^{M} → red	25	35, 139, 140
Orange^{M}-HgI_2 → α-HgI_2	Tetragonal → monoclinic	Orange → yellow^{HT}	127	140,141

Table 5.24 Aqueous solubility of mercury(II) iodide as a function of temperature (adapted from Ref. 4).

T *(K)*	T *(°C)*	C_{HgI_2} *(mol dm^{-3})*	*Molality (mol kg^{-1})*	*Ref.*
291 ± 2	17.85 ± 2	7.4×10^{-5}		144
290.65	17.50	8.87×10^{-5}		145
295.15	22.00	1.18×10^{-4}		145
295.65	22.50		2.2×10^{-4}	146
298.15	25.00	$\sim1.3\times10^{-4}$		68
298.15	25.00	9.77×10^{-5}		147
298.15	25.00	$1.05\pm0.055\times10^{-4}$		148
298.15	25.00	1.3×10^{-4}		83
373.15	100.00		4.0×10^{-3}	149
469	195.85		8.1×10^{-2}	90
502	228.85		0.21	90
514	240.85		0.25	90

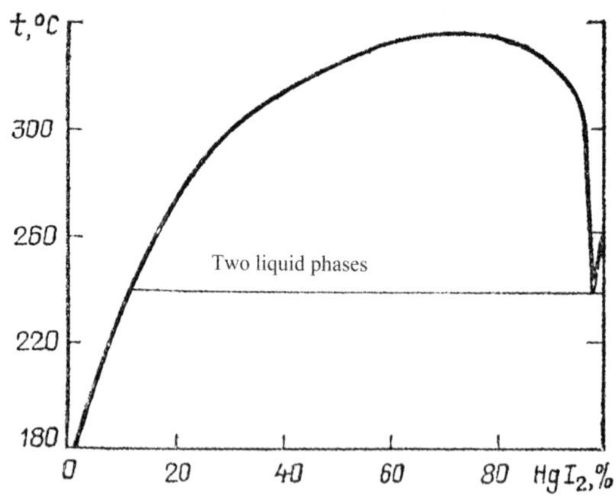

Figure 5.6 HgI_2–H_2O phase diagram.
Reproduced with permission from Ref. 90. Copyright © 1937 Verlag GmbH & Co. KGaA, Weinheim.

structure contains almost linear (178.3°) I–Hg–I chains and has an orthorhombic structure. Under pressure, both the orange and metastable yellow forms of HgI_2 transform into the red HgI_2 structure, although by different mechanisms.[138] The crystal structure of the stable high-temperature yellow form is monoclinic.[139]

5.2.8.2 *Orange HgI_2*

Three forms of orange HgI_2 are built from Hg_4I_{10} supertetrahedra. Two forms are polytypic structures[141] and the third form is created from interpenetrating three-dimensional networks of Hg_4I_{10} supertetrahedra. The two polytypic structures are built on Hg_4I_{10} supertetrahedra linked at each corner.

Experimental values for the solubility of HgI_2 in water are listed in Table 5.24. A critical assessment of the literature data was performed on the solubility of mercury(II) iodide in water.[4] Its solubility in the temperature interval 288–323 K can be determined with the equation

$$\ln m_{HgI_2} = 7.608 - 49.576\left(\frac{T}{100}\right) \text{mol kg}^{-1} \tag{5.44}$$

and in the temperature interval 463–513 K using

$$\ln m_{HgI_2} = 10.751 - 64.134\left(\frac{T}{100}\right) \text{mol kg}^{-1} \tag{5.45}$$

The physical properties of HgI_2 are given in Table 5.25. The temperature coefficients of the HgI_2 solubility curves, in coordinates of $\log m$ *versus* $1/T$, up to the transition temperature from the tetragonal red to monoclinic yellow modification at 400 K (127 °C), do not differ much between themselves. This is due to the small change in dissolution enthalpy, 41 kJ mol^{-1}, for the tetragonal red modification and 52 kJ mol^{-1} for the monoclinic yellow modification.[4] At 514 K (241 °C), the HgI_2–H_2O system goes through a transition from solid liquid to liquid–liquid. As can be seen from the HgI_2–H_2O phase diagram[90] (Figure 5.6), the system achieves full miscibility at 611 K (338 °C)

Table 5.25 Physical properties of HgI_2.

Property	*α-HgI_2 Red*	*β-HgI_2 Yellow*	*Ref.*
Molecular weight (g mol^{-1})	454.40		
Melting point (K)		530.15	113
$\Delta H^{\alpha\rightarrow\beta}$ (kJ mol^{-1})	2.9 ± 0.02		142
	2.85 ± 0.02		143
$\Delta H^{\beta\rightarrow \text{melt}}$ (kJ mol^{-1})		18.8	150
ΔS_{melt} (J mol^{-1} K^{-1})	35.98		
$\Delta H^{\text{sublimation}}$ (kJ mol^{-1})	85.8		102
	85.5		104
Boiling point (K)		624	36
Density (g cm^{-3})	6.30	6.38	151
$\Delta H^{\circ}_{f,298.15}$ (kJ mol^{-1})	–105.4	–102.9	130
$\Delta G^{\circ}_{f,298.15}$ (kJ mol^{-1})	–101.7		130
$\Delta S^{\circ}_{298.15}$ (J mol^{-1} K^{-1})	180.0		130
Crystal structure	See text		

and 77% HgI_2. The $HgCl_2$–H_2O, $HgBr_2$–H_2O and $Hg(CN)_2$–H_2O systems do not feature areas of liquid–liquid immiscibility.[90]

Calculation of the solubility product of HgI_2 by Clever *et al.*[4] and Yatsimirsky and Shutov[152] are in very close agreement at 298.15 K. The solubility product is $K_s = 2.9\times10^{-29}$ mol³ kg⁻³. The relationship between the solubility product and temperature can be calculated using the equation

$$\log K_s = -3.276 - 182.765 / \left(\frac{T}{100}\right) \text{mol}^3\text{kg}^{-9} \tag{5.46}$$

within an accuracy of 15%.[4] Mercury(II) iodide is moderately soluble in organic solvents also is also soluble in dioxane, chloroform and aqueous KI solution. The stability constant of mercury complexes formed *via* the reaction

$$Hg^{2+} + 4X^- \rightleftharpoons [HgX_4]^{2-} \tag{5.47}$$

increases with increasing deformability of anions in the sequence $X^- = Cl^-$, Br^-, I^- ($\beta_4 = 1.3\times10^{15}$, 9.2×10^{20} and 5.6×10^{29}, respectively). Formation constants of the complex $HgX_m^{(2-m)-}$ for equations (5.22)–(5.29) are given in Table 5.17.

5.2.9 Mixed Mercury(II) Halides

Equilibrium constants of the halide exchange reaction:[1]

$$HgX_{2(aq)} + HgY_{2(aq)} \rightleftharpoons 2HgXY_{(aq)} \tag{5.48}$$

where X, Y = Cl^-, Br^- and I^-, are listed in Table 5.26.

Equilibrium constants (*K*) of the exchange reaction at 298 K:[1,2,152,158–161]

$$HgBr_4^{2-} + nI \rightarrow [HgBr_{4-n}I_n]^{2-} + nBr^- \tag{5.49}$$

with $n = 1$, 2, 3 and 4 are given in Table 5.27.

HgI_2 is an important commercial material. It is used in X-ray and γ-ray detectors, metal halide lamps, electrochemical experiments, production of electrolytes needed to obtain high-purity mercury and analytical chemistry (*e.g.*, saturated $K_2[HgI_4]$) and $Ba[HgI_4]$ solutions).

5.2.10 Mercury(II) Cyanide – $Hg(CN)_2$

The physical properties of $Hg(CN)_2$ are given in Table 5.28. Mercury(II) cyanide decomposes at 693 K (420 °C) into mercury and cyanogen. Early attempts to obtain $Hg_2(CN)_2$ only succeeded in producing $Hg(CN)_2$ and

Table 5.26 Equilibrium constants and properties of mixed mercury(II) halides.

Mixed halide	*LogK*[153]	*LogK*[154]	*Density (g cm⁻³)*[155]	*Color*
HgBrI	1.07 ± 0.08	1.10 ± 0.20		Yellow–orange[156]
HgClI	1.35 ± 0.17	1.75 ± 0.20		
HgClBr	1.14 ± 0.11	2.0 ± 0.5	5.72 ± 0.19	White[157]

Table 5.27 Equilibrium constant in mixed halide exchange reaction (5.49).[1,2,152–161]

n	*Equilibrium constant*
1	1.6×10^3
2	3.1×10^5
3	2.4×10^7
4	6.0×10^8

Table 5.28 Physical properties of $Hg(CN)_2$.

Property	*Value*	*Ref.*
Molecular weight (g mol^{-1})	252.62	
Melting point (K)	320 C (decomp.)	165
Density (g cm^{-3})	3.996 (25 °C)	166
$\Delta H^{\circ}_{f,298.15}$ (kJ mol^{-1})	263.6	130
Crystal structure	Tetragonal	163,164
Distances (nm):		
d(Hg–C)	0.199	163,164
d(Hg–N)	0.270	163,164

metallic mercury as a result of a disproportionation reaction. Some researchers believe that $Hg_2(CN)_2$ can be obtained in non-aqueous solutions at low temperatures.[162] $Hg(CN)_2$ forms white tetragonal crystals with C–Hg–C bond angles of 171°.[163,164] It exhibits high water solubility, 93 g kg^{-1} H_2O at 287 K and 539 g kg^{-1} H_2O at 373 K.

Aqueous solutions of mercury(II) cyanide contain fully undissociated molecules $Hg(CN)_2$. The equilibrium constant for the dissociation of $Hg(CN)_2$:

$$Hg(CN)_2 \rightleftharpoons Hg^{2+} + 2CN^- \tag{5.50}$$

is 2.9×10^{-35}. The equilibrium constant of the reaction

$$Hg(CN)_2 \rightarrow Hg^{2+} + 2CN^- \tag{5.51}$$

is 1.9×10^{14}.[1,97] The equilibrium constants of exchange reactions of the type

$$Hg(CN)_2 + Hg(X)_2 \rightarrow 2HgX(CN) \tag{5.52}$$

where $X^- = Cl^-$, Br^- and I^- have been published,[167,168] and for $X^- = I^-$ is 0.11[167] or 0.14.[168]

The equilibrium constants of cyanide complexes depend on their composition,[1,169] as seen in the following equations:

$$Hg(CN)_2Cl^- + CN^- \rightarrow [Hg(CN)_3Cl]^{2-} \quad K = 3.3\times10^3 \tag{5.53}$$

$$Hg(CN)_3^- + Br^- \rightarrow [Hg(CN)_3Br]^{2-} \quad K = 4.2 \tag{5.54}$$

$$Hg(CN)_2 + Cl^- \rightarrow [Hg(CN)_2Cl]^- \quad K = 0.5 \tag{5.55}$$

$$Hg(CN)_3^- + Cl^- \rightarrow [Hg(CN)_3Cl]^{2-} \quad K = 0.3 \tag{5.56}$$

The resulting solutions that contain cyanide complexes, including misaligned ones, are electrically conductive and are used in engineering. Electrically conductive solutions can also be obtained by introducing excess sodium cyanide or other alkali metal cyanides:

$$Hg(CN)_2 + 2NaCN \rightleftharpoons 2Na^+ + Hg(CN)_4^{2-} \tag{5.57}$$

as background salts {$K_2[Hg(CN)_4]$, $Li_2[Hg(CN)_4]$, *etc.*}. Mercury(II) cyanide can be obtained from mercury oxide and $KFe^{II}Fe^{III}(CN)_6$ according to

$$3HgO + KFe^{II}Fe^{III}(CN)_6 + 3H_2O \rightarrow 3Hg(CN)_2 + Fe(OH)_2 + Fe(OH)_3 + K^+ + OH^- \tag{5.58}$$

when the mixture is heated at 363 K (90 °C) for several hours.[3] $Hg(CN)_2$ can be isolated from the solution using standard operations. Mercury(II) cyanide is highly soluble in various organic solvents.

5.2.11 Mercury(I) Dithiocyanate – $Hg_2(SCN)_2$

Mercury(I) dithiocyanate has an orthorhombic crystal structure with $a = 1.571$ nm, $b = 0.643$ nm and $c = 0.638$ nm.[170] It is poorly soluble in water. When analyzing the experimental data concerning the solubility of mercury(I) thiocyanate in water, one should take into account both the disproportionation reaction of Hg_2^{2+} ions to Hg^0 and Hg^{2+} and the formation of complexes $HgSCN^+$, $Hg(SCN)_2$, $Hg(SCN)_3^-$ and $Hg(SCN)_4^{2-}$. According to experimental data from the literature,[1–4] the solubility of $Hg_2(SCN)_2$ in water at 298.15 K is 2.7×10^{-7} mol dm^{-3}. The solubility product of $Hg_2(SCN)_2$ in water at 298.15 is $K_s = 3.2 \times 10^{-20}$ (Ref. 4) to 3.0×10^{-20}.[160]

$Hg_2(SCN)_2$ is obtained *via* an exchange reaction by mixing a weakly acidic solution of $Hg(NO_3)_2$ after contact with metallic mercury (in less than stoichiometric proportions) with KSCN solution. This first reaction produces a dark green or dark gray precipitate, which, upon stirring in the dark, converts within a few days to $Hg_2(SCN)_2$. $Hg_2(SCN)_2$ is a white precipitate that is sensitive to light. $Hg_2(SCN)_2$ is then isolated by filtering, rinsing a few times in boiling distilled water and drying in vacuum.[171] Mercury(I) thiocyanate disproportionates in KSCN solution through the reaction

$$Hg_2(SCN)_2 \rightleftharpoons Hg^0 + Hg(SCN)_2 \tag{5.59}$$

5.2.12 Mercury(II) Dithiocyanate – $Hg(SCN)_2$

The physical properties of mercury dithiocyanate are given in Table 5.29. $Hg(SCN)_2$ has a greater solubility in water than $Hg_2(SCN)_2$. As with

Table 5.29 Physical properties of $Hg(SCN)_2$.

Property	*Value*	*Ref.*
Molecular weight (g mol^{-1})	316.74	
Melting point (K)	438 (decomp.)^a	27
Density (g cm^{-3})	3.71	174
	3.73	175
$\Delta G^\circ_{f,298.15}$ (kJ mol^{-1})	–253.1^a	176
Crystal structure	Monoclinic	175

^{a}Decomposition begins at 110 °C and is spontaneous at 165 °C (438 K).

Table 5.30 Solubility of divalent ions at 293 K in$[Hg(SCN)_4]$.[2–4,158,159]

Divalent ion	*Solubility at 293 K (mol dm^{-3})*
Zn^{2+}	1.75×10^{-4}
Cd^{2+}	19.0×10^{-4}
Cu^{2+}	1.82×10^{-4}
Co^{2+}	5.37×10^{-4}
Pb^{2+}	9.72×10^{-3}
Mn^{2+}	0.660

$Hg_2(SCN)_2$, to determine the solubility of mercury(II) dithiocyanate in water one should take into account the formation of thiocyanate complexes of mercury(II). The solubility of mercury(II) thiocyanate in water is 1.74×10^{-3} mol m^{-3} at 293 K[172] and 2.2×10^{-3} mol dm^{-3} at 298 K. The solubility product is 2.15×10^{-8}.[4]

Tetrathiocyanatomercurates(II) of alkali metals, $Me_2[Hg(SCN)_4]$, are fairly soluble in water and ethanol and have high electric conductivity. $Me_2[Hg(SCN)_4]$ is obtained by dissolving $Hg(SCN)_2$ in a boiling solution of aqueous KSCN. Mercury sulfide is produced as the solution cools, is filtered off and the filtrate is stripped in the normal way. Dazzling white crystals of $Me_2[Hg(SCN)_4]$ are produced.[173] Divalent ions of heavy metals Zn^{2+}, Cd^{2+}, Cu^{2+}, Co^{2+}, Pb^{2+} and Mn^{2+} combine with$[Hg(SCN)_4]^{2-}$ anions to create poorly soluble $Me[Hg(SCN)_4]$ salts, the solubilities of which at 293 K is given in Table 5.30.

Mercury thiocyanate complexes of Zn,[177] Co[178] and other metals[179] have been studied in detail.

5.3 Oxygen Compounds of Mercury(I) and Mercury(II)

Mercury reacts with oxygen to form mercury(I) oxide (Hg_2O), mercury(II) oxide (HgO) and mercury peroxide (HgO_2).

5.3.1 Mercury(I) Oxide – Hg_2O

Hg_2O is a thermally unstable compound and decomposes into HgO and Hg when exposed to light. Hg_2O forms black crystals with $d_4 = 9.8$ g cm^{-3}. Mercury(I) oxide is obtained by exposing mercury to water vapor at temperatures below 373 K. When hydroxide ions are added to solutions of mercury(I) salts, a number of successive reactions take place in which the resulting hydroxides and oxides of mercury(I) were observed as short-lived intermediates:

$$Hg_2^{2+} + 2OH^- \rightarrow Hg_2(OH)_2 \tag{5.60}$$

$$Hg_2(OH)_2 \rightarrow Hg_2O + H_2O. \tag{5.61}$$

The resulting mercury(I) oxide disproportionates in the presence of water through the reaction

$$Hg_2O_{sol} \rightarrow HgO_{sol} + Hg_{liq}. \tag{5.62}$$

A disproportionation reaction inn solid Hg_2O has been observed at 373 K. The heat of disproportionation is 35.56 kJ mol^{-1}.[180] The hydrolysis reaction

$$Hg_2^{2+} + H_2O \rightleftharpoons Hg_2(OH)^+ + H^+ \tag{5.63}$$

has an equilibrium constant of 1×10^{-5}. Hg_2O is poorly soluble in water, $K_s = 1.6\times10^{-23}$, but fairly soluble in nitric acid. The standard electrode potential of the Hg_2O/Hg half-reaction in alkaline solutions:

$$Hg_2O + H_2O + 2e \rightarrow 2Hg + 2OH^- \tag{5.64}$$

is +0.123 V (*versus* normal hydrogen potential). Numerous claims about having obtained black Hg_2O from solutions of mercury(I) salts and alkali are not proven. X-ray diffraction, magnetic susceptibility testing and heat of formation measurements were used to prove that the product of the reaction is a mixture of metallic mercury and mercury(II) oxide. A study of the Hg_2^{2+}/OH^- reaction equilibrium:

$$Hg_2OH^+ \underset{k_{-1}}{\overset{k_1}{\rightleftharpoons}} Hg^0 + HgOH^+ \tag{5.65}$$

has shown that, when $K_{disprop} = k_1/k_{-1} = 5.5\times10^{-9}$ M and $k_{-1} = 9.0\times10^7$ mol^{-1} s^{-1}, the disproportionation reaction rate constant $k_1 = 0.495$ s^{-1}.[162]

5.3.2 Mercury(II) Oxide – HgO

HgO exists in two modifications, yellow and red. Thermodynamic and structural values of the red and yellow modifications of HgO are given in Ref. 181 and references contained therein. Both modifications are orthorhombic crystals. The crystals have tetrahedral valence angles Hg–O–Hg and O–Hg–O of 109° and 179°, respectively, in z-shaped chains –Hg–O–Hg–O–. The Hg–O bond length is 0.203 nm and the shortest Hg–O– distance is 0.282 nm.[182–184] Figure 5.7 illustrates the geometry of the Hg–O bonding in HgO.

The density of the yellow modification is $d_{yellow} = 11.03$ g cm^{-3} and that of the red modification is $d_{red} = 11.14$ g cm^{-3}.[182–184] HgO decomposes at 773 K (500 °C) with a heat of +180.9 kJ mol^{-1}. Red HgO turns black when heated but

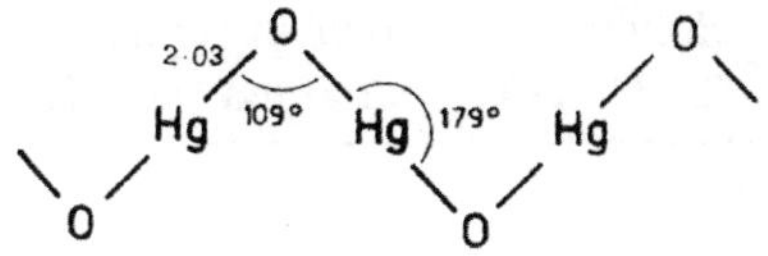

Figure 5.7 Characteristic structure of HgO.
Reproduced with permission from Ref. 43.

returns to its original color as it cools. Yellow HgO turns red when heated. At the atomic level, HgO crystals appear as endless chains of –O–Hg–O–Hg–O, with an –Hg–O–Hg– angle of 109° and an –O–Hg–O– angle of 180°.[185]

Red HgO is synthesized by either a dry or a wet method. The dry method consists of oxidation of metallic mercury with oxygen or ozone at 573–673 K or careful heating of $Hg_2(NO_3)_2$ and $Hg(NO_3)_2$ at 623–673 K. In the wet method, mercury(II) oxide (HgO) is precipitated from hot mercury(II) salt solutions with the help of hydroxides of alkali or alkaline earth metals. The resulting hydroxide of mercury(II), $Hg(OH)_2$, on the introduction of alkali metal hydroxides, immediately decomposes into HgO and Hg_2O. Mercury(II) oxide can also be obtained *via* anodic dissolution of mercury in a solution of hydroxides. Standard half-reaction potentials are as follows:

$$HgO_{red} + H_2O + 2e^- \rightleftharpoons Hg + 2OH^- \quad E^o = +0.0981\ V^{22,23} \tag{5.66}$$

$$HgO + 2H_2O + 2e^- \rightarrow Hg + 2OH^- \quad E^o = +0.0966\ V^{22,23} \tag{5.67}$$

$$Hg(OH)_2 + 2H^+ + 2e^- \rightarrow Hg + 2H_2O \quad E^o = +1.034\ V^{23} \tag{5.68}$$

$$2HgO + 4H^+ + 2e^- \rightarrow Hg_2^{2+} + 2H_2O \quad E^o = +1.065\ V^{23} \tag{5.69}$$

$$2Hg(OH)_2 + 4H^+ + 2e^- \rightarrow Hg_2^{2+} + 4H_2O \quad E^o = +1.279\ V^{23} \tag{5.70}$$

Yellow HgO is obtained by exposing mercury(II) solutions to alkali metal hydroxides. Mercury oxide powders exhibit maxima in IR absorption at $v = 491$ and 595 cm^{-1} and become phosphorescent in the spectral range 2.0–4.5 eV. The equilibrium constant for the reaction

$$HgO + H_2O \rightleftharpoons Hg(OH)_2 \tag{5.71}$$

is $\sim 10^{-2}$. Mercury(II) hydroxide begins to precipitate at pH ≈ 2; complete precipitation occurs as pH $\approx$ 5–12. The hydrolysis reactions are

$$Hg^{2+} + H_2O \rightarrow Hg(OH) + H^+ \tag{5.72}$$

$$Hg(OH)^+ + H_2O \rightleftharpoons Hg(OH)_2 + H^+ \tag{5.73}$$

$$Hg(OH)_2 + OH^-(OH)_3^- \tag{5.74}$$

$$HgO_{red} + H_2O(OH)_3^- \tag{5.75}$$

Table 5.31 Solubility of HgO in aqueous solution at 25 °C (293 K) (adapted from Ref. 186).

Solubility ($\times 10^{-4}$ mol L^{-1})		*Ref.*
Red	Yellow	
2.26 ± 0.03	2.36 ± 0.03	186
–	2.41	187
2.34	–	188
2.37	2.39	189
2.33	–	190
2.25	2.37	191

Table 5.32 Solubility of $Hg(OH)_2$ in water–organic media.[192]

Component	*Solubility*, S *($\times 10^{-9}$ mol L^{-1})*	K_s
CH_3OH	0.981	3.77×10^{-27}
C_2H_5OH	1.02	4.42×10^{-27}
C_3H_7OH	1.12	5.61×10^{-27}

Data for the solubility of HgO in water are given in Table 5.31. The solubility product of $Hg(OH)_2$ at 291 K is $K_s = 4 \times 10^{-26}$ mol dm^{-3}.[3] The solubility of mercury(II) oxide in water depends on the size of the particles. Water–organic media (water–methanol, water–ethanol, water–2-propanol) have little effect on the solubility of mercury(II) hydroxide.[192] Thus, the solubility of $Hg(OH)_2$ in water, determined using a radioactive tracer technique, is 1.35×10^{-9} mol L^{-1} ($K_s = 9.84 \times 10^{-27}$) at 293.65 K. For a 1:1 solvent:solute ratio, the values are as given in Table 5.32.[192]

Mercury(II) hydroxide exhibits amphoteric properties. In acidic solutions it ionizes through the following reactions:

$$Hg(OH)_2 \rightarrow H^+ + HHgO_2^- \tag{5.76}$$

$$HHgO_2^- \rightarrow H^+ + HgO_2^{2-} \tag{5.77}$$

$$Hg(OH)_2 \rightarrow HgOH^+ + OH^- \tag{5.78}$$

$$Hg(OH)_2 \rightarrow Hg^{2+} + 2OH^- \tag{5.79}$$

$$HgO_{solid} + H^+ \rightarrow HgOH^+ \tag{5.80}$$

$$HgO_{solid} + 2H^+ \rightarrow Hg^{2+} + H_2O \tag{5.81}$$

In alkaline solutions, the reactions are as follows:

$$HgO_{solid} + OH^- \rightarrow HHgO_2^- \tag{5.82}$$

$$HgO_{solid} + 2OH^- \rightarrow HgO_2^{2-} + H_2O \tag{5.83}$$

Table 5.33 Thermodynamic properties of organic mercury compounds R_2Hg and RHgX.

Compound[a]	*Name*	ΔH^0_{form} *(kJ mol⁻¹)*	*Ref.*
$(CH_3)_2Hg$ (l)	Dimethylmercury	+59.8	1
CH_3HgCl (c)	Methylmercury chloride	–119.6	212
CH_3HgBr (c)	Methylmercury bromide	–87.8	212
CH_3HgI (c)	Methylmercury iodide	–44.3	212
C_2H_5HgCl (c)	Ethylmercury chloride	–142.2	213
C_2H_5HgBr (c)	Ethylmercury bromide	–108.8	213
C_2H_5HgI (c)	Ethylmercury iodide	–66.9	213
$C_{10}H_{22}Hg$	Diphenylmercury	282.8	1
C_6H_5ClHg	Phenylmercury chloride	-2.3 ± 9.6	214

[a]l = liquid, c = crystalline.

The equilibrium constants of reactions (5.78)–(5.81) are described by the following equations at 298.15 K

$$K_{5.78} = \frac{x_{HgOH^+} x_{OH^-}}{x_{Hg(OH)_2}} = 7.1 \times 10^{-12}\,\mathrm{mol\,dm^{-3}} \tag{5.84}$$

$$K_{5.79} = \frac{x_{Hg^{2+}} x^2_{OH^-}}{x_{Hg(OH)_2}} = 2.2 \times 10^{-23}\,(\mathrm{mol\,dm^{-3}})^2 \tag{5.85}$$

$$K_{5.80} = \frac{x_{HgOH^+}}{x_{Hg^+}} = 0.17 \tag{5.86}$$

$$K_{5.81} = \frac{x_{Hg^{2+}}}{x^2_{H^+}} = 53\,(\mathrm{mol\,dm^{-3}})^{-1} \tag{5.87}$$

This is why the solubility of HgO depends on pH. At pH 10.4, the solubility of HgO_{yellow} is 4.64×10^{-4} mol dm^{-3}. The solubility of HgO as a function of acidity and alkalinity has been reported.[193] Mercury(II) hydroxide starts to precipitate at pH ≈ 2 and stops at pH ≈ 5–12.[194] $Hg(OH)_2$ dissolves in concentrated alkaline solutions.

The structure of the Pourbaix diagram of equilibrium in the mercury–water system at 298.15 K is discussed in Ref. 195 and in acidic solutions in Ref. 196. The solubility of mercury(II) oxide in aqueous solutions of salts, as shown in Table 5.31, increases considerably.

Mercury(II) hydroxide dissolves in concentrated alkaline solutions and in HCl and HNO_3, but does not dissolve in alcohols. For practical purposes, data on the thermal stability of mercury(II) oxide are important. HgO starts to decompose at 903 K (603 °C).[46,180] At red heat, HgO completely sublimes into the gas-phase constituents Hg and O_2. The following equations were obtained for the dissociation pressure of HgO:[180]

$$\log P_{\Sigma Hg+O_2} = 10.9518 - \frac{5273.5}{T} + 1.75\log T - 0.001033T\ \mathrm{Pa} \tag{5.88}$$

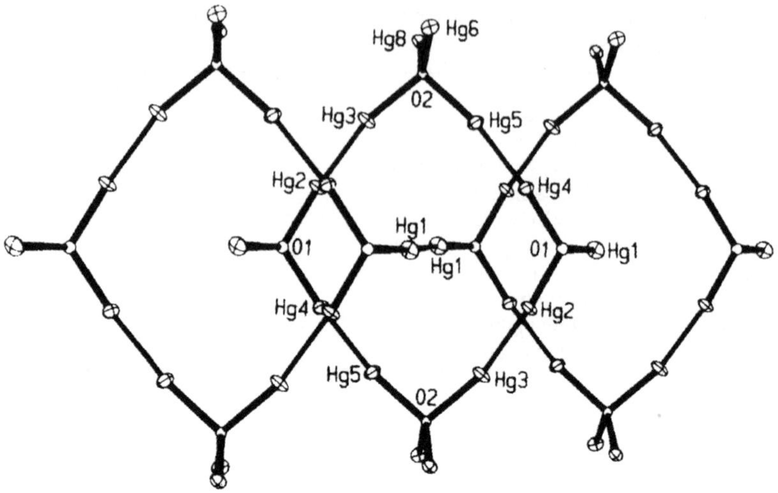

Figure 5.8 Linkage of –O–Hg(I)–Hg(I)–O chains. Reproduced with permission from Ref. 197.

$$\log P_{O_2} = 206.07966 - \frac{27569}{T} - 57.581\log T \text{ Pa} \quad (5.89)$$

$$\log P_{Hg} = 66.71495 - \frac{10529.8}{T} - 16.61\log T \text{ Pa} \quad (5.90)$$

Mercury peroxide, HgO_2, is obtained from the reaction between the yellow modification of mercury oxide with a 30% solution of H_2O_2 at 258 K or by adding H_2O_2 and K_2CO_3 to an alcoholic solution of $HgCl_2$. If a dry synthesis method is used, mercury oxide is fused, *i.e.*, melted, with an alkali metal peroxide to obtain the colorless compound M_2HgO_2 (where M = alkali metal), which decomposes into the original components when it comes into contact with water. M_2HgO_2 contains structural fragments $[O–Hg–O]^{2-}$. Mercury peroxide is barely stable and explodes upon heating or impact.

Mercury can form mixed valence compounds with halides, chalcogenides and other anions. Appendix III gives examples of mixed valence compounds. An interesting example is the crystal structure of an oxide–bromide of mercury(I, II), $Hg_8O_4Br_3$.[197] As can be seen in Figure 5.8, five atoms of Hg(I) (Hgl–Hg5) and three atoms of Hg(II) (Hg6–Hg8) are positioned asymmetrically. The atoms of mercury(I) form three pairs with interatomic Hg–Hg spacings of 0.2517(2)–0.2557(3) nm. These distances are somewhat greater than the spacings found in Hg(I) compounds.

5.3.3 Mercury(I) Nitrate Dihydrate – $Hg_2(NO_3)_2{\cdot}2H_2O$

Mercury(I) nitrate dihydrate forms monoclinic, colorless crystals.[198] The crystals contain Hg–Hg–OII$_2$ chains with a bond angle of 167.5°. A density of 4.785 g cm^{-3} was reported by Grdenić.[198] Dry mercury(II) nitrate, obtained by

Potts and Allred,[199] decomposes into nitrogen oxide and a yellow product at temperatures above 373 K. The infrared spectra of the latter showed that NO_3 groups form dual coordination bonds with mercury. Studies of the Hg_2^{2+}–NO_3–H_2O system suggests a weak complex with constant $K_1 \approx 1^{200}$ is developed. Mercury(I) nitrate is obtained by exposing excess metallic mercury to moderately concentrated nitric acid for several days, according to the equation

$$6Hg + 8HNO_3 \rightarrow 3Hg_2(NO_3)_2 + 2NO + 4H_2O \tag{5.91}$$

The resulting colorless short monoclinic crystals of $Hg_2(NO_3)_2{\cdot}2H_2O$ are separated. Mercury(I) nitrate can also be obtained by heating metallic mercury in excess in moderately dilute HNO_3. $Hg_2(NO_3)_2{\cdot}2H_2O$ melts at 343 K (70 °C); it hydrolyzes in the presence of excess water and produces the basic salt $Hg_2(OH)(NO_3)$. Acidic solutions of $Hg_2(NO_3)_2$ are stable if they are not exposed to air or oxygen.

5.3.4 Mercury(II) Nitrate – $Hg(NO_3)_2$

Depending on the experimental conditions, mercury(II) nitrate forms an octahydrate, $Hg(NO_3)_2{\cdot}8H_2O$, a monohydrate, $Hg(NO_3)_2{\cdot}H_2O$, and a hemihydrate, $Hg(NO_3)_2{\cdot}0.5H_2O$. Mercury(II) nitrate hemihydrate is the most easily obtained salt. $Hg(NO_3)_2{\cdot}0.5H_2O$ melts at 352 K (79 °C). Its crystals are colorless, with a density of 4.30 g cm^{-3}, and are sensitive to light. Attempts to obtain dry salt using a thermal method resulted in a basic salt, $Hg_3O_2(NO_3)_2$. Dry $Hg(NO_3)_2$ is produced *via* a reaction between N_2O_4 and HgO.[3] $Hg(NO_3)_2$ is slightly volatile in vacuum.

Mercury(II) nitrate exists in the form of almost completely undissociated molecules in aqueous solutions. It exhibits considerable solubility in dilute HNO_3 or acetone. Nitrates are widely used for the synthesis of complex mercury(II) compounds. Mercury(II) nitrate is normally obtained by dissolving metallic mercury or mercury oxide in an excess of nitric acid:

$$3Hg + 8HNO_3 \rightarrow 3Hg_2(NO_3)_2 + 2NO + 4H_2O \tag{5.92}$$

Crystals of mercury(II) nitrate monohydrate are obtained by evaporating the solution with subsequent crystallization. $Hg(NO_3)_2$ solutions are only stable in the presence of a certain amount of nitric acid, which prevents hydrolysis. $Hg(NO_3)_2$ quickly hydrolyzes in excess water and produces a precipitate of $Hg_3O_2(NO_3)_2{\cdot}H_2O$ or, when boiled in dilute solutions, forms mercury(II) oxide (HgO).[3] There is experimental proof of complex formation[200] occurring through the reactions

$$Hg(NO_3)_2 + NO_3^- \rightleftharpoons Hg(NO_3)_3^- \tag{5.93}$$

$$Hg(NO_3)_3^- + NO_3^- \rightleftharpoons Hg(NO_3)_4^{2-} \tag{5.94}$$

5.3.5 Mercury(I) Perchlorate – $Hg_2(ClO_4)_2$

Mercury(I) perchlorate has a molecular mass of 600.086 g mol^{-1} and forms two hydrates: $Hg_2(ClO_4)_2{\cdot}4H_2O$ and $Hg_2(ClO_4)_2{\cdot}2H_2O$ (at temperatures above 309 K). $Hg_2(ClO_4)_2$ has an extremely high solubility in water. In such solutions, the perchlorate dihydrate, $Hg_2(ClO_4)_2{\cdot}2H_2O$, is the stable phase upon contact with a solution. Such solutions exhibit a tendency to hydrolyze (a 0.2 M solution has pH 2.1). $Hg_2(ClO_4)_2$ is completely dissociated in aqueous solutions. There is no information about the formation of dry $Hg_2(ClO_4)_2$. The hydrate $Hg_2(ClO_4)_2{\cdot}n\ H_2O$ is obtained by dissolving mercury(I) carbonate in perchloric acid or by electrolytic dissolution of metallic mercury in perchloric acid solutions of specific concentrations.[200]

5.3.6 Mercury(II) perchlorate – $Hg(ClO_4)_2$

Mercury(II) perchlorate also forms crystalline hydrates, $Hg(ClO_4)_2{\cdot}H_2O$, $Hg(ClO_4)_2{\cdot}2H_2O$ and $Hg(ClO_4)_2{\cdot}6H_2O$.

Mercury(II) perchlorate crystallizes into hexagonal crystals of hexahydrate, which cannot be dehydrated by thermal means. $Hg(ClO_4)_2$ also exhibits considerable water solubility: 2.980 kg $Hg(ClO_4)_2$ per kilogram H_2O at 298.15 K. Aqueous solutions of $Hg(ClO_4)_2$ are highly acidic [0.5 M $Hg(ClO_4)_2$ is hydrolyzed by 37%].[3] The main product occurring during the hydrolysis of aqueous solutions is a precipitate of $Hg_3O_2(ClO_4)_2$.

Perchlorate solutions exhibit good electrical conductivity and contain hydrated ions Hg^{2+} and $HgOH^+$. $Hg(ClO_4)_2$ is obtained by dissolving mercury(II) carbonate and oxide in perchloric acid with gentle heating. When $Hg(ClO_4)_2$ solution contacts metallic mercury, the equilibrium

$$Hg(ClO_4)_2 + Hg \overset{H_2O}{\rightleftharpoons} Hg_2(ClO_4)_2 \tag{5.95}$$

is shifted to the right.

5.4 Organometallic Mercury Compounds

5.4.1 Organometallic Mercury(I) Compounds

Only a limited number of organometallic compounds of mercury(I) are known. Mercury(I) halides and perchlorates form several complexes with weak bases:

2-ethylpyridine, $[Hg_2(C_7H_9N)_4]X_2$
4-cyanopyridine, $[Hg_2L_2']X_2$
3-chloropyridine, $[Hg_2L_2'']X_2$
4-benzoylpyridine, $[Hg_2L_2''']X_2$

where $L' = C_6H_4N_2$, $L'' = C_6H_6NCl$, $L''' = C_{12}H_9ON$, and $X^- = Cl^-$, Br^-, I^-, ClO_4^-, NO_3^-.[201–203]

The complexes formed by dinitrate and diperchlorate with mercury(I), 1,10-phenathroline, $[Hg(C_{12}H_8N_2)_2]X_2$, and 2,2′-bipyridyl, $[Hg(C_{10}H_8N_2)_2]X_2$, have been briefly described.[204,205] The ability of mercury(I) to form stable covalent complexes with bases was first demonstrated by Wirth and Davidson.[206] In complexes formed by mercury(I) with 4-cyanopyridine and 3-chloropyridine, which share the common structure of the complex $Hg_2L_2(ClO_4)_2$, the ligands are coordinated *via* a base nitrogen atom in approximately the axial position of the dimer Hg_2^{2+} with weak interaction between mercury and perchlorate ion.[201] This structure is characteristic of all the complexes of the type.

An important factor for the formation of stable mercury(I) complexes with various nitrogen-containing donors is basicity (effective base strength). As it turns out, only ligands with low basicity (based on replaced atoms of pyridine in third and fourth positions) are able to build stable complexes, $[Hg_2L_2](ClO_4)_2$ and $[Hg_2L_4](ClO_4)_2$, whereas the base strength, characteristic of 4-benzylpyridine (pK_a 3.35) and pyridine (pK_a 5.21), is the critical factor.[201] $Hg_2(ClO_4)_2$ forms stable bidentate complexes with 1,8-naphthopyridine (pK_a 3.36) and 5-nitro-1,10-phenanthroline (pK_a 3.55), which have melting points above 673 K.[201] More basic ligands result in disproportionation of mercury(I) complexes due to the high affinity of mercury(II) ions toward nitrogen-containing ligand donors. Interesting results were obtained from the reactions between mercury(I) and diphenyltrifluorophosphine, which creates complexes with the structural formula $[Hg–Hg–P(CF_3)\ Ph_2]^{2+}$ and trifluorophosphine $[Hg–Hg–PF_3]^{2+}$.[207,208]

5.4.2 Organometallic Mercury(II) Compounds

A large number of different metallorganic mercury(II) compounds have been used for the synthesis of different classes of organic compounds.[3,209–211] Organomercury(II) compounds fall into two classes: (a) R_2Hg and $R'HgR$ and (b) RHgX, where R and R' are organic radicals and $X^- = Cl^-$, Br^-, I^-, ClO_4^-, NO_3^-, SO_4^{2-}, *etc.* If $X^- = Cl^-$, Br^-, I^-, CN^-, SCN^- or OH^- ions with polarized electron shells (soft acids), engaging in covalent bonding with mercury ions, then organomercury(II) compounds with these ions exhibit the properties of non-polar covalent compounds that have higher solubility in organic solvents than in water. Organomercury(II) compounds with anions that have poorly deformable electron shells (hard bases), such as SO_4^{2-}, CO_3^{2-}, PO_4^{3-}, ClO_4^- and NO_3^-, are salt-like and heteropolar ($[RHg]_m^+\ X_n^-$).

The solvation energy of organomercuric compounds with covalent bonding is much greater compared with the hydration energy of mercury(II) chlorides, bromides, iodides, *etc.*, which is why they are readily extractible with organic solvents. Thus, even inert solvents (benzene, toluene) can be used to extract mercury from neutral aqueous solutions.[215, 216] Some physical properties of organic mercury(II) compounds are given in Table 5.17. Additional thermodynamic data can be found elsewhere.[217,218]

The properties of metallorganic mercury compounds depend on the nature of R radicals. There are three types of compounds. Type one compounds are mercury compounds with aliphatic or aromatic hydrocarbon groups, type two are mercury compounds with radicals of the type

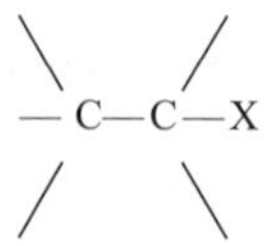

where $X^- = OH^-$, Cl^-, Br^-, *etc.*, and type three are organic halocarbon compounds (CF_3, C_2F_5, C_6Cl_5).[2,3,209–211] Many known synthesis reactions can be used to produce organometallic mercury(II) compounds:

(a) Grignard reactions:

$$RMgX + HgX_2 \rightarrow RHgX + MgX_2 \qquad (5.96)$$

$$RHgX + RMgX \rightarrow R_2Hg + MgX_2 \qquad (5.97)$$

(b) Amalgamation reactions:

$$2RI + Na(Hg)_x \rightarrow R_2Hg + 2NaI \qquad (5.98)$$

$$2RI + Cd(Hg)_x \rightarrow R_2Hg + 2CdI_2 \qquad (5.99)$$

$$2RBr + K(Hg)_x \rightarrow R_2Hg + 2KI \qquad (5.100)$$

where R = alkyl (C_nH_{2n+1}), vinyl (C_nH_{2n-1}) and aryl (C_6H_5 and benzene derivatives).

(c) Substitution reactions:

$$RH + HgX_2 \rightleftharpoons RHgX + HX \qquad (5.101)$$

$$C_6H_6 + Hg(OCOCH_3)_2 \rightarrow C_6H_6 - HgOCOCH_3 + CH_3COOH; \qquad (5.102)$$

$$RMe + HgX_2 \rightarrow RHgX + MeX \qquad (5.103)$$

$$RMe + RHgX \rightarrow R_2Hg + MeX \qquad (5.104)$$

where Me = Li, Mg, Al, Tl, Zn, Cd, Pb, *etc.*

(d) Mercurization reactions:

$$C_6H_5OCOOH + Hg \rightarrow C_6H_5HgOCOOH \qquad (5.105)$$

$$RX + Hg \rightarrow RHgX \qquad (5.106)$$

where X = Br^-, I^-, N_2^+ and $NHNH_2$.

(e) Addition reactions:

$$HC{\equiv}CH + HgCl_2 \rightarrow CClH{=}CHHgCl \qquad (5.107)$$

Organomercury(II) compounds RHgX and R_2Hg are non-linear. Dialkyl and diaryl mercury compounds are highly volatile, toxic, colorless liquids or low-melting solids. Owing to the low polarity of the C–Hg bond and the low oxygen affinity, organomercury(II) compounds resist oxygen contained in air and water. However, owing to the poor stability of the C–Hg bond and poor reactivity, organomercury(II) compounds break down when exposed to light, irradiation or heat (thermal decomposition) and produce free radicals. In recent years, the ability of organomercury(II) compounds to generate free radicals of organic intermediates has been used to perform various organic syntheses and construct organic molecules with predetermined properties.[219] The properties of free radicals and the methods of their production and application have been reported.[219–222] The properties (spectral characteristics, fluorescence) of the excited atoms of mercury, their dimers and trimers and various mercury complexes with NH_3, H_2O, H_2, rare gases (HgNe, HgAr, HgKr), butylamine, aliphatic alcohols, *etc*., have been studied.[223]

Organomercury(II) complexes may serve as a base for the production of electrolytes needed to obtain high-purity mercury by electrolytic refining. Organomercury(II) compounds are commonly used to produce a broad class of organometallic compounds:

$$(C_6H_5)_2Hg + 2Na \rightarrow 2C_6H_5Na + Hg \tag{5.108}$$

$$(C_2H_5)_2Hg + 2Na \rightarrow 2C_2H_5Na + Hg \tag{5.109}$$

$$(CH_3)_2Hg + Mg \rightarrow (CH_3)_2Mg + Hg \tag{5.110}$$

For practical purposes, it is interesting to look at the compounds of mercury(I) acetate:

$$Hg_2Ac_{2(cryst)} \rightarrow Hg_2^{2+} + Ac^- \tag{5.111}$$

with $K_s = 2.4 \times 10^{-10}$, which dissolve in excess sodium acetate and acetic acid through the reaction

$$Hg_2Ac_2 + Ac^- \overset{H_2O}{\rightleftharpoons} HgAc_3^- + Hg \tag{5.112}$$

The solubility of $Hg_2Ac_{2(crystal)}$ in water, measured in grams per kilogram H_2O, is 100 at 298 K and 1000 at 373 K. Consecutive acetate complexes formation constants are[1,224] $K_1 = 3.6 \times 10^5$, $K_2 = 2 \times 10^9$, $K_3 = 1.9 \times 10^{13}$ and $K_4 = 1.2 \times 10^{11}$.

In oxalate solutions, mercury(II) forms complexes:

$$Hg_2^{2+} + 2C_2O_4^{2-} \rightarrow Hg_2(C_2O_4)_2^{2-} \tag{5.113}$$

with equilibrium constant $K = 9.2 \times 10^6$. Very valuable properties are offered by mercury(II) ion complexes with glycinate (Gly) ions ($HgGly_2$ $K = 1.5 \times 10^{19}$), ethylenediamine, pyridine, citrate, ethylenediaminetetraacetate (Y^{4-}, HY^{3-}), *etc.*[160]

In conclusion, it will be observed that the chemistry of mercury(I) and mercury(II) compounds, which extends to a wide range of reactions, is both interesting and intricate and, depending on the nature of the ligands and their ratio to mercury, allows one to obtain mercury compounds with different chemical and physical properties.

References

1. L. G. Hepler and G. Olofsson, *Chem. Rev.*, 1975, **75**, 585.
2. H. L. Roberts, *Adv. Inorg. Chem. Radiochem.*, 1968, **11**, 309.
3. B. J. Aylett, *Comprehensive Inorganic Chemistry, Group IIB*, Pergamon Press, New York, 1973, vol. 3, pp. 187–328.
4. H. W. Clever, S. A. Johnson and M. E. Derrick, *J. Phys. Chem. Ref. Data*, 1985, **14**, 631.
5. N. V. Sidgwick, *Chemical Elements and Their Compounds*, Oxford University Press, London, 1950, vol. 1.
6. H. C. Moser and A. F. Voigt, *J. Am. Chem. Soc.*, 1957, **79**, 1837.
7. I. Sanemasa and T. Hirata, *Bull. Chem. Soc. Jpn.*, 1977, **50**, 3255.
8. T. Y. Toribara, C. P. Schlelds and L. Koval, *Talanta*, 1970, **17**, 1025.
9. M. A. Thompson, J. C. Sullivan and E. Deutch, *J. Am. Chem. Soc.*, 1971, **93**, 5667.
10. I. Sanemasa, *Bull. Chem. Soc. Jpn.*, 1975, **48**, 1795.
11. D. N. Glew and D. A. Hames, *Can. J. Chem.*, 1971, **49**, 3114.
12. E. Onat, *J. Inorg. Nucl. Chem.*, 1974, **36**, 2029.
13. S. S. Choi and D. G. Tuck, *J. Chem. Soc.*, 1962, 4080.
14. J. C. Pariaud and P. Archinard, *Bull. Soc. Chim. Fr.*, 1952, 454.
15. A. Stock, F. Cucuel, F. Gerstner, H. Köhle and H. Lux, *Z. Anorg. Allg. Chem.*, 1934, **217**, 241.
16. E. K. Mroczek, in *Proceedings of the 19th Stanford Geothermal Reservoir Engineering Workshop*, Stanford University, Stanford, CA, 18–20 January 1994, SGP-TR-147.
17. V. I. Sorokin and T. P. Dadze, in *Fluids in the Crust: Equilibrium and Transport Properties*, ed. K. I. Shmulovich, B. W. D. Yardley and G. G. Gonchar, Chapman & Hall, London, 1994, p. 57.
18. V. P. Chviruk and N. V. Koneva, *Ukr. Khim. Zh.*, 1975, **41**, 1162.
19. V. P. Chviruk, N. V. Koneva and M. I. Ovrutsky, *Ukr. Khim. Zh.*, 1972, **38**, 275.
20. V. P. Chviruk and N. V. Koneva, *Zh. Prikl. Khim.*, 1978, **51**, 743.
21. V. P. Chviruk, N. V. Koneva and A. Yu. Noel, *Zh. Prikl. Khim.*, 1974, **11**, 2546.
22. W. M. Latimer, *Oxidation States of Elements and Their Potentials in Aqueous Solutions*, Foreign Literature Publishing, Moscow, 1954.
23. G. Milazzo and S. Caroli, *Tables of Standard Electrode Potentials*, Wiley, New York, 1978.
24. L. F. Kozin, *Electrodeposition and Dissolution of Polyvalent Metals*, Naukova Dumka, Kiev, 1979.
25. S. Rondinini, A. Cavadore, P. Longhi and T. Mussini, *J. Chem. Thermodyn.*, 1988, **20**, 711.
26. C. E. Vanderzee and J. A. Swanson, *J. Chem. Thermodyn.*, 1974, **6**, 827.

27. M. Simon, P. Jönk, G. Wühl-Couturier and S. Halbach, *Mercury, Mercury Alloys and Mercury Compounds, Ullman's Encyclopedia of Industrial Chemistry*, Wiley-VCH, Weinheim, 2006, pp. 559–594.
28. R. A. Feldhoff, *Pharm. Ztg.*, 1930, **75**, 11.
29. R. Hirayama, *Jpn. Pat.*, 174 101, 1946.
30. N. P. Chopey, *Chem. Eng. (N.Y.)*, 1961, **68**, 120.
31. E. L. Cadmus, Process for production of mercuric chloride and red oxide of mercury, *US Pat.*, 3 424,552, 1969.
32. Y. V. Karyankin and J. J. Angelov, Chisty Klimicheskie Reativy (Pure Chemical Reagents), State-Owned Scientific–Technical Publisher for Chemical Literature, Moscow, 1955, p. 457.
33. J. P. Ponpon, R. Stuck, P. Siffert, B. Meyer and C. Schwab, *IEEE Trans. Nucl. Sci.*, 1975, **22**(1), 182–191.
34. N. L. Skinner, C. Ortale, M. M. Schieber and L. Van den Berg, *Nucl. Instrum. Methods Phys. Res. A*, 1989, **283**(2), 119.
35. G. A. Jeffrey and M. Vlasse, *Inorg. Chem.*, 1967, **6**, 396–399.
36. D. C. Bebout, *Encyclopedia of Inorganic Chemistry, Encyclopedia of Inorganic and Bioinorganic Chemistry*, Wiley, Chichester, 2006 pp. 1–15.
37. J. N. Blocher and E. H. Hall, *J. Phys. Chem.*, 1959, **63**, 127.
38. G. G. Koerber and T. deVries, *J. Am. Chem. Soc.*, 1952, **74**, 5008.
39. F. Ebert and H. Woitinek, *Z. Anorg. Allg. Chem.*, 1933, **210**, 269.
40. E. Dorm, *J. Chem. Soc. D*, 1971, 466.
41. L. M. Volkova and S. A. Magarill, *J. Struct. Chem.*, 1999, **40**, 262.
42. D. Grdenić and C. Djordjević, *J. Chem. Soc.*, 1956, 1316.
43. D. Grdenić, *Q. Rev. Chem. Soc.*, 1965, **19**, 303.
44. M. Hostettler and D. Schwarzenbach, *C. R. Chim.*, 2004, **8**, 147.
45. S. A. Polyshchuk, M. G. Khmeliova, G. M. Zadneprovsky, T. A. Kaidalova and N. V. Kuptsova, *J. Less-Common Met.*, 1970, **21**, 63.
46. Ya. I. Gerasimov, A. E. Krestovnikov and S. I. Gorbov, *Chemical Thermodynamics in Nonferrous Metallurgy*, Metallurgiya, Moscow, 1973, vol. 5.
47. V. P. Glushko, ed, *Thermal Constants of Substances*, Moscow, Nauka, 1972, Issue 6, Part 1, 370 p.
48. *JANAF Thermochemical Tables*, J. Phys. Chem. Ref. Data, Vol. 14, Supplement 1, 1985.
49. D. T. Meshri, in *Kirk-Othmer Encyclopedia of Chemical Technology*, ed. M. Grayson, Wiley, New York, 1980, vol. 3.
50. Y. Marcus, *J. Phys. Chem. Ref. Data*, 1980, **9**, 1307.
51. C. Bárta and C. Barta Jr., Growth of Acousto-Optic Materials, *Materials Science Forum*. Ed. N. B. Singh and D. J. Todd, 1991, **61**, 93.
52. M. Bass (ed.), *Handbook of Optics*, McGraw-Hill, New York, 1995.
53. N. Gupta, *Proc. SPIE*, 2005, **5953**, 1.
54. R. J. Havighurst, *J. Am. Chem. Soc.*, 1926, **48**, 2113.
55. N. V. Belov and I. M. Rumanova, *Trudy Inst. Krist. Akad. Nauk SSSR*, 1949, **5**, 57.

56. V. Subramanian and K. Self, *Acta Crystallogr., Sect. B*, 1980, **36**, 2132.
57. O. Bonnin, K. O. Awitor, J. P. Badaud, L. Bernard, B. Coupat, J. P. Fournier and P. Verdier, *J. Phys. III*, 1997, **7**, 311.
58. A. Ditte, *Ann. Chim. Phys.*, 1881, **22**, 551.
59. S. M. Aslanov and V. P. Blidin, *Russ. J. Inorg. Chem.*, 1959, **4**, 748.
60. M. Étard, *Ann. Chim. Phys.*, 1894, **2**, 503.
61. R. D. Eddy and A. W. C. Menzies, *J. Phys. Chem.*, 1940, **44**, 207.
62. V. A. Mikhailov and E. F. Grigor'eva, *J. Struct. Chem., (USSR)*, 1968, **9**, 687.
63. S. M. Aslanov, *Russ. J. Inorg. Chem.*, 1963, **8**, 772.
64. F. Flöttmann, *Fresenius' Z. Anal. Chem.*, 1928, **73**, 1.
65. H. G. Greenish and F. A. U. Smith, *Pharm. J.*, 1903, **71**, 881.
66. J. S. Laird, *J. Phys. Chem.*, 1920, **24**, 736.
67. V. I. Tikhomirov, *Zh. Russ.-Khim. O-va. Chast. Khim.*, 1907, **39**, 731.
68. H. Morse, *Z. Phys. Chem. Stoechiom. Verwandschaftsl.*, 1902, **41**, 709.
69. H. W. Foote, *Am. Chem. J.*, 1903, **30**, 339.
70. M. S. Sherrill, *Z. Phys. Chem. Stoechiom. Verwandschtsl.*, 1903, **43**, 705.
71. Y. Osaka, *Mem. Coll. Sci. Eng. Kyoto Imp. Univ.*, 1908, **1**, 290.
72. W. Herz and G. Anders, *Z. Anorg. Chem.*, 1907, **52**, 164.
73. W. Herz and W. Paul, *Z. Anorg. Chem.*, 1913, **82**, 431.
74. E. Moles and M. Marquina, *An. R. Soc. Esp. Fis. Quim.*, 1924, **22**, 551.
75. A. Benrath, *Z. Anorg. Allg. Chem.*, 1927, **163**, 396.
76. A. Benrath and G. Ammer, *Z. Anorg. Allg. Chem.*, 1929, **177**, 129.
77. H. Bassett and H. H. Croucher, *J. Chem. Soc.*, 1930, 1784.
78. H. C. Thomas, *J. Am. Chem. Soc.*, 1939, **61**, 920.
79. P. Laurent, P. Hagenmuller and Dag-Quoc-Quan, *C. R. Acad. Sci.*, 1955, **241**, 1044.
80. V. P. Blidin, *Zh. Neorg. Khim.*, 1957, **2**, 1149.
81. L. S. Lilich, *Russ. J. Inorg. Chem.*, 1959, **4**, 67.
82. K. P. Anderson, A. L. Cummings, J. L. Bells and K. J. Walker, *J. Inorg. Nucl. Chem.*, 1974, **36**, 1837.
83. Ya. D. Fridman, S. D. Gorokhov, T. V. Fokina and N. V. Dolgashova, *Sov. J. Coord. Chem.*, 1977, **3**, 1424.
84. P. A. Meerburg, *Chem. Weekbl.*, 1908, **5**, 429.
85. P. A. Meerburg, *Z. Anorg. Chem.*, 1908, **59**, 136.
86. F. A. H. Schreinemakers, *Chem. Weekbl.*, 1910, **7**, 197.
87. C. Tourneux, *Ann. Chim. (Paris)*, 1919, **11**, 225.
88. F. A. H. Schreinemakers and J. C. Thonus, *Proc. K. Ned. Akad. Wet.*, 1912, **15**, 472.
89. S. Toda, *Mem. Coll. Sci. Kyoto Imp. Univ.*, 1922, **4**, 305.
90. A. Benrath, F. Gjedebo, B. Schiffers and H. Wunderlich, *Z. Anorg. Allg. Chem.*, 1937, **231**, 285.
91. L. F. Kozin, *Physicochemistry and Metallurgy of High-Purity Mercury and Its Alloys*, Naukova Dumka, Kiev, 1992.
92. L. G. Sillen, *Acta Chem. Scand.*, 1949, **3**, 539.
93. Y. Marcus, *Acta Chem. Scand.*, 1957, **11**, 599.

94. L. Ciavatta and H. Grimaldi, *J. Inorg. Nucl. Chem.*, 1968, **30**, 197.
95. P. B. Bentham, C. G. Romak and H. F. Shurvell, *Can. J. Chem.*, 1985, **63**, 2303.
96. J. H. R. Clarke and L. A. Woodward, *Trans. Faraday Soc.*, 1965, **61**, 207.
97. R. Arnek, *Ark. Kemi*, 1965, **24**, 531.
98. Yu. N. Kukushkin, *Chemistry of Coordination Compounds*, Vysshaya Shkola, Moscow, 1985.
99. S. Sjoberg, *Acta Chem. Scand.*, 1977, **A31**, 705.
100. I. Persson, M. Landgren and A. Marton, *Inorg. Chim. Acta*, 1986, **116**, 135.
101. I. Eliezer and S. Adida, *J. Phys. Chem.*, 1973, **77**, 87.
102. R. W. Kiser, J. G. Dillard, and D. L. Dugger, in *Mass Spectrometry in Inorganic Chemistry*, ed. J. L. Margrave, American Chemical Society, Washington, DC, 1968, vol. 72, pp. 153–180.
103. D. Cubicciotti, H. Eding and J. W. Johnson, *J. Phys. Chem.*, 1966, **70**, 2989.
104. F. D. Rossini, D. D. Wagman, W. H. Evans, S. Levine and I. Jaffe, *Natl. Bur. Stand. Circ.*, 1952, **500**.
105. J. W. Johnson, W. J. Silva and D. Cubicciotti, *J. Phys. Chem.*, 1966, **70**, 1169.
106. V. I. Vasilev, E. V. Kozlovskii and A. A. Mokeev, *Zh. Neorg. Khim.*, 1981, **26**, 1213.
107. H. Oppermann, Ch. Barta and J. Trnka, *Z. Anorg. Allg. Chem.*, 1989, **568**, 187.
108. M. C. Morris, H. F. McMurdie and E. Evans, *NBS Monogr.*, 1976, **25**, 29.
109. V. I. Pakhomov, *Zh. Neorg. Khim.*, 1990, **35**, 2476.
110. A. I. Brodskiy, *Electrochemistry and Thermodynamics of Solutions*, Kiev, 1974, vol. 1.
111. H. Oppermann, *Z. Anorg. Allg. Chem.*, 1989, **576**, 229.
112. C. Guminski, *J. Phase Equilib.*, 2000, **21**, 539.
113. C. Guminski, *J. Phase Equilib.*, 1997, **18**, 206.
114. C. Guminski, *J. Phase Equilib.*, 1994, **15**, 101.
115. R. Dworsky and K. Komarek, *Monatsh. Chem.*, 1970, **101**, 976.
116. J. S. Kim, S. B. Trivedi, J. Soos, N. Gupta and W. Palosz, *Proc. SPIE*, 2007, **6661**, 66610-B2.
117. S. R. Gupta, G. J. Hills and D. J. G. Ives, *Trans. Faraday Soc.*, 1963, **59**, 1886.
118. W. Leuschke and K. Schwabe, *J. Electroanal. Chem.*, 1970, **25**, 219.
119. D. J. Knuteson, N. B. Singh, M. Gottlieb, D. Suhre, N. Gupta, A. Berghmans, D. Kahler, B. Wagner and J. Hawkins, *Opt. Eng.*, 2007, **46**, 064001.
120. L. M. Volkova and S. A. Magarill, *J. Struct. Chem.*, 1999, **40**, 262.
121. F. M. G. Johnson, *J. Am. Chem. Soc.*, 1911, **23**, 777.
122. E. B. R. Prideaux, *J. Chem. Soc.*, 1910, **97**, 2032.
123. A. E. Brodsky, *Z. Elektrochem. Angew. Phys. Chem.*, 1929, **35**, 833.
124. A. J. Read, PhD Thesis, University of New Zealand, 1960.

125. M. Pernot, *C. R. Acad. Sci.*, 1932, **195**, 238.
126. H. J. V. Tyrell and J. Richards, *J. Chem. Soc.*, 1953, 3812.
127. A. Zajdler and D. M. Czakis-Sulikowska, *Zesz. Nauk Politech. Lodz. Chem.*, 1974, **30**, 21.
128. A. B. Garrett, *J. Am. Chem. Soc.*, 1939, **61**, 2744.
129. M. Iwamoto, S. N. Johnson and H. Miyamoto, Mercury in Liquids, in *Compressed Gases, Molten Salts and Other Elements*, Solubility Data Series, Pergamon Press, New York, 1986, vol. 29.
130. D. D. Wagman, W. H. Evans, V. B. Parker, I. Halow, S. M. Bailey and R. H. Schumm, *Nat. Bur. Stand. Tech. Note*, 1969, No. 270–4.
131. V. P. Vasilev, E. V. Kozlovskii and A. A. Mokeev, *Zh. Neorg. Khim.*, 1982, **27**, 632.
132. H. Braekken, *Z. Kristallogr.*, 1932, **81**, 152.
133. N. B. Singh, R. H. Hopkins, R. Mazelsky and M. Gottlieb, *J. Cryst. Growth*, 1987, **85**, 240.
134. M. Kars, T. Roisnel, V. Dorcet, A. Rebbah and O.-D. L. Carlos, *Acta Crystallogr.*, 2012, **E68**, i11.
135. H. Oppermann and W. Ludwig, *Z. Anorg. Allg. Chem.*, 1990, **590**, 161.
136. M. Piechotka and E. Kaldis, *J. Less-Common Met.*, 1986, **115**, 315.
137. S. C. Hansen and D. Kobertz, in *Solid State Transformations*, J. Čermák and I. Stloukal, Trans Tech Publications, Zurich, 2008, vol. 138, pp. 29–42.
138. J. B. Newkirk, *Acta Met.*, 1956, **4**, 316.
139. M. Hostettler, H. Birkedal and D. Schwarzenbach, *Helv. Chim. Acta*, 2003, **86**, 1410.
140. M. Hostettler, H. Birkedal and D. Schwarzenbach, *Chimia*, 2001, **55**, 541.
141. M. Hostettler, H. Birkedal and D. Schwarzenbach, *Acta Crystallogr. B*, 2002, **58**, 903.
142. B. Baranowski, M. Friesel and A. Lunden, *Z.* Phys. Chem., 1988, **269**, 585.
143. Y. F. Nicolan, M. Dupuy and Z. Kabsch, *Nucl. Instrum. Methods Phys. Res.*, 1989, **A283**, 149.
144. N. A. Tananaev and A. T. Pilipenko, *Zh. Prikl. Khim.*, 1937, **10**, 549.
145. E. Bourgoin, *Bull. Soc. Chim. Fr.*, 1884, **42**, 620.
146. S. M. Naudé, *Z. Phys. Chem. Stoechiom. Verwandschaftsl.*, 1927, **125**, 98.
147. J. A. de Bruijn, *Recl. Trav. Chim. Pays-Bas*, 1941, **60**, 309.
148. G. Biedermann and L. G. Sillen, *Sven. Kem. Tidskr.*, 1949, **61**, 63.
149. S. Hietanen and L. G. Sillen, *Sven. Kem. Tidskr.*, 1947, **59**, 22.
150. *Gmelins Handbuch, Mercury*, Verlag Chemie, Weinheim, 1967, vol. 34, Pt. B2, p. 820.
151. R. J. Havighurst, *Am. J. Sci.*, 1925, **10**, 15.
152. K. B. Yatsimirsky and A. A. Shutov, *Zh. Neorg. Khim.*, 1952, **26**, 842.
153. T. G. Spiro and D. N. Hume, *J. Am. Chem. Soc.*, 1961, **83**, 4305.
154. Y. Marcus, *Acta Chem. Scand.*, 1957, **11**, 610.

155. R. P. Rastogi, B. L. Dubey and N. D. Agrawal, *J. Inorg. Nucl. Chem.*, 1975, **37**, 1167.
156. E. S. Nozhko and V. D. Tishina, *Zh. Prikl. Khim.*, 1983, **56**, 1381; *J. Appl. Chem. USSR*, 1983, **56**, 1298.
157. M. A. Beg and A. Jain, *Polyhedron*, 1992, **11**, 2775.
158. I. M. Korenman, F. P. Sheyanova and M. A. Potapova, *Zh. Obshch. Khim.*, 1956, **26**, 2114.
159. I. M. Korenman, F. P. Sheyanova and M. N. Barishnikova, *Zh. Obshch. Khim.*, 1956, **26**, 365.
160. L. G. Sillen and A. E. Martell, *Stability Constants of Metal-Ion Complexes*, Special Publication No. 17, Chemical Society, London, 1964, vol. 1, and 1971, vol. 2.
161. V. P. Vasiliev, *Thermodynamic Properties of Electrolyte Solutions*, Vysshaya Shkola, Moscow, 1982.
162. B. M. Chadwick and A. G. Sharpe, *Adv. Inorg. Chem. Radiochem.*, 1966, **8**, 83.
163. J. Hvoslef, *Acta Chem. Scand.*, 1958, **12**, 1568.
164. R. C. Seccombe and C. H. L. Kennard, *J. Organomet. Chem.*, 1969, **18**, 243.
165. P. Kocovsky, G. Wang and V. Sharma, *e-EROS Encyclopedia of Reagents for Organic Synthesis*, Wiley, Chichester, 2001.
166. W. Biltz, *Z. Anorg. Allg. Chem.*, 1928, **170**, 161.
167. M. T. Beck and F. Gaizer, *J. Inorg. Nucl. Chem.*, 1964, **26**, 1755.
168. J. S. Coleman, R. A. Penneman, L. H. Jones and I. K. Kressin, *Inorg. Chem.*, 1968, **7**, 1174.
169. L. Newman and D. N. Hume, *J. Am. Chem. Soc.*, 1961, **83**, 1795.
170. K. Huttner and S. Knappe, *Z. Anorg. Allg. Chem.*, 1930, **190**, 27.
171. D. K. Breitinger and W. A. Herrmann, *Synthetic Methods of Organometallic and Inorganic Chemistry. Copper, Silver, Gold, Zinc, Cadmium, Mercury*, Georg Thieme, New York, 1999.
172. A. N. Kirgintsev, L. N. Trushnikova and V. G. Lavrentev, *Solubility of Inorganic Substances in Water*, Khimiya, Leningrad, 1972.
173. G. Brauer (ed.), *Inorganic Synthesis*, Mir, Moscow, 1985, vol. 4, pp. 1144–1157.
174. H. Puff and H. Becker, *Acta Crystallogr.*, 1965, **18**, 299.
175. A. L. Beauchamp and D. Goutier, *Can. J. Chem.*, 1972, **50**, 977.
176. L. G. Sillen and A. E. Martell, *Spec. Publ. Chem. Soc.*, No. 17, 1964.
177. D. Xu, W.-T. Yu, X.-Q. Wang, D.-R. Yuan, M.-K. Lu, P. Yang, S.-Y. Guo, F.-Q. Meng and M.-H. Jiang, *Acta Crystallogr. C: Crystal Struct. Commun.*, 1999, **55**, 1203.
178. J. W. Jeffrey and K. M. Rose, *Acta Crystallogr. B*, 1968, **24**, 653.
179. G. A. Bowmaker, A. V. Churakov, R. K. Harris, J. A. K. Howard and D. C. Apperley, *Inorg. Chem.*, 1998, **37**, 1734.
180. G. B. Taylor and G. A. Hullet, *J. Am. Chem. Soc.*, 1922, **44**, 1443.
181. C. Guminski, *J. Phase Equilib.*, 1999, **20**, 85.
182. K. Aurivillius, *Acta Chem. Scand.*, 1944, **8**, 523.

183. K. Aurivillius, *Acta Chem. Scand.*, 1956, **10**, 852.
184. K. Aurivillius and O. Carlson, *Acta Chem. Scand.*, 1957, **11**, 1069.
185. R. Gillespie, *Molecular Geometry*, Mir, Moscow, 1975.
186. K. Aurivillius and O. van Heidenstam, *Acta Chem. Scand.*, 1961, **15**, 1993.
187. W. Ostwald, *Z. Phys. Chem.*, 1900, **34**, 69.
188. G. A. Hulett, *Z. Phys. Chem.*, 1901, **37**, 406.
189. K. Schick, *Z. Phys. Chem.*, 1903, **42**, 172.
190. G. Fuseya, *J. Am. Chem. Soc.*, 1920, **42**, 368.
191. A. B. Garrett and A. E. Hirschler, *J. Am. Chem. Soc.*, 1938, **60**, 299.
192. A. Cecal and G. I. Ciuchi, *Rev. Roum. Chim.*, 1989, **34**, 953.
193. A. S. Kertes, *Solubility Data Series, Vol. 23, Copper, Silver, Gold, Cadmium, Mercury Oxides and Hydroxides*, Pergamon Press, New York, 1986.
194. V. A. Nazarenko, V. P. Antonovich and E. M. Nevskaya, *Hydrolysis of Metal Ions in Dilute Solutions*, Atomizdat, Moscow, 1979.
195. M. Pourbaix, *Atlas d'Equilibres Electrochimique*, Gauthier-Villars, Paris, 1963.
196. A. Ballester, E. Otero and F. Gonzales, *Rev. Met.*, CENIM, 1988, **24**, 16.
197. C. Stålhandske, *Acta Chem. Scand.*, 1987, **41A**, 576.
198. D. Grdenić, *Acta Crystallogr. B*, 1975, **31**, 2174.
199. R. A. Potts and A. L. Allred, *Inorg. Chem.*, 1966, **5**, 1066.
200. A. F. Cotton and G. Wilkinson, *Advanced Inorganic Chemistry*, Wiley, New York, 1988.
201. D. L. Kepert and D. Taylor, *Aust. J. Chem.*, 1974, **27**, 1199.
202. D. L. Kepert, D. Taylor and A. H. White, *Inorg. Chem.*, 1972, **11**, 1639.
203. D. L. Kepert, D. Taylor and A. H. White, *J. Chem. Soc., Dalton Trans.*, 1973, 893.
204. G. Anderegg, *Helv. Chim. Acta*, 1959, **42**, 344.
205. R. C. Elder, J. Halpern and J. S. Pond, *J. Am. Chem. Soc.*, 1967, **89**, 6874.
206. T. H. Wirth and N. Davidson, *J. Am. Chem. Soc.*, 1964, **86**, 4314.
207. P. A. W. Dean and D. G. Ibbott, *Inorg. Nucl. Chem. Lett.*, 1975, **11**, 119.
208. P. A. W. Dean and D. G. Ibbott, *Can. J. Chem.*, 1976, **54**, 177.
209. L. G. Makarova and A. N. Nesmeyanov, *Methods of Element-Organic Chemistry, Mercury*, Nauka, Moscow, 1965.
210. W. Pryor, *Free Radicals*, Atomizdat, Moscow, 1970.
211. A. N. Nesmeyanov and N. A. Nesmeyanov, *Introduction to Organic Chemistry*, Khimiya, Moscow, 1974, vol. 2.
212. K. Hartley, H. O. Pritchard and H. A. Skinner, *Trans. Faraday Society*, 1950, **46**, 1019.
213. K. Hartley, H. O. Pritchard and H. A. Skinner, *Trans. Faraday Society*, 1951, **47**, 254.
214. C. L. Chernick, H. A. Skinner and I. Wadsö, *Trans. Faraday Soc.*, 1956, **52**, 1088.
215. Yu. A. Zolotov, B. Z. Ioffa and L. K. Chuchalin, *Extraction of Halogenide Complexes of Metals*, Nauka, Moscow, 1973.

216. Yu. A. Zolotov and N. M. Kuzmin, *Concentration of Traces of Elements*, Khimiya, Moscow, 1982.
217. J. D. Cox and G. Pilcher, *Thermochemistry of Organic and Organometallic Compounds*, Academic Press, New York, 1970.
218. D. M. Fairbrother and H. A. Skinner, *Trans. Faraday Soc.*, 1956, **52**, 956.
219. J. Barluenga and M. Yus, *Chem. Rev.*, 1988, **88**, 487.
220. V. D. Pokhodenko, A. A. Beloded and V. G. Koshechko, *Oxidation–Reduction Reactions of Free Radicals*, Naukova Dumka, Kiev, 1977.
221. V. D. Pokhodenko and M. F. Guba, *Bull. Ukr. Acad. Sci.*, 1988, **5**, 24.
222. D. C. Nonhebel and J. C. Walton, *Free-Radical Chemistry*, Mir, Moscow, 1977.
223. A. B. Callear, *Chem. Rev.*, 1987, **87**, 335.
224. D. Banerjee and I. P. Singh, *Z. Anorg. Allg. Chem.*, 1964, **331**, 225.

CHAPTER 6

Electrochemical Properties of Mercury

6.1 Kinetics and Mechanism of Discharge and Ionization of Mercury in Simple Electrolytes

Mercury is an ideal electrode material. It exhibits a positive electrode potential and a high overpotential for releasing hydrogen. It is liquid at room temperature and can be easily purified. Therefore, it is a favorable material for the construction of precise electrode devices with continuously renewed surfaces of mercury. The fresh mercury surface ensures reproducible thermodynamic parameters for equilibrium in systems such as Hg/Hg_2X_2 and Hg/HgX_2, and also kinetic characteristics of mercury discharge and ionization.[1–13] The relationship of equilibria in Hg/Hg_2X_2 ($X = ClO_4^-$, NO_3^-, SO_4^{2-}) and Hg/HgX_2 ($X = Cl^-$, Br^-, I^-, CN^-, SCN^- and others) systems have been discussed in the literature[1–6,11,14–19] and also in Chapter 5. In solutions of perchloric and nitric acids, the equilibrium in the Hg/HgX_2 system:

$$Hg + Hg(ClO_4)_2 \rightleftharpoons Hg_2(ClO_4)_2 \tag{6.1}$$

is shifted towards the formation of monovalent mercury ions. The standard electrode potential of the reduction of Hg_2^{2+}:

$$Hg_2^{2+} + 2e \rightarrow 2Hg \tag{6.2}$$

is $E^0_{Hg_2^{2+}/Hg} = 0.7960 \pm 0.0005$ V (*versus* NHE)[5] in solutions of perchloric acid. According to Wanderzee and Swanson,[6] $E^0_{Hg_2^{2+}/Hg} = 0.7965 \pm 0.001$ V (*versus* NHE). A similar value for $E^0_{Hg_2^{2+}/Hg}$ in the $Hg/Hg_2(ClO_4)_2$ system was obtained

Mercury Handbook: Chemistry, Applications and Environmental Impact
By Leonid F Kozin and Steve Hansen

Published by the Royal Society of Chemistry, www.rsc.org

by Choudary and Prasad.[11] The potential of the electrode $Hg/Hg_2(ClO_4)_2$ is described by the relationship

$$E_t = 0.80252 - 2.51\times10^{-4}(T-273) - 1.0668\times10^{-6}(T-273)^2$$

$E^0_{Hg_2^{2+}/Hg^0} = 0.795578$ V in the temperature range 5–35 °C (278–308 K).

Numerous investigators have studied the kinetics and mechanism of electrode reactions of mercury(I) in solutions of perchloric acid.[1–5,7–10] The cathodic process of electroreduction of mercury(I) ions consists of two single-electron reactions:

$$Hg_2^{2+} + e \xrightarrow{k_1} Hg^+ + Hg \tag{6.3}$$

$$Hg^+ + e \underset{k_{a_2}}{\overset{k_{k2}}{\rightleftharpoons}} Hg^0 \tag{6.4}$$

that are described by a single kinetic equation:

$$i = k_{k_1}\left[Hg_2^{2+}\right]\exp\left(\frac{-\alpha_1 FE}{RT}\right) - k_{a_2}\exp\left(\frac{\beta_1 FE}{RT}\right) \tag{6.5}$$

where α and β are transfer coefficients ($\alpha_1 = 0.4$; $\beta_1 = 1.6$) and z_1 is the number of electrons at the limiting stage ($z_1 = 1$). For the proposed kinetic equation, the order of the cathodic reduction process in eqn (6.2) for Hg_2^{2+} ions is $n = 1$. However, upon processing the data collected in the experiments, it was found that, depending on the current density and concentration of Hg_2^{2+} ions, the order of the reactions changes from 2 at high concentrations $\{[Hg_2^{2+}] = (2.3–5.3)\times10^{-3}\ M\}$ to 0.65 at lower concentrations $\{[Hg_2^{2+}] = (1.0–0.59)\times10^{-3}\ M\}$. These data demonstrate the complexity of the electrode processes of mercury. Formation of the intermediate ion Hg^+ behaves like radicals[12] and their mutual interaction follows the second-order dimerization

$$Hg^+ + Hg^+ \xrightarrow{k_{dimerization}} Hg_2^{2+} \tag{6.6}$$

with a high rate constant, $k_{dimerization} = (8.0 \pm 1.0)\times10^9\ L\ mol^{-1}\ s^{-1}$.[13] This is why this reaction can complicate the course of reaction (6.2).

The kinetics of the electroreduction of Hg_2^{2+} ions at a mercury electrode in 0.1–1.0 M perchloric acid solution were studied using an advanced galvanostatic double-pulse method[7] and also by a quasi-equilibrium method.[2,8] The rate of electrode reduction given in eqn (6.2) is described by the current density (current per unit surface area)[2] and is given by

$$g = k'\left\{C_a\left(\frac{C_1}{C_0}\right)^{\alpha}\exp\left(-\frac{2\alpha\eta F}{RT}\right) - C_0^{1-\alpha}C_1^{\alpha}\exp\left[-\frac{2(\alpha-1)\eta F}{RT}\right]\right\} \tag{6.7}$$

where c_0 and c_a are the concentration of Hg_2^{2+} ions in solution and on the surface of a mercury drop, c_1 is the concentration of mercury in the metallic phase, k'_0 is the rate constant at the standard potential, η is the overpotential and α is the transfer coefficient.

For cathodic electroreduction of Hg_2^{2+} ions with kinetic control, the current is given by the equation

$$i = \frac{4\pi F a^2 k_0' C_0^{1-\alpha} C_1^{\alpha} \exp\left(-\frac{2\alpha\eta F}{RT}\right)\left[1 - \exp\left(\frac{2\eta F}{RT}\right)\right]}{1 + \frac{a k_0'}{2D}\left(\frac{C_1}{C_0}\right)^{\alpha} \exp\left(-\frac{2\alpha\eta F}{RT}\right)} \tag{6.8}$$

where a is the radius of the mercury drop. The transfer coefficient α and rate constant k_0' are determined by using equation (6.9) for characteristic transfer time τ:

$$\frac{1}{\tau} = \frac{M\left(k_0'\right)^2}{\rho D} C_0^{1-2\alpha} C_1^{2\alpha} \exp\left(-\frac{4\alpha\eta F}{RT}\right)\left[1 - \exp\left(\frac{2\eta F}{RT}\right)\right] \tag{6.9}$$

where M is molecular mass, ρ is density and D is the diffusion coefficient.

Table 6.1 gives the transfer coefficients, α, apparent (k_0') and standard (k_0) rate constants of electroreduction of Hg_2^{2+} ions at a mercury electrode in solutions of perchloric acid, according to Bindra *et al.*[2] Table 6.1 also gives apparent transfer coefficients, α', which at $\eta = 0$ agree well with the value[9] of $\alpha = 0.40$ for a two-step cathodic reaction:

$$Hg_2^{2+} + e \rightarrow Hg_2^{+} \tag{6.10}$$

$$Hg_2^{+} + e \rightarrow 2Hg^0 \tag{6.11}$$

with $z = 1$ and first-order for Hg_2^{+} ions.

It should be noted that Hg_2^{+} ions are reactive and have a short lifetime as they follow the disproportionation reaction

$$Hg_2^{+} + Hg_2^{+} \rightarrow Hg_2^{2+} + 2Hg \tag{6.12}$$

at a high rate. The disproportionation reaction rate constant is $2k_{\text{disproptionation}} = (1.4 \pm 0.2) \times 10^{10}$ L mol^{-1} s^{-1}.[13]

The rate of mercury discharge and ionization at a stationary mercury electrode is limited by the mass transfer and, takes place with a very low overpotential.[10] The resulting cathodic reaction follows eqn (6.2), whereas the anodic process is a single-electron reaction:

$$Hg^0 \rightarrow Hg^{+} + e \tag{6.13}$$

Table 6.1 Kinetic characteristics of cathodic reduction of Hg_2^{2+} ions in solutions of perchloric acid.

Composition of solution	*A*	$k_0^1 \times 10^2$ *(cm s^{-1})*	k_0 *(cm s^{-1})*	α' *at* $\eta = 0$
1×10^{-3} M Hg_2^{2+}: 1 M $HClO_4$	0.32	2.0	2.37	0.37
1×10^{-3} M Hg_2^{2+}: 0.1 M $HClO_4$	0.28	2.5	1.06	0.38
1×10^{-3} M Hg_2^{2+}: 0.01 M $HClO_4$	0.34	0.76	1.92	0.37
5×10^{-4} M Hg_2^{2+}: 0.1 M $HClO_4$	0.26	1.1	1.11	0.37
5×10^{-3} M Hg_2^{2+}: 0.1 M $HClO_4$	0.27	2.2	1.13	0.38

with the associated dimerization reaction, eqn (6.6), of Hg^+ ions. There is virtually no oxidation of Hg to Hg^{2+} ions at the liquid mercury electrode.

The electrode process of electroreduction of mercury(II) ions is also complex. The standard electrode potential of the reduction reaction

$$2Hg^{2+} + 2e \rightarrow Hg_2^{2+} \tag{6.14}$$

in perchloric acid solution is $E^0_{Hg^{2+}/Hg_2^{2+}} = 0.9119 \pm 0.0003$[14] and 0.913 ± 0.003 V (*versus* NHE).[15] According to Dobosh,[16] the standard electrode potential $E^0_{Hg^{2+}/Hg_2^{2+}} = 0.920$ V (*versus* NHE). The standard electrode potential for a two-step reaction:

$$Hg^{2+} + 2e \rightarrow Hg \tag{6.15}$$

as was shown in Chapter 5, is $E^0_{Hg^{2+}/Hg^0} = 0.854$ V (*versus* NHE).[17]

Taking into account the literature values[6,15] for $E^0_{Hg_2^{2+}/Hg^0}$ and $E^0_{Hg^{2+}/Hg_2^{2+}}$, the Luther equation gives $E^0_{Hg^{2+}/Hg^0} = 0.8547$ V (*versus* NHE). Similar values of the electrode potentials $E^0_{Hg^{2+}/Hg_2^{2+}}$, $E^0_{Hg_2^{2+}/Hg^0}$ and $E^0_{Hg^{2+}/Hg^0}$ are also observed in nitric acid solutions.[4,16–18] The zero charge potential of mercury is –0.193 V (*versus* NHE).[19] Therefore, the surface of a mercury electrode at applied potentials is positively charged in simple electrolytes.[20–23]

Polarization curves for mercury(II) ion reduction and mercury oxidation in nitric acid electrolyte on liquid mercury electrodes. In this case, electroreduction of mercury ions also takes place at low polarization and is reversible according to eqn (6.14). In the presence of mercury(II) ions, metallic mercury is not formed at moderate polarization since eqn (6.1) reaches equilibrium very fast. Therefore, Hg_2^{2+} ions are the only product of electroreduction. Studies of the electroreduction of Hg^{2+} ions in a solution composed of $1\,M\,KNO_3 + 0.01\,M\,HNO_3$ conducted using polarography, chronopotentiometry and cyclic voltamperometry methods on a gold electrode established that Hg^{2+} ions are reduced reversibly according to eqn (6.14).[20] The cyclic voltamperometric diagram is characterized by one wave. Analysis of the polarographic diagrams in coordinates $\log[i/(i_d - i)^2] - E$ resulted in a straight line with a slope of 0.029 V at 298 K, which is typical for two-electron processes. It has also been found that with increase in voltamperometric polarization rate the electrode reaction starts to become irreversible.

Studies of the electroreduction of Hg^{2+} to Hg° in a 0.1 M solution of HNO_3 at a glassy carbon electrode found a single peak that was interpreted as an integral one and was attributed to the two-step reactions $Hg^{2+} \rightarrow Hg_2^{2+}$ and $Hg_2^{2+} \rightarrow Hg^0$ of reduction of Hg^{2+} to Hg^0 at potentials too close to form two independent peaks.[23]

Figure 6.1 shows a voltamperometric curve for electroreduction of Hg^{2+} ions at a glassy carbon electrode.[22] Investigation of the influence of potential sweep rate (v, V s^{-1}) revealed that at $v \leq 0.02$ V s^{-1} the electrode process takes place in

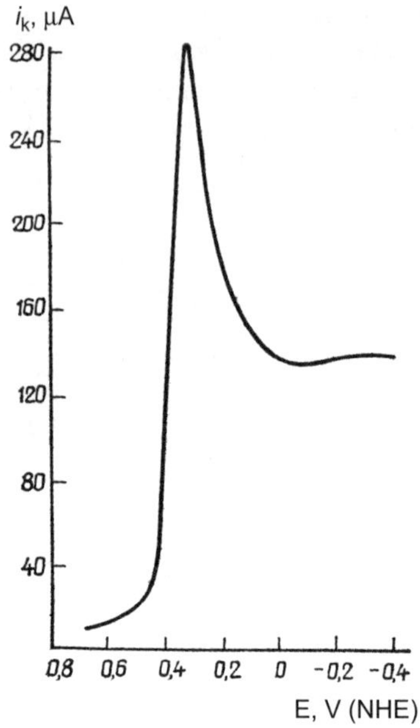

Figure 6.1 Voltamperometric diagram for reduction of Hg^{2+} ions in a solution composed of 0.1 M KNO_3 + 0.01 M HNO_3 at a glassy carbon electrode. $[Hg^{2+}] = 2.29 \times 10^{-3}$ mol; potential sweep rate $v = 0.02$ V s^{-1}. Reproduced with permission from Ref. 22.

a reversible fashion, whereas an increase in the sweep rate causes kinetic control of the process (the same as in the study by Torsi and Mamantov[20]) and at $v \geq 0.4$ V s^{-1} electroreduction of Hg^{2+} ions is irreversible.[22] On the other hand, by taking into account the difference between standard potentials $E^0_{Hg^{2+}/Hg_2^{2+}}$ and $E^0_{Hg_2^{2+}/Hg^0}$, one could anticipate two waves or two distinct peaks. Whether reaction (6.14) is a single-stage or one-step type, it should be accompanied by a chemical reproportionation reaction, eqn (6.16), in the surface layer of the mercury electrode:

$$Hg^{2+} + Hg \rightleftharpoons Hg_2^{2+} \tag{6.16}$$

Voltamperometric studies at a rotating disk electrode made of glassy carbon showed that the shape of the voltamperometric diagrams depends on whether the electrode is coated with mercury or not. If the surface of the electrode is not covered with mercury, reduction of Hg^{2+} ions is manifested by a single elongated wave ($\Delta E \approx 0.5$ V), whereas if it is covered with mercury, two distinct S-shaped waves of different heights are found, as shown in Figure 6.2. Vetter[9] suggested that a cathodic processing treatment of the electrode surface only partially covers it with a mercury deposit. Mercury is deposited only on active centers of the electrode in the form of fine droplets that act as small mercury electrodes at the corresponding polarization. Those centers are the locations where the reproportionation reaction takes place, generating Hg_2^{2+} ions. These Hg_2^{2+} ions are immediately reduced to Hg^0, producing the first reversible S-shaped wave with a half-wave potential $E_{\frac{1}{2}}=0.427$ V (*versus* SCE). The second wave with $E_{\frac{1}{2}}=0.297$ V (*versus* SCE) is responsible for reduction of Hg^{2+} to Hg^0 on the bare parts of the electrode surface.

We believe that this interpretation of the interesting experimental data is insufficient since it does not match the electrode potentials of eqns (6.14) and (6.2). The first wave corresponds to eqn (6.14) of electroreduction of Hg^{2+} to Hg_2^{2+}, concurrently with the Hg^{2+} ion reproportionation eqn (6.16), and results in the generation of Hg_2^{2+} ions. Transformation of Hg^{2+} ions in the course of reaction (6.16) causes a decrease in the height of wave 1 at $E_{\frac{1}{2}}=0.427$ V (*versus* SCE). Generation of Hg_2^{2+} ions according to reaction (6.16) increases the height of wave 2. Analysis of Figure 6.2 shows that the height of wave 1 decreases by 58% due to the consumption of Hg^{2+} ions and the second wave increased by the same amount.

Taking into account reproportionation, the electrode process can be represented as follows:

$$3Hg^{2+} + Hg + 2e \rightarrow 2Hg_2^{2+} + 4e \rightarrow 4Hg^0 \quad \text{(with } 1Hg^0 \text{ returned from the product to the reactant side)} \tag{6.17}$$

Different heights of the first and second waves cannot be attributed to different diffusion coefficients for Hg^{2+} and Hg_2^{2+} ions. These diffusion coefficients, equal to 8.2×10^{-10} and $9.2\times10^{-10}\,m^2\,s^{-1}$, respectively,[24] are fairly close to

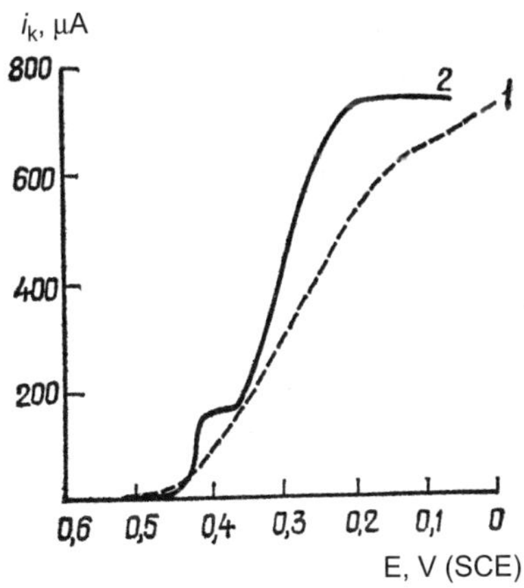

Figure 6.2 Voltamperometric diagrams obtained for reduction of Hg^{2+} at a rotating disk-shaped electrode. $[Hg^{2+}]=2.29\times10^{-3}$ mol; electrode rotation speed $\omega=1500$ rpm. (1) Completely cleaned from traces of metallic mercury deposits during ionization at $E=0.8$ V (*versus* SCE) for 5 min; (2) electrode partially covered with mercury deposits.
Reproduced with permission from Ref. 22.

each other. In fact, as can be seen in ref. 3, polarographic waves of the electroreduction of Hg^{2+} and Hg_2^{2+} ions and metallic mercury oxidation, i_k, that corresponds to the reaction $Hg^{2+} \rightarrow Hg_2^{2+^+}$, is just slightly greater than i_k of the reaction $Hg_2^{2+} \rightarrow Hg^0$. Comparison between the electroreduction of ions Hg^{2+} and Hg_2^{2+} was first made by Kolthoff and Miller.[24]

Ref. 3 also demonstrates that the maximum current of mercury ionization in acidic solutions of easy-soluble salts of mercury(I) and -(II) is virtually impossible to reach. This shape of the curves is typical for reversible electrochemical reactions.[25–28] The results in ref. 3 also show that the discharge and ionization in simple electrolytes takes place with very low polarization. According to Bindra *et al.*,[3] potentials of dropping mercury electrodes in solutions of simple salts of mercury(I) and -(II) are defined by the Nernst equations:

$$E = E^0_{Hg_2^{2+}/Hg^0} + \frac{RT}{2F}\ln\left[Hg_2^{2+}\right]_0 \tag{6.18}$$

$$E = E^0_{Hg^{2+}/Hg^0} + \frac{RT}{2F}\ln\left[Hg^{2+}\right]_0 \tag{6.19}$$

When Hg_2^{2+} ions are being reduced, the current is limited by their diffusion to the electrode and is described by

$$i_k = 0.627zFD^{\frac{1}{2}} m^{\frac{2}{3}} t^{\frac{1}{6}}\left(\left[Hg_2^{2+}\right] - \left[Hg_2^{2+}\right]_0\right) \tag{6.20}$$

while the half-wave potential follows the equation[3]

$$E_{\frac{1}{2}} = E^0_{Hg_2^{2+}/Hg} + \frac{RT}{2F}\ln\left(\frac{\left[Hg_2^{2+}\right]}{2}\right) \tag{6.21}$$

where z is the number of electrons involved in the electrode reaction, $z = 2$, F is Faraday's constant, D is the diffusion coefficient of mercury ions, m is the mass of the dropping mercury per unit time and t is the dropping time.

According to Geyrovsky and Kuta,[3] the equations shown above are also valid for the reduction of mercury(II) ions. The mercury(II) ions react immediately with metallic Hg on its surface according to eqn (6.16).

The half-wave reduction potentials of mercury(II) ions in non-complexing media (ClO_4^-, NO_3^-) on a rotating disk electrode depend on the nature of the electrode material and the solubility, S, of metallic mercury.[23] It was found that $E_{\frac{1}{2}}$ at an Hg^{2+} concentration of 2×10^{-4} mol L^{-1} for graphite ($S_{Hg} = 0\%$), platinum ($S_{Hg} = 0.090\%$), gold ($S_{Hg} = 0.136\%$) and mercury ($S_{Hg} = 100\%$) is 0.060, 0.220, 0.395 and 0.410 V (*versus* SCE), respectively. It was shown that some intermetallics with Hg were formed on the surface of platinum electrodes.[29] Pt_2Hg and PtHg were found at partial coverages (less than a monolayer of Hg) of the platinum electrode surface. $PtHg_2$ formed as a result of deposition of two monolayers of mercury. Oxidation of the surface and subsurface of the intermetallic compounds produces mercury(II) ions, whereas the oxidation of volumetric mercury (deposition of more than 50 monolayers of Hg atoms on platinum surfaces) produces only mercury(I) ions.

In sulfuric acid solution (1 M H_2SO_4), the electroreduction of Hg^{2+} ions was studied with the help of a rotating disk electrode with ring and voltammetry using a linear potential sweep.[34] The anodic dissolution of metallic mercury was studied with the help of the polarographic method ($Na_2SO_4 + H_2SO_4$, $NaClO_4$ up to $\mu = 0.1$).[35] It was established that the rate of the electrode process depends on the potential and, at $E < 0.4$ V (*versus* SCE), eqn (6.22) occurs on the glassy carbon electrode:[34]

$$2Hg^{2+} + 2e \rightarrow Hg_2^{2+} \tag{6.22}$$

whereas at $E_k = 1.4$ V (*versus* SCE), only Hg_2^{2+} ion oxidation, the reaction

$$Hg_2^{2+} \rightarrow 2Hg^{2+} + 2e \tag{6.23}$$

occurs at the ring-disk electrode. The oxidation current at the ring-disk electrode depends on the electrode preparation conditions. In the course of

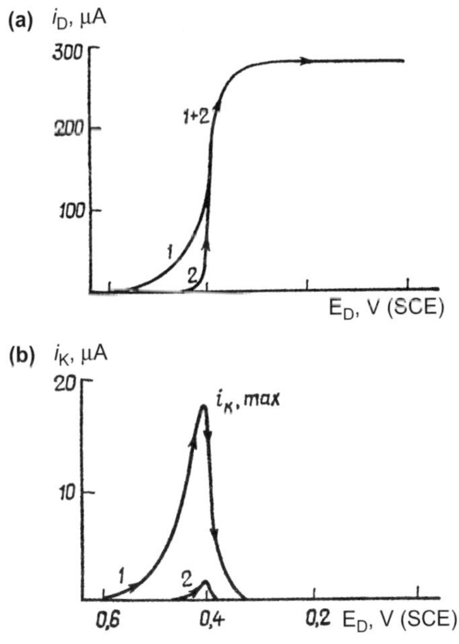

Figure 6.3 (a) Polarization *i–E* curves for electroreduction of 9.9×10^{-4} M Hg^{2+} in 1 M H_2SO_4 on a glassy carbon disk electrode and (b) oxidation of Hg_2^{2+} intermediates on a ring electrode, obtained with (1) one pulse on a freshly treated electrode and (2) two subsequent pulses. $E_k = 1.48$ (*versus* SCE), $\nu = 0.01$ V s^{-1}, $\omega = 251$ rad s^{-1}.
Reproduced with kind permission from Elsevier Science © P. Kiekens, R. M. H. Verbeeck, H. Donche and E. Temmerman, *Electroanal. Chem.*, 1983, **147**, 235. Ref. 34.

cathodic polarization of the disk electrode, the current at the ring first increases with decreasing disc potential, then drops to zero (see Figure 6.3). This is because the formation of Hg(I) ions on the surface of the disc or in the adjacent layer is limited to the Hg^{2+} ion reduction potential region on the disc. In the case of a shift of the disk (glassy carbon) electrode potential towards the cathodic region, the electroreduction of Hg^{2+} ions occurs as a one-step reaction, eqn (6.15), and the electrochemical reaction does not produce any Hg^{2+} ions exhibiting the properties of intermediates. Figure 6.3 shows that, indeed, Hg(I) ions are generated, according to eqn (6.22), on the surface of the disc at the wave base potentials. However, it will be observed that some amount of Hg_2^{2+} ions may also be generated *via* reproportionation, eqn (6.16). It is also found that subsequent scanning without reactivation of the electrode surface produces much smaller quantities of Hg(I) ions.

From Figure 6.4, it follows that the ring electrode current, i_k, is a function of the disk electrode current, i_D, at potentials that are more positive than the peak current potential at the ring-disk electrode. The relation between i_k and i_D is $i_k/i_D = N_{exp}$, where N_{exp} is the experimental electrode efficiency factor.[1] At $E<0.4$ V (*versus* SCE) i_D increases, whereas i_k decreases; i_k/i_D decreases and reaches zero at $E=0.35$ V (*versus* SCE).

The maximum Hg(I) oxidation current at the ring-disk electrode as a function of the electrode rotation speed is shown in Figure 6.5.[34] The shape of

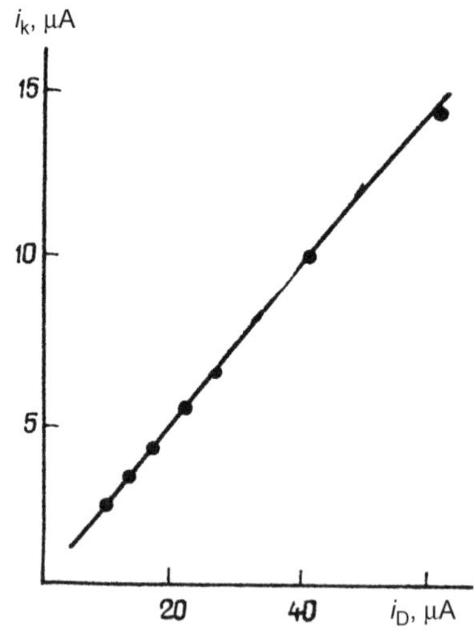

Figure 6.4 Ring-disk electrode current, i_k, as a function of disk-shaped electrode current, i_D, of Hg(II) reduction, given potentials that are more positive than the Hg_2^{2+} ion oxidation peak potential at the ring; $N_{exp}=0.24$, where $N_{exp}=i_k/i_D$.
Reproduced with kind permission from Elsevier Science © P. Kiekens, R. M. H. Verbeeck, H. Donche and E. Temmerman, *Electroanal. Chem.*, 1983, **147**, 235. Ref. 34.

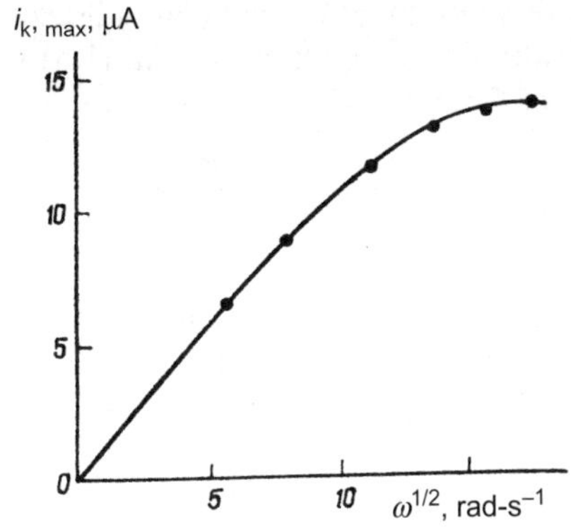

Figure 6.5 $i_{k\ max}$ of oxidation of Hg_2^{2+} ions at the ring-disk electrode as a function of rotation speed $\omega^{\frac{1}{2}}$. $[Hg^{2+}] = 9.9 \times 10^{-4}$ M in 1 M H_2SO_4; $E_k = 1.4$ V (versus SCE).
Reproduced with kind permission from Elsevier Science © P. Kiekens, R. M. H. Verbeeck, H. Donche and E. Temmerman, *Electroanal. Chem.*, 1983, **147**, 235. Ref. 34.

the $i_{k\ max}$–$\omega^{\frac{1}{2}}$ curve is typical for an electrode reaction at a disk electrode, which is controlled partly by mass transfer and partly by charge transfer at potentials corresponding to the maximum current at the ring-disk electrode. The subsequent beginning of direct reduction of Hg(II) to metallic mercury is preceded, as seen from Figure 6.3, by the transition of the reaction occurring at the disk electrode into the diffusion-controlled region (i_D = constant) at $E > 0.35$ V (*versus* SCE).

Kiekens *et al.*[34] believe that at high i_D values, Hg(I) disproportionation according to eqn (6.24) is likely to occur:

$$Hg_2^{2+} \rightleftharpoons Hg^0 + Hg^{2+} \tag{6.24}$$

At potentials that are more negative than 0.4 V and at i_{max} at the ring-disk electrode, seen in Figure 6.5, direct reduction of mercury(II) to metallic mercury via eqn (6.15) will gradually replace the single-electrode transfer based on the reaction

$$Hg^{2+} + e \rightarrow Hg^+ \tag{6.25}$$

Therefore, the electroreduction of mercury(II) ions may be considered as an electrode process which, depending on electrode potential, occurs via consecutive reactions (6.14) and (6.24) and parallel reactions (6.15), (6.25), (6.14) and (6.24). This is why the total number of electrons involved in the cathodic process is non-integral and varies from $z = 1.65$ to 1.85.[34] It has been demonstrated that under diffusion-controlled conditions, electroreduction of Hg(II) proceeds with two electrons ($z = 2$) and ultimately produces the metal. The calculated value of the Hg(II) diffusion coefficient is $(6.9 \pm 0.1) \times 10^{-6}$ cm^2 s^{-1}.

In cases of anodic dissolution of mercury in sulfuric acid solutions under conditions where the solubility product of mercury(I) sulfate (Hg_2SO_4) cannot be reached, the function E *versus* $\ln i$ should be described by the equation[35]

$$E = E^0_{Hg_2^{2+}/Hg^0} + \frac{RT}{2F}\ln i_A \tag{6.26}$$

It follows from eqn (6.26) that the anodic current should not depend on the concentration of sulfate ions. However, experiments[35] have shown that anodic current increases with increasing concentration of sulfate ions in the solution, which contradicts eqn (6.26). It follows[35] that the anodic current does not reach the limiting diffusion value, which is observed when a deposit forms on the electrode surface. The electrochemical behavior of such systems has been addressed.[36,37]

It follows[35] that the anodic current is a linear function of the concentration of sodium sulfate (a), while at the same time the mercury electrode potential is shifted towards less positive values (b). The electrode behavior of mercury is due to the formation of a complex of mercury(I) with sulfate ions $[Hg_2SO_4]$ the stability constant of which is 1.5×10^2 L mol^{-1}.[35] The expected values of i_A and E for the studied sodium sulfate solutions, obtained with the help of a new numerical method developed by Kikuchi and Murayama.[35] As can be seen, there is a very good correlation between the calculated and experimental values for mercury oxidation currents and mercury electrode potentials in solutions with different sodium sulfate concentrations.

Table 6.2 gives some kinetic parameters for the electroreduction of mercury ions in different solutions. It can be seen that the exchange currents reach high values in non-complexing media. In solutions containing ligands that form complex mercury(II) compounds, the exchange currents and consequently the rate constants become smaller. If inert materials, such as glassy carbon, are used as the electrode material, the exchange current is also greatly reduced.

6.2 Kinetics and Mechanism of Discharge and Ionization of Mercury in Complex-forming Media

Mercury(I) halides, except fluorides, are not dissociated in aqueous solutions, yet, it follows from Chapter 5 that they dissolve in the presence of excess alkali metal halides and ammonium in solution, producing complex ions HgX_2^- and HgX_4^{2-}. On the whole, mercury(I) compounds are fairly stable in aqueous solutions and become involved in disproportionation reactions only in the presence of ligands. Equation (6.27) illustrates a disproportionation reaction of mercury(I).

$$Hg_2X_2 + zX^- \underset{}{\overset{k_{\text{disproportionation}}}{\rightleftharpoons}} [HgX_{z+2}]^{z-} + Hg^0 \tag{6.27}$$

Table 6.2 Rate constants, exchange currents and transfer coefficients during electroreduction of mercury(I) and -(II) ions and mercury oxidation.

Background (mol)	*Electrode*	*Method*[a]	T *(K)*	*Transfer coefficient* α	β	k_S *(cm s⁻¹)*	i_0 *(A cm⁻²)*	*Ref.*
Simple electrolytes								
$(0.59–3.5)\times10^{-3}$ Hg_2^{2+} 1 $HClO_4$	Hg_{liq}	DPGST	298	(0.3)[b]	0.7 ± 0.03	–	16.5 ± 1.5	31
$(0.6–3.4)\times10^{-3}$ Hg_2^{2+} 1.0 $HClO_4$	Hg_{liq}	DPGST	234.35	(0.3)	0.7 ± 0.03	–	14 ± 1.5	31
1.8×10^{-3} Hg_2^{2+} 1.0 $HClO_4$	Hg_{sol}	DPGST	234.35	(0.3)	0.7 ± 0.03	–	10 ± 1	31
$(0.59–5.3)\times10^{-3}$ Hg_2^{2+} 1.0 $HClO_4$	Hg_{liq}	DACPMM	298	0.75	(0.25)	$\geq7.2\times10^{-2}$	≥5	38
$(0.5–2.0)\times10^{-3}$ Hg_2^{2+} 1.0 $HClO_4$	Hg_{liq}	FR	298	0.86	–	~1.5[c]	0.17–0.69	33
1.0×10^{-3} Hg_2^{2+} 0.98 $HClO_4$	Hg_{liq}	DPGST	298	0.24	(0.76)	5.2×10^{-3}	0.25	32
3.5×10^{-4} Hg_2^{2+} 1.1 $HClO_4$	Hg_{liq}	FR	298	0.30	(0.70)	1.4[c]	5.9[§]	33
9.9×10^{-4} Hg_2^{2+} 1.0 H_2SO_4	Glassy carbon	VDEK	298	0.54–0.60[d]	0.40–0.44	$\sim10^{-7}$[d]	–	34
4.7×10^{-4} Hg_2^{2+} 0.2 $HClO_4$	Hg_{liq}	FR	296.5 ± 0.5	0.28	(0.72)	0.36	1.9[§]	33
$(0.02–5.15)\times10^{-3}$ Hg_2^{2+} 1 $HClO_4$	Hg_{liq}	PRM	298	0.32 ± 0.02	(0.68 ± 0.02)	0.019 ± 0.02		39
2.0×10^{-3} Hg^{2+} 1 $HClO_4$	Hg_{liq}	FR	298	(0.53)	(0.57)	3.5×10^{-3}	0.68	33
Complex-forming media								
1×10^{-5} Hg^{2+}, 0.1 CN^- 0.9 NaCl	Hg_{liq}	PPST	285	0.30 ± 0.03	0.67 ± 0.05	3.9×10^{-5}	0.03	40
$(0.83–1.04)\times10^{-3}$ Hg^{2+} 4.5 NH_4Br, 3.5 HBr	Hg_{liq}	XP	298	–	–		1.5×10^{-4}	42
$(1\times10^{-5}–0.5)$ Hg^{2+} 4.5 NH_4Br 3.5 HBr	Hg_{liq}	DPGST	298	0.5	0.5			47

[a]DPGST, double-pulse galvanostatic method; DACPM, direct and alternating current polarization method; FR, Faradaic rectification; VDEK, voltammetric; PRM, pulse relaxation method; PPST, pulse potentiostatic method; XP, Chronopotentiometry method.
[b]Expected transfer coefficients from the relationships $\alpha-\beta=1$ and $\partial \ln i_0/\partial \ln[Hg_2^{2+}]=\beta/2$.
[c]Standard rate constant.
[d]Kinetic parameters of the first step $Hg^{2+}+e\rightarrow Hg^{+}$.

Therefore, the electrochemical behavior of mercury in simple electrolytes ($HClO_4$, $NaClO_4$, H_2SO_4, HNO_3, *etc.*) and complex-forming electrolytes (Cl^-, Br^-, I^-, CN^-, OH^-, SCN^-, CH_3COO^-, *etc.*) is different on mercury and on inert electrode materials.[1–4,7–9,30–44]

In the presence of low concentrations of anions (Cl^-, Br^-, I^-) that form insoluble compounds with mercury(I), the anodic polarization of mercury electrodes is manifested by the limiting current and is dependent on the anion concentration and passivation phenomena.[1,36,43] The passivation of mercury anodes is presumably due to a film of poorly soluble salts of univalent mercury, which covers the electrode.[45] The passivation of mercury electrodes can be greatly reduced by introducing ammonium ions into the solution.[46] The kinetic regularities of discharge–ionization of mercury in this case are of considerable practical interest. Halide–ammonium electrolytes are used for electrochemical processes with liquid mercury and amalgam electrodes to produce high-purity metals. From a theoretical viewpoint, the process of mercury discharge–ionization in a complex electrolyte with several ligands is very interesting.

References

1. L. F. Kozin, *Electrostatic Precipitation and Dissolution of Multivalent Metals*, Naukova Dumka, Kiev, 1989.
2. P. Bindra, A. R. Brown, N. Fleischmann and D. Pletcher, *J. Electroanal. Chem.*, 1975, **58**, 39.
3. Ya. Geyrovsky and Ya. Kuta, *Basic Polarography*, Mir, Moscow, 1965.
4. P. Wrona and Z. Galus, in *Encyclopedia of Electrochemistry of the Elements*, ed. A. J. Bard, Marcel Dekker, New York, 1982, vol. 9, pt. A, pp. 1–227.
5. S. E. S. Wakkad and T. M. Salem, *J. Phys. Chem.*, 1950, **54**, 1371.
6. C. E. Wanderzee and J. A. Swanson, *J. Chem. Thermodyn.*, 1974, **6**, 827.
7. M. Kogoma, T. Nakayama and S. Aoyaqui, *J. Electroanal. Chem.*, 1972, **34**, 123.
8. P. Bindra, A. P. Brown and M. Fleischmann, *J. Electroanal. Chem.*, 1975, **58**, 31.
9. K. J. Vetter, *Electrochemical Kinetics; Theoretical and Experimental Aspects*, Academic Press, New York.
10. L. F. Kozin, N. I. Ivanchenko and A. A. Nikitin, *Zhn. Prikl. Khim.*, 1982, **55**, 1001.
11. C. K. Choudary and B. Prasad, *J. Indian Chem. Soc.*, 1982, **59**, 555.
12. R. S. Eachus, M. C. R. Symons and J. K. Yandell, *Chem. Commun.*, 1969, **17**, 979.

13. M. Farraggi and A. Amozig, *Int. J. Radiat. Phys. Chem.*, 1972, **4**, 353.
14. S. Papoff, J. A. Riddick, V. I. Wirth and L. D. Orgh, *J. Am. Chem. Soc.*, 1931, **53**, 1195.
15. S. R. Carter and R. Robinson, *J. Chem. Soc.*, 1972, **1**, 267.
16. D. Dobosh, *Electrochemical Constants*, Mir, Moscow, 1980.
17. V. M. Latimer, *The Oxidation States of the Elements and Their Potentials in Aqueous Solutions*, Izd-vo Inostr. Lit., Moscow, 1954.
18. S. Hietanen and L. G. Sillen, *Ark. Kemi*, 1956, **10**, 103.
19. E. I. Hrusheva and V. E. Kazarinov, *Electrochemistry*, 1986, **22**, 1262.
20. G. Torsi and G. Mamantov, *J. Electroanal. Chem.*, 1971, **32**, 465.
21. M. S. Shuman, *Anal. Chem.*, 1969, **41**, 142.
22. R. G. Dhaneshwar and A. V. Kulkarni, *J. Electroanal. Chem.*, 1979, **99**(2), 207.
23. M. Stulikova, *J. Electroanal. Chem.*, 1973, **48**, 33.
24. I. M. Kolthoff and C. S. Miller, *J. Am. Chem. Soc.*, 1941, **63**, 2732.
25. K. P. Butin, R. D. Rackimov and I. V. Novikova, *Vestnik MGU, Ser. 2*, 1990, **31**, 574.
26. A. G. Stromberg, *Zh. Fiz. Khim.*, 1953, **27**, 1287.
27. A. G. Stromberg, *Zh. Fiz. Khim.*, 1955, **29**, 409.
28. A. G. Stromberg, *Zh. Fiz. Khim.*, 1955, **29**, 2152.
29. M. Z. Hassan, D. F. Unterekcr and S. Bruckenstein, *J. Electroanal. Chem.*, 1973, **42**, 161.
30. R. Tammamushi, *Kinetic Parameters of Electrode Reactions of Metallic Compounds*, Butterworths, London, 1975.
31. H. Gerisher and M. Krause, *Z. Phys. Chem.*, 1958, **14**, 184.
32. H. Matsuda, S. Oka and P. Delahay, *J. Am. Chem. Soc.*, 1959, **81**, 5077.
33. H. Imai and P. Delahay, *J. Phys. Chem.*, 1962, **66**, 1108.
34. P. Kiekens, R. M. H. Verbeeck, H. Donche and E. Temmerman, *Electroanal. Chem.*, 1983, **147**, 235.
35. K. Kikuchi and T. Murayama, *Bull. Chem. Soc. Jpn.*, 1988, **61**, 4269.
36. L. F. Kozin, *Physico-Chemical Basics of Amalgam Metallurgy*, Nauka, Alma-Ata, 1964.
37. K. Kikuchi and T. Murayama, *Bull. Chem. Soc. Jpn.*, 1988, **61**, 3159.
38. H. Gerisher, *Z. Elektrochem.*, 1951, **55**, 98.
39. W. D. Weir and C. G. Enke, *J. Phys. Chem.*, 1967, **71**, 280.
40. M. Perez-Fernandez and H. Gerisher, *J. Phys. Chem.*, 1960, **64**, 477.
41. N. Tanaka and T. Murayama, *Z. Phys. Chem. (Frankfurt)*, 1959, **21**, 146.
42. N. Tanaka and T. Murayama, *Z. Phys. Chem. (Frankfurt)*, 1958, **14**, 370.
43. L. F. Kozin and Ya. P. Suchkov, *Ukr. Khim. Zh.*, 1991, **57**, 612.

44. J. Schwade and A. Chon Dien, *Electrochim. Acta*, 1962, **7**, 1.
45. M. T. Kozlovskii, A. I. Zebreva and B. P. Gladyshev, *Amalgams and Their Application*, Nauka, Alma-Ata, 1971, pp. 81–82.
46. L. F. Kozin and A. A. Nikitin, *Izv. Akad. Nauk Kazakh., Ser. Khim.*, 1970, **1**, 36.
47. L. F. Kozin, *Ukr. Khim. Zh.*, 1991, **57**, 585.

CHAPTER 7
Lighting

7.1 Introduction

Mercury plays an extremely important role in discharge lighting. It is used as a light source and as a buffer gas to carry current through a discharge. Many types of discharge light sources contain mercury, including fluorescent, high-pressure mercury, ultra-high-performance (UHP) mercury, high-pressure sodium and metal halide lamps. Numerous monographs[1–4] have been devoted to the physics of lamp discharge and operation, so only a very brief explanation of the mechanism of the emission of light is given here. The emphasis is on the role of mercury in the discharge and mercury-containing materials used in these lamps. Mercury is the essential element necessary for producing light in fluorescent, high-pressure mercury and UHP lamps. Mercury is used in other lamp types in conjunction with other metals or metal halides. Mercury(II) iodide is added to some lamps.

Fluorescent lamps involve low-pressure discharges and may be classified as compact, linear, cold cathode or germicidal. High-pressure and UHP (*i.e.*, ultra-high-pressure) lamps contain mercury at pressures from a few to several hundred atmospheres and emit light in the ultraviolet, visible and infrared regions. High-pressure sodium lamps contain a binary Na–Hg or a multi-component amalgam. Sodium radiation is the dominant emission in these discharges. Metal halide lamps may contain mercury or may be mercury free. Metal halide lamps may contain an assortment of halides in addition to mercury or a mercury replacement such as zinc iodide. The metal halides emit important spectral lines and contribute substantially to the efficiency of converting electrical energy to light and to the quality of the emitted light.

The science of quantifying light quality is a complex subject. Although discussions of the topics involved in light source quality have been reported,[5]

Mercury Handbook: Chemistry, Applications and Environmental Impact
By Leonid F Kozin and Steve Hansen

Published by the Royal Society of Chemistry, www.rsc.org

Table 7.1 Mercury contents in different discharge light sources.[7–9]

Lamp type	*Mercury content range (mg)*	*Efficacy (lm W^{-1})*	*CCT (K)*	*CRI*
Mercury vapor	25–225	40–60		
Ultra-high-pressure	2.5–10	60–75	7000–8000	57
High-pressure sodium	20–145	100–120	2200–2700	20–70
Quartz metal halide[a]	2.5–225	60–120	2700–6500	50–95
Ceramic metal halide	2.5–30	60–120	2700–6500	70–95
T-5 linear fluorescent	1–3	61–100	4000–5000	62
Compact fluorescent	1–3	61–100	4000–5000	62

[a] May be higher for very high wattage lamps.

several key parameters are used to describe light quality and lamp operating temperature. These include the color rendering index (CRI) and the correlated color temperature (CCT). Typical mercury contents, lamp performances and light quality measures, namely CRI, CCT and efficacy, found in discharge lighting applications, according to lamp type, are given in Table 7.1.

7.1.1 Lamp Color and Quality Measurements

A measure of a lamp's color quality is defined as the color rendition index (CRI). CRI is defined on a scale from 0 and 100, where the CRI for sunlight and a halogen lamp is defined as 100. The correlated color temperature (CCT) is the measure used to describe the relative color appearance of a white light source. The CCT of a light source is the color temperature of a Planckian (black-body) radiator that best approximates it and indicates whether a light source appears more yellow/gold/orange or more blue, in terms of the range of available shades of 'white.' CCT is given in kelvin. A CCT of 2700 K is 'warm' and 5000 K is 'cool.'[6] CCTs over 6000 K are bluish white in color whereas CCTs around 2700 K appear slightly yellowish and are typical color temperatures of an incandescent lamp. Halogen lamps have a CCT of about 3000 K and sunlight about 5000– 6500 K.

7.2 Fluorescent Lighting

Fluorescent lamps contain a small mercury dose that emits UV radiation at 185.0 and 253.7 nm. The UV radiation is converted into visible light by a phosphor. The ideal mercury vapor pressure in a fluorescent lamp is 0.8 Pa (6 mTorr) and, for pure mercury, occurs close to 40 °C (313 K). Only around 50 μg of the mercury dose is in the discharge as a vapor during operation.[10] It should be mentioned that an auxiliary or buffer gas is also necessary for the operation of fluorescent lamps. An optimum pressure for an auxiliary gas such as argon is between 200 and 300 Pa.[11] A schematic showing the operation of fluorescent lamps is shown in Figure 7.1.

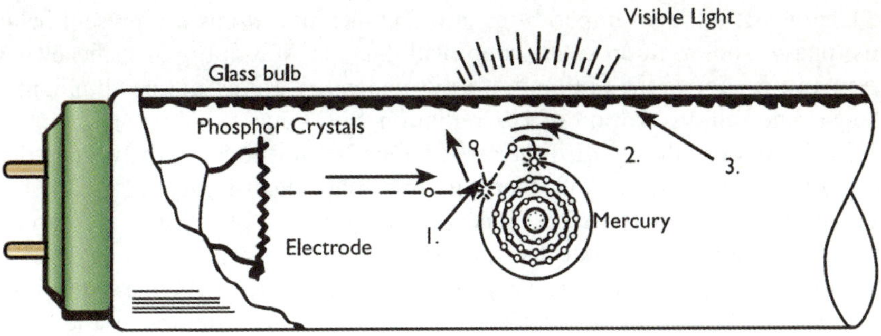

Figure 7.1 Schematic illustration of a fluorescent lamp. Courtesy of Philips Lighting.

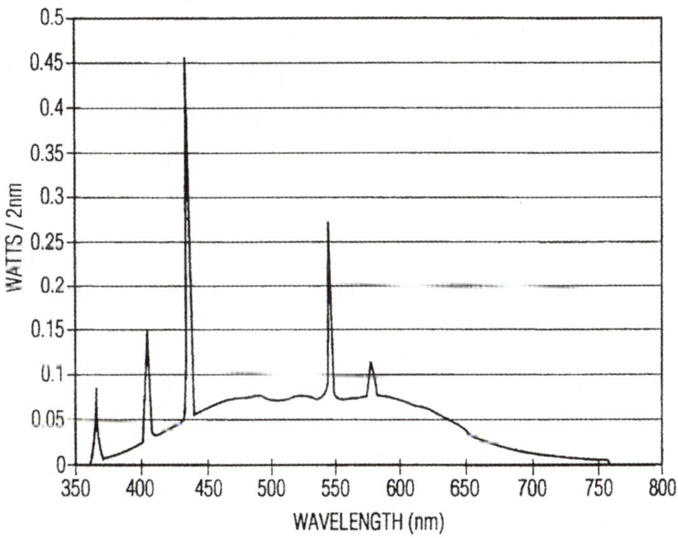

Figure 7.2 Spectrum of a high-CRI fluorescent lamp. Reproduced from Ref. 12.

Fluorescent lamp operation consists of the following steps:

1. Power through the electrodes heats them and gives off electrons (the electrodes are covered with an emission mix that assists in extracting electrons).
2. The electrons excite mercury atoms.
3. The mercury atoms fluoresce and create UV radiation (185 and 254 nm).
4. The phosphorescent wall coating converts UV into visible light.

The spectrum of a commercial fluorescent lamp is given in Figure 7.2. The lumen output of fluorescent lamps is strongly dependent on the mercury vapor pressure, and the latter is strongly dependent on the temperature, as described in Chapter 1. At low temperatures, too few mercury atoms are present in the gas phase and a fluorescent lamp will have a sub-optimum efficacy. At temperatures above the optimum pressure, mercury atoms will be abundant in the gas and self-absorption of UV radiation will occur.

The luminous efficacy of fluorescent lamps has improved greatly since they were first introduced in 1938. Not only have phosphors improved greatly and contributed to the increased efficiency, but also the use of krypton or xenon as the fill gas and the recent use of high-frequency operation have enabled fluorescent lamps to become very efficient light sources. At the same time, mercury has been more accurately dosed into fluorescent lamps by the use of amalgams, glass capsules and other devices.

7.2.1 Mercury Content in Fluorescent Lamps

The mercury vapor required in the gas phase for maximum light output at the operating temperature is small relative to a typical mercury dose. For example, a typical 4 ft T8 lamp requires only about 50 μg of Hg and a 15 W CFL requires only about 10 μg. Mercury consumption over the life of the lamp demands that much larger doses are used. The relative importance of these consumption reactions is shown in Table 7.2.[13]

Successful approaches that permit mercury reduction due to consumption include coatings on the glass bulb wall, coatings on the phosphor particles, additives to the phosphor layer[14,15] and coatings on top of the phosphor layer. Materials used in coatings as barriers to mercury consumption reactions include Al_2O_3, Y_2O_3 and other metal oxides in both particulate and vitreous forms. Industry has made substantial progress in reducing the mercury content in fluorescent lamps over the last 30 years. The reduction in mercury dosed into fluorescent lamps is provided in Table 7.3 for the 4 ft linear lamp types.[16]

7.2.2 Amalgam-controlled Mercury Vapor Pressure

Amalgams have one of two purposes in fluorescent lamps: one is to regulate the mercury vapor pressure and the other is to make handling and dosing mercury safer and more accurate than handling liquid mercury. Mercury vapor pressure control, *i.e.*, regulation, of the mercury vapor pressure over a wide temperature range, generally between 60 and 140 °C, is required in fluorescent lamps operating in outer jackets or in enclosed fixtures and in high-power germicidal

Table 7.2 Mercury consumption mechanisms.[13]

Component consuming Hg	*Factor affecting consumption*	*Relative importance*
Glass without coating	Glass type, wall loading, phosphor properties	High
Phosphor layer	Phosphor type, grain size, additives	Medium to low
Emission material	Cathode shield, electrical factors	Medium to low
Glass stems	Glass type	Low

Table 7.3 Mercury dose per lamp in 4 ft linear fluorescent lamps using the best available technology.

Year	*mg Hg (using best available technology)*
1984	64
1986	43
1988	40
1990	33
1992	33
1994	33
1996	16
1998	10
2000	5
2006	4

lamps with high lamp wall temperatures. Various amalgams have been used to regulate mercury vapor pressure, two of the most successful being Pb–Bi–Sn–Hg and Bi–In–Hg. The use of Pb–Bi–Sn–Hg has been steadily declining because of its lead content. However, its main advantage is its relatively low cost compared with Bi–In–Hg. Indium metal is significantly more expensive then either bismuth or tin.

Bi–In–Hg amalgams were reported in 1977.[17] Early compositions of Bi–In–Hg amalgams contained between about 3 and 12 wt% of mercury and 20–30 wt% of indium. Indium dramatically reduced the mercury vapor pressure of these amalgams between about 50 and 140 °C. Recently, Hein and Raiser[18] and Hellebreker and Kaldenhoven[19] suggested the use of Bi–Sn–In–Hg (where the indium content is between 3 and 5 wt%) as a lower indium modification of Bi–In–Hg. Vapor pressure curves for these amalgams are shown in Figure 7.3.

Many other amalgams have been developed to regulate the mercury vapor pressure in a fluorescent lamp. Examples of amalgams regulating the mercury vapor pressure are given in Table 7.4.

7.2.3 Temperature-controlled Amalgams

Temperature-controlled amalgams have no significant effect on the vapor pressure of mercury. They only provide a mechanism for administering (dosing) lamps with a precise quantity of mercury.

7.2.3.1 Zn–Hg Amalgams

Zinc–mercury amalgams[27] were introduced as a means of delivering a solid mercury dose into a fluorescent lamp. The shape of the equilibrium phase diagram[28,29] (Figure 7.4) favors a high mercury vapor pressure at room temperature. When the mercury vapor pressure above a mercury–zinc pellet

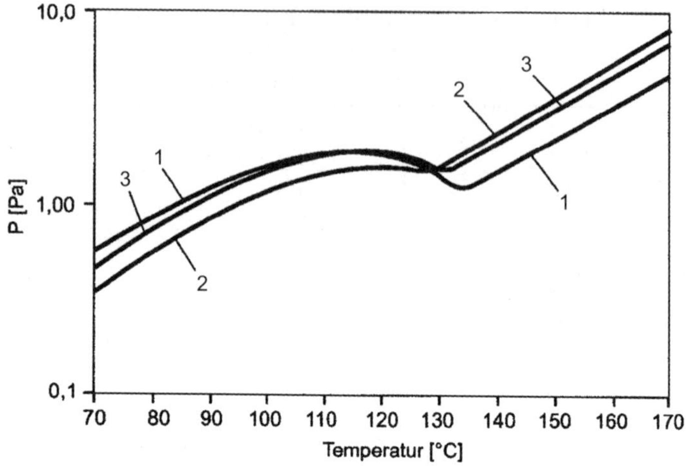

Figure 7.3 Vapor pressure of Bi–Sn–In–Hg amalgams.[18] Compositions (wt%): 1, 58 Bi–38 Sn–3 In–0.7 Hg; 2, 57 Bi–37 Sn–5.5 In–1.2 Hg; 3, 58 Bi–37 Sn–4 In–1 Hg.
Reproduced from Ref. 18.

Table 7.4 Regulating amalgams used in fluorescent lamps.

Amalgam	*Ref.*
In–Hg	20
In–Sn–Hg	21
Pb–Bi–In–Hg	22
Bi–In–Hg	23
Pb–Bi–Sn–Hg	24
Pb–Bi–Ag–Hg	25
In–Sn–Zn–Hg	26

containing 50 wt% Hg was measured,[30] its vapor pressure was found to be about 95% of the vapor pressure of pure mercury at temperatures well above the peritectic point. The surprisingly high vapor pressure was explained by the presence of a metastable mercury-rich liquid in the manufactured product.[27]

The mercury vapor pressure over the temperature range 5–70 °C is expressed by the equation

$$\log(p_{\mathrm{Zn-Hg}}, \mathrm{Pa}) = -\frac{3214}{T} + 7.293 \tag{7.1}$$

Zinc is used only as a carrier to hold mercury until it can be released inside a fluorescent lamp. The mercury vapor pressure above a equimolar Zn-Hg amalgam is given in ref. 30. As indicated by the phase diagram, the γ phase is reported to melt at 42.9 °C. However, in actual tests by differential scanning calorimetry, alloys of predominantly γ phase melt at 65–75 °C.

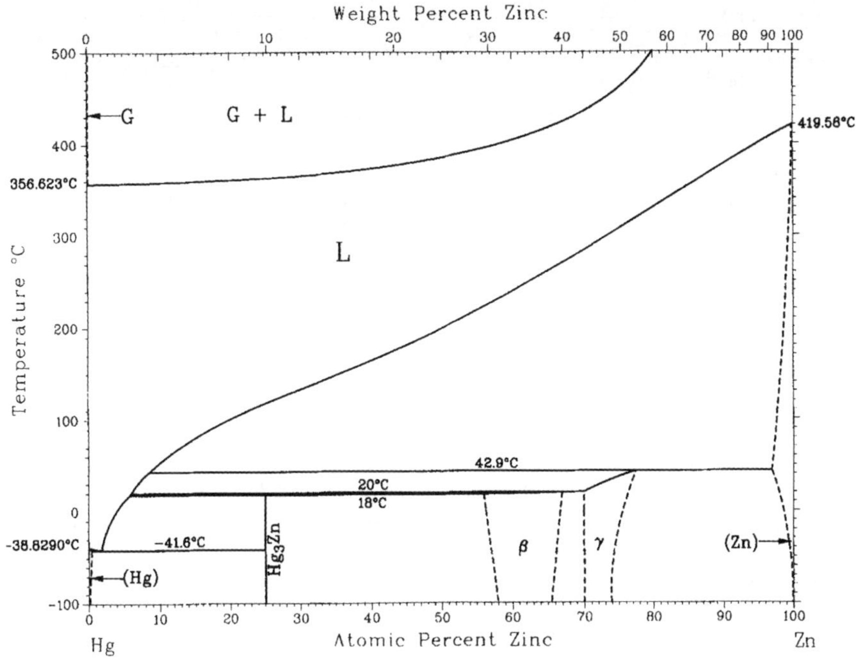

Figure 7.4 Zn Hg equilibrium phase diagram.
Reproduced with kind permission from ASM International.[28]

7.2.3.2 *Sn–Hg Amalgams*

Sn–Hg amalgams also exhibit a high mercury vapor pressure at room temperature. Although measurements of the mercury vapor pressure above liquid Sn–Hg are plentiful, vapor pressure measurements above the solid are more scarce.[31] Vapor pressure measurements of an Sn–Hg amalgam are given in and replace with ref. 31.

Sn–Hg amalgams have been known almost since antiquity and are a key component of dental amalgam. A thermodynamic assessment of the Sn–Hg phase diagram[32] is shown in Figure 7.5. Sn–Hg amalgams are useful in fluorescent lighting applications because they can have mercury contents between about 10 and 50 wt%.[33]

7.2.4 Mercury Dispensers

7.2.4.1 *Bi–Sn–Hg Amalgams*

Bismuth–tin–mercury amalgams have been employed as a method to reduce mercury vapor pressure, similarly to Bi–In–Hg, and as a mercury dispenser for cold cathode fluorescent lamps.[37] A thermodynamic model of the Bi–Sn–Hg phase diagram has been calculated based on experimental data.[38–42] Figure 7.6 gives the calculated liquidus projection. It predicts the

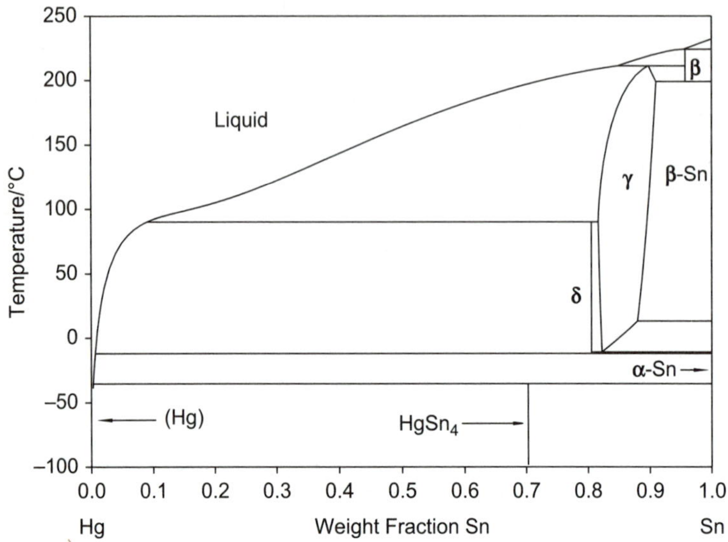

Figure 7.5 Calculated Sn–Hg phase diagram. The β-phase has been simplified to a stoichiometric compound ($HgSn_{38}$).
Reproduced with kind permission from Springer Science + Business Media.[32]

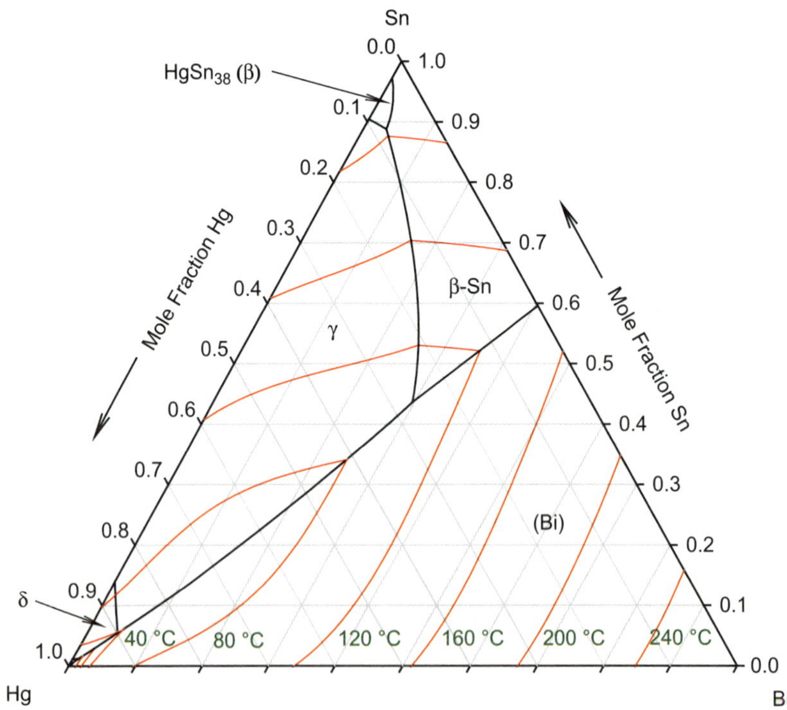

Figure 7.6 Calculated Bi–Sn–Hg liquidus projection.[44] Courtesy of APL Engineered Materials, Inc.

decomposition of the β-phase ($HgSn_{38}$) from the binary Sn–Hg phase diagram into γ- and β-Sn. The vapor pressure above one composition of Bi–Sn–Hg has been measured.[43]

7.2.4.2 Ti–Hg Amalgams

Intermetallic compounds composed of titanium and mercury were developed to introduce mercury into fluorescent lamps.[34–36] The amalgams operate by releasing mercury at very high temperatures from the compounds Ti_3Hg and Zr_3Hg. Della Porta and Rebaudo[34,35] reported the synthesis of γ-Ti_3Hg. Fine titanium powder passing through a 400-mesh screen is placed in a crucible with mercury and heated at about 800 °C for 3 h. The resulting alloy consists essentially of γ-Ti_3Hg. Zirconium may also be used to form a zirconium–mercury amalgam.[36]

7.3 Measurement of Mercury Vapor Pressure of Fluorescent Lamp Amalgams

Several methods are available for quantifying the temperature dependence of the mercury vapor pressure in the region of operation of fluorescent lamps. These methods include direct static measurement, atomic absorption spectrometry (AAS) and Knudsen effusion mass spectrometry. Amalgam vapor pressures at room temperature are fairly low and require relatively sophisticated equipment and careful experimental work to be measured correctly.

7.3.1 Vapor Pressure Measurement System

The vapor pressure measurement system (VPMS) is a static measurement.[45–47] Total degassing of the sample and the apparatus is a prerequisite for a static measurement of its vapor pressure. The pressure is measured by a capacitance diaphragm absolute gauge. Calibration of the capacitance diaphragm absolute gauge is performed at equally spaced pressures from 0 to 1300 Pa. The sample is immersed in a constant-temperature bath that allows adjustment of the sample temperature. Temperatures in the range from 223 K (–50 °C) to about 400 K (127 °C) are generally possible. The sensor temperature is kept constant and constrains the upper temperature limit of the vapor pressure measurement to 5 K below the sensor temperature.

The vapor pressure of the Zn–Hg amalgam discussed in Section 7.2.3.1 was measured by the VPMS while decreasing the temperature from 70 to 40 °C.[48] The vapor pressure of Zn–Hg obtained during the descending temperature profile is expressed by the equation

$$\log(p, \mathrm{Pa}) = \frac{-3202.4}{T} + 10.143 \tag{7.2}$$

and was in excellent agreement with AAS measurements.[30]

7.3.2 Atomic Absorption Spectrometry

AAS can be used to measure the vapor pressure of a fluorescent lamp amalgam.[17,30] The 254 nm radiation from mercury is used to measure the absorbance of mercury vapor from an amalgam. Measurement temperatures are limited by the dimensions of the absorption cell. The quartz absorption cell must be at a higher temperature than the sample to prevent condensation of mercury vapor on the window and it must be aligned by minimizing the absorbance reading. Absorbance (A) readings of the mercury vapor in the cell are typically recorded to a precision of ± 0.0001 absorbance units (± 0.001 absorbance units above $A = 1.000$) as the temperature is increased.[30] The zero absorbance value can be checked with the cell removed from the path. Absorbance readings may be converted into pressure by use of mercury as a standard.

7.3.3 Knudsen Effusion Mass Spectrometry

Measurement of the vapor pressure of pure mercury by AAS at temperatures above 80 °C is difficult owing to high absorbance values. One means of overcoming this obstacle is to use Knudsen effusion mass spectrometry (KEMS). Vapor pressure measurements at temperatures up to 200–300 °C were made by Hilpert on a specially designed Knudsen effusion system.[49–53] An advantage of the KEMS method is the measurement of all gaseous species. Hilpert measured the vapor pressure of magnesium and calcium amalgams in a KEMS system.

A quadrupole mass filter and Knudsen cell operating under high vacuum were used. A pressure of 10^{-5} Pa in the effusion orifice corresponds to a pressure smaller than 10^{-9} Pa in the ion source. Two orifice diameters were used in the measurement of the vapor pressures of MgHg and $MgHg_2$. The following least-squares equations were obtained for Hg vapor pressures:

$$\log(p_{\mathrm{MgHg}}, \mathrm{Pa}) = -(4845.0 \pm 43.1)\,/\,T + (9.882 \pm 0.092) \qquad (7.3)$$

$$\log(p_{\mathrm{MgHg_2}}, \mathrm{Pa}) = -(3400.7 \pm 31.6)\,/\,T + (10.087 \pm 0.045) \qquad (7.4)$$

7.4 High-pressure Mercury Lamp

The high-pressure mercury discharge lamp has been extensively discussed by Elenbaas.[4] A continuous transition from a low- to a high-pressure discharge occurs in a mercury discharge. The luminous efficacy exhibits a maximum for a given tube diameter and current at the pressure used for fluorescent lamp operation. As pressure is increased, the luminous efficacy passes through a minimum before it begins to increase with increasing pressure. The high-pressure mercury arc operates at a pressure of between 2 and 5 atm.

The core of the high-pressure mercury arc itself operates at a temperature between 5000 and 7000 K. Some of the important spectral lines of mercury in the discharge are $\lambda = 491.6$, 546.1, 577.0, 579.0 and 643.8 nm.

7.5 Ultra-high-performance Lamps

Ultra-high performance (UHP) lamps are ultra-high-pressure mercury lamps used commercially in video and projection applications. UHP lamps contain up to 10 mg of mercury and operate at pressures of ~20–30 MPa (~200–300 atm). These lamps produce a luminous efficacy of about 60 lm W^{-1}. Their electrode gap is very short, typically on the order of 1.0–1.3 mm.[3] An approximately 2 mm thick quartz envelope is used to contain the extraordinary pressures obtained in UHP lamps. The coldest point on the quartz envelope (the cold spot temperature) must be above 1100 K (823 °C) to evaporate completely all of the mercury in the arc tube. A schematic of the arc tube is shown in Figure 7.7.

The spectrum of UHP lamps is comprised of atomic emission from mercury, molecular emission from the Hg_2 dimer and electron-atom bremsstrahlung radiation. The 185 and 253.7 nm resonance lines of mercury (both UV lines) are totally absorbed within the plasma.[3] The mercury plasma exhibits a high electrical resistance that results in high operational voltages and large power dissipation into a very small volume. The spectrum of a UHP is shown in Figure 7.8.

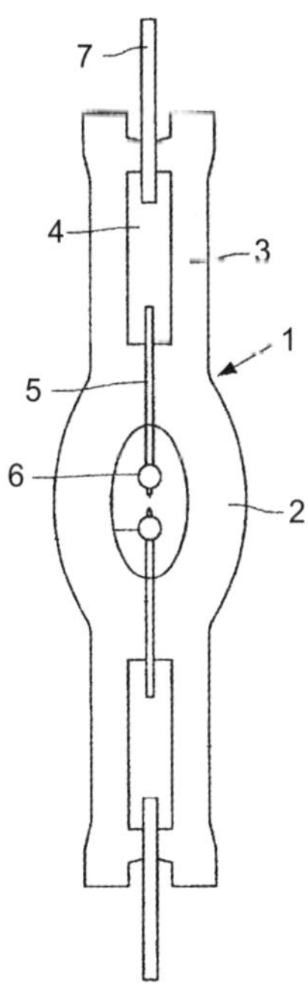

Figure 7.7 Schematic diagram of a 50 W UHP lamp[54] operating at ~200 bar. The CCT is between 7000 and 8000 K, the Hg content is 6 mg, the wall loading is 130 W cm^{-2}, the luminous efficacy is 58 lm W^{-1} and the electrode gap is 1.2 mm. 1, Lamp; 2, elliptical lamp envelope; 3, cylindrical quartz parts; 4, molybdenum foil; 5, electrodes; 6, electrode coils; 7, current supply wires. Reproduced from Ref. 54.

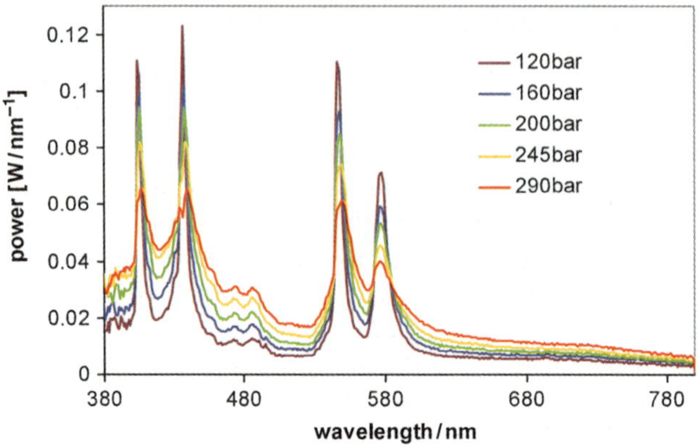

Figure 7.8 Spectra of a 120 W UHP lamp at different mercury pressures. The lamp has an arc gap of 1 mm. The 290 bar spectrum has the highest continuum radiation and the lowest overall peak heights.
Reproduced with permission from Ref. 8.

7.6 High-pressure Sodium Lamps

High-pressure sodium (HPS) discharge lamps are highly efficient light sources.[1,55] High-pressure sodium lamps are characterized by a long service life of 20 000 h or more[2,56] and are capable of providing a maximum luminous efficacy of 100–120 lm W^{-1}.

Sodium amalgams, *i.e.*, sodium and mercury alloys, are used as the light-emitting media in these lamps. Spheres of sodium amalgam[57,58] are obtained from high-purity sodium and mercury but may also contain a third element such as thallium, indium or cesium to improve color rendition. A typical high-pressure sodium lamp design is shown in ref. 1.

The arc tube is mounted in an external vacuum outer jacket. A dose of ~25 mg of sodium amalgam is placed into a 400 W lamp discharge tube and xenon is injected until its pressure reaches (2.4–2.7) $\times 10^3$ Pa. Lamps of lower power receive smaller amounts of amalgam.

A special high-voltage pulse starts a discharge in the xenon gas and leads to evaporation of mercury and sodium. Mercury vapor acts as a buffer gas. The ionization potential of sodium is $E^{u}_{Na} = 5.14$ eV, the excitation potential and the deepest light emission levels of sodium, $E^{b}_{Na} = 2.09$ eV, are significantly smaller than those of mercury: $E^{u}_{Hg} = 10.39$ eV; $E^{b}_{Hg} = 4.89$ and 6.71 eV at the principal quantum number of $n = 6$ of the outer shell of an excited mercury atom.[59] Since the sodium ionization and excitation potentials are considerably smaller than those of mercury, most atoms of sodium in the vapor are ionized and radiate light. Owing to their high excitation potential, the atoms of mercury are virtually not involved in the emission, even under considerably higher pressure. Therefore, the vapor of mercury controls the lamp current and power.

In effect, only atoms of sodium in the gaseous phase are responsible for light and color parameters of the high-pressure lamps. The main excitation levels of sodium atoms are shown in ref. 1 and 60. Broadened resonance lines of sodium ($\lambda = 589.59$ and 589.99 nm) are responsible for the yellow–orange color emitted by HPS lamps.[59] Spectra from a high-pressure sodium lamp are shown in ref. 3.

HPS lamps with a color rendering index of about 30[2,56] are not used for indoor illumination where high CRI is necessary. This is why improvement of the spectral distribution of the emitted light is of critical importance for many researchers. It has been found that the addition of lithium, thallium or indium to mercury lamps causes emission of red ($\lambda = 671$ nm), green ($\lambda = 535$ nm), blue ($\lambda = 451$ nm) and violet ($\lambda = 410$ nm) light.[59] It is possible to obtain a continuum spectrum similar to that of white light (daylight) using additives to sodium amalgam. Some physicochemical and spectral properties of mercury and sodium and also some promising additives[2,61–65] for improved color rendering and other parameters of HPS lamps are given in Table 7.5.

Thorough studies of the discharge physics of sodium atoms demonstrated that by altering the sodium vapor pressure it is possible to control the discharge parameters and improve color rendering.[66] It has been found that the maximum luminous efficacy is reached at a sodium vapor pressure of 2.7×10^3 to 2.7×10^4 Pa (20–200 Torr of mercury).[2,56,59–62,67] The discharge temperature in a 400 W HPS lamp is 2300–2770 K depending on the specifics of the lamp design. The cold spot temperature of the discharge tube is about 930–970 K. Thermal properties of sodium amalgams in a broad range of temperatures have been studied.[68–73]

Data analysis shows that negative deviations from an ideal solution are observed even at high temperatures, although the deviation from ideality

Table 7.5 Physicochemical and spectral characteristics of various elements.[2,61–65]

Element[a]	T_{melt} *(K)*	T_{evap} *(K)*	*Ionization potential (eV)*	*Excitation potential (eV)*	*Characteristic radiation wavelength (nm)*	*Vapor pressure, (Pa) (T = 900 K)*
Sodium (3)	371	1151.2	5.14	2.10	589.59 588.99	5.07×10^3
Mercury (6)	234.32	629.81	10.43	4.89 6.71	253.05 184.95	1.05×10^8
Thallium (6)	577	1748	6.11	3.74	535.05 377.57	2.41
Indium (5)	429.78	2297	5.78	3.02	451.13 410.18	3.07×10^{-3}
Potassium (4)	336.4	1032	4.34	1.61 1.62	769.90 764.49	2.43×10^4
Rubidium (5)	312.7	959.2	4.18	1.56 1.59	794.76 780.03	5.40×10^4
Cesium (6)	301.6	943	3.89	1.39 1.45	894.35 852.11	6.55×10^4

[a] Values in parentheses are the principal quantum numbers for the outer shells of excited atoms.

becomes smaller with increase in temperature. The composition of sodium amalgam commonly used in HPS lamps is 78–85 at.% sodium and 15–22 at.% mercury.[2,61,62,66,67,73]

Sodium vapor pressures[56,60,67,71,73] may be calculated under conditions similar to the operating conditions of an HPS lamp from thermodynamic data for the Na–Hg system.

Various metals are added to sodium amalgam to improve the color rendition and enhance the sodium vapor pressure in HPS lamps.[67,73–76] The thermodynamic properties of ternary amalgam systems Na–Me–Hg, where Me = In, Tl, Cs or Rb, have been studied.[59,73,74,77–82]

Values of sodium activity in the sodium–mercury binary system. Component activities in the sodium–thallium–mercury ternary system for the three *T–c* planes of the Gibbs triangle are given in ref. 55. It can be seen that both the sodium and mercury activities exhibit negative deviations from ideal solutions at virtually all concentrations, whereas the thallium activity at certain concentrations shows positive deviations. Apparently, the sodium activity in the ternary sodium–thallium–mercury system is higher than that in the binary sodium–mercury system. This is responsible for an increase in the vapor pressure in the Na–Tl–Hg system. The pressures of sodium and mercury vapor are higher in the ternary than in the binary system. This pattern also holds for the sodium vapor pressure at 873–973 K.[55] The vapor pressure increase is favorable for application of sodium–thallium amalgams in HPS lamps.

The activity of sodium in the sodium–indium–mercury system at 733 K is known. The pressures are lower than the pressure of sodium above a sodium–mercury melt at 773 K based on calculations[71] and shown in Table 7.6. A temperature increase from 873 to 973 K has virtually no effect on the sodium vapor pressure. It should be noted that there is a close relation between the pressure of sodium and mercury vapors in the discharge tube, voltage applied to the lamp, light intensity and patterns of the sodium discharge lines.

To determine possible combinations of thallium and indium in ternary amalgams, Dergacheva *et al.*[74] studied the electrical and spectral characteristics of HPS lamps with various media in the discharge tube. Table 7.6 gives the electrical and spectral characteristics of lamps with thallium and indium doses

Table 7.6 Electrical and light characteristics of HPS lamps with Tl and In added to the discharge.[74]

Amalgam	*Composition (at.%)*	*Voltage (V)*	*Current ($A\ cm^{-2}$)*	*Power (W)*	*Light flux (lm)*	*Luminous efficacy ($lm\ W^{-1}$)*
Na–Hg	78–22	120	4.9	400	40 000	100
Na–Tl–Hg	50–25–25	136	3.7	420	41 320	98
Na–Tl–Hg	40–12–48	130	3.0	373	32 220	86
Na–Tl–Hg	20–16–64	107	4.5	390	19 990	50
Na–In–Hg	12–48–40	100	2.6	310	13 950	45
Na–In–Hg	8–32–60	105	2.5	270	10 780	40
Na–Tl–In–Hg	63–7–13–17	103	4.6	375	38 070	101

Table 7.7 Effect of thallium on CCT.

Amalgam	*Composition (at.%)*	*CCT (K)*
Na–Hg	–	2673
Na–Tl–Hg	40:12:48	2523
Na–Tl–Hg	50:25:25	3273

in the discharge tube in the form of three- and four-component amalgams. The changes in the lamp current and power are insignificant. The light flux and efficiency are very sensitive to the sodium content since a decrease in Na content reduces the sodium vapor pressure.

Decreases in the sodium concentration in a ternary amalgam from 50 to 20 at.% reduce the luminous efficacy by almost half. For the same reason, low light flux and low luminous efficacy are observed in lamps with high concentrations of indium (see Table 7.6). Use of a quaternary amalgam as the emission source with the sum of thallium and indium no greater than 20–25 at.% and sodium content no less than 50 at.% makes it possible to develop lamps with sufficiently high light flux and efficiency. The drawbacks of sodium lamps with additions of other components, *e.g.*, with a quaternary amalgam, are more difficult ignition and complications with stabilization of parameters. Increasing the indium and thallium content in the amalgam by more than 25 at.% cause HPS lamps to ignite and operate in a stable fashion only at 380 V.

Of all compositions surveyed, the best results were obtained with Na–Tl–Hg (50:25:25 at.%). A series of lamps filled with this composition demonstrated stable characteristics after 100 h of operation. Introduction of thallium in the sodium–mercury discharge causes pronounced changes in the emission spectrum. Thallium lines and broadened sodium lines appear. A faint peak at $\lambda = 451$ nm has been noted in the spectra of lamps containing indium. This small peak at 451 nm has almost no impact on the light radiation power in the blue region. The CCT change caused by introduction of thallium in the discharge process is demonstrated in Table 7.7.

Cesium is often added to sodium mercury amalgams. A detailed discussion of the effects of cesium on high-pressure sodium discharges is can be found elsewhere.[55] Light and electrical characteristics of Na–Cs–Hg lamps are given in Table 7.8. A high-CRI lamp was reported by Gottschling *et al.*[83] The lamp contained 3 mg of sodium, 2 mg of mercury and 0.5 mg of cesium with 30 kPa of xenon fill gas and had a CRI of 70. The lamp, operated at 70 W, had 1000 current pulses per second superimposed in-phase upon a supply voltage in the form of a 1 kHz square-wave current of 0.1 A. The CCT was 3600 K and the efficiency was ~ 75 lm W^{-1}.

7.7 Metal Halide Lamps

Metal atoms introduced into high-pressure mercury lamps give additional spectral lines at visible wavelengths. Generally, metals are introduced as

Table 7.8 Light and electrical characteristics of lamps containing Na–Cs–Hg amalgam.

Alloy composition							
Hg	*Na*	*Cs*	*Voltage (V)*	*Current (A)*	*Power (W)*	*Light flux (lm)*	*Luminous efficacy (lm W⁻¹)*
0.3	0.60	0.10	104	4.5	404	29,570	73
0.3	0.55	0.15	122	4.5	405	29,420	72
0.3	0.50	0.20	117	4.2	413	26,710	64
0.3	0.40	0.30	44	5.7	250	13,600	54
0.3	0.30	0.40	150	3.2	378	14,300	37
0.2	0.65	0.15	109	4.6	425	20,300	45
0.2	0.50	0.30	143	3.5	376	24,230	65
0.2	0.40	0.40	87	4.9	380	12,240	32
0.2	0.725	0.075	71	5.1	333	32,890	98
0.1	0.825	0.075	123	4.0	422	27,470	65

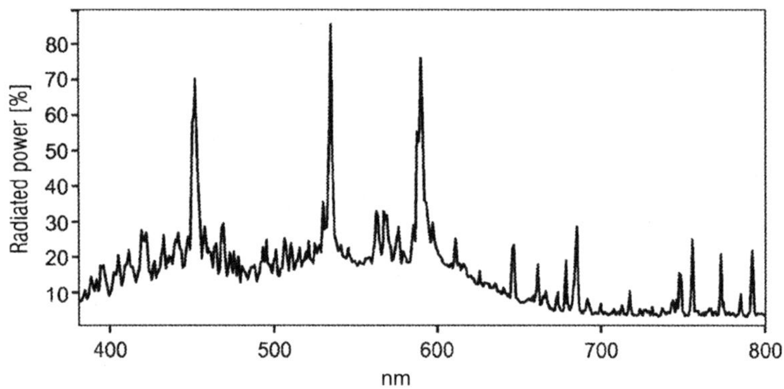

Figure 7.9 Spectrum of a 73 W metal halide lamp containing TmI_3 46.4 wt%, GdI_3 23.2 wt%, NaI 19.6 wt%, TlI 5.8 wt%, InI 5.0 wt% and Hg 1–10 mg cm^{-3}, with the following lamp characteristics after 100 h: CCT 6000 K, CRI 81 and efficacy 92 lm W^{-1}.
Reproduced from Ref. 84.

iodides, bromides or even chlorides in some cases because of their much higher vapor pressure. Many metal halide lamps contain rare earth metal halides because of their rich spectral features in the visible range.[84] Other metal halide lamps rely upon calcium iodide as an important part of the dose.[85] Figure 7.9 shows a spectrum of a metal halide lamp with a high CCT. It should be noted that mercury is present as a buffer gas. It also contributes several lines to the visible spectrum. Note the large continuum background in the 400–650 nm range.

The spectrum from a metal halide lamp containing calcium iodide[85] is shown in Figure 7.10; spectral features and efficiency of the 340 W lamp are as follows: luminous efficacy 105 lm W^{-1}, CRI 90 and CCT 3860 K. The fill gas is a neon–argon Penning mixture with 98.0–99.5% neon.

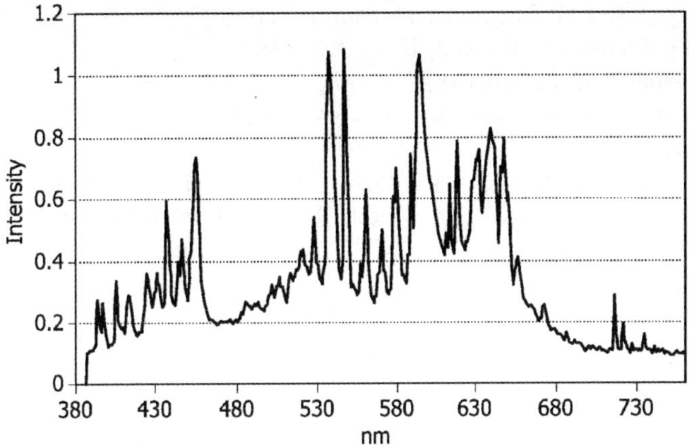

Figure 7.10 Spectrum of a 340 W metal halide lamp containing CaI_2–NaI–TlI–CeI_3–InI and ~5.3 mg Hg.
Reproduced from Ref. 85.

References

1. J. J. De Groot and J. A. J. M. Van Vliet, *The High Pressure Sodium Lamp*, Philips Technical Library, Kluwer Technische Boeken, Deventer, 1986.
2. J. F. Waymouth, *Electric Discharge Lamps*, MIT Press, Cambridge, MA, 1971.
3. P. Flesch, *Light and Light Sources, High Intensity Discharge Lamps*, Springer, Berlin, 2010.
4. W. Elenbaas, *The High Pressure Mercury Vapor Discharge*, North-Holland, Amsterdam, 1951.
5. Various articles in *Light Sources 2012, Proceedings of the 13th International Symposium on the Science and Technology of Lighting, Troy, NY*, IOP Publishing, Bristol, 2012.
6. Eden Park Illumination, *Microplasma Planar Lighting*, 2012, http://edenpark.com/wp-content/uploads/Microplasma_Planar_Lighting-by_Eden_Park_Illumination.pdf, (last accessed 20 June 2012).
7. Environment Canada, *Fluorescent Lamps*, 2012, http://www.ec.gc.ca/mercure-mercury/default.asp?lang=En&n=2486B388-1 (last accessed 7 March 2013).
8. G. Derra, H. Moench, E. Fischer, H. Giese, U. Hechtfischer, G. Heusler, A. Koerber, U. Niemann, F.-C. Noertemann, P. Pekarski, J. Pollmann-Retsch, A. Ritz and U. Weichmann, *J. Phys. D: Appl. Phys.*, 2005, **38**, 2995.
9. European Environmental Bureau, *Mercury-Containing Lamps Under the Spotlight*, Publication No. 2008/010, European Environmental Bureau, Brussels, 2008.
10. G. G. Lister, J. E. Lawler, W. P. Lapatovich and V. A. Godyak, *Rev. Mod. Phys.*, 2004, **76**, 541.

11. C. Meyer and H. Nienhuis, *Discharge Lamps*, Philips Technical Library, Kluwer Technische Boeken, Deventer, 1988.
12. C. Trushell and L. Magean, *US Pat.*, 5 612 590, 1997.
13. D. C. Nesting, M. Oomen and P. Walitsky, *J. Illum. Eng. Soc.*, 2001, Winter, 76.
14. V. D. Hildenbrand, C. J. M. Denissen, A. J. H. P. Van der Pol, A. H. C. Hendriks, C. van der Marel, J. H. M. Snijders, Y. Tamminga, H. H. Brongersma and M. M. Viitanen, *J. Electrochem. Soc.*, 2003, **150**, H147.
15. I. Snijkers-Hendrickx, *et al.*, in *Proceedings of the 11th International Symposium on the Science and Technology of Light Sources*, 2007, pp. 361–369.
16. R. Devonshire, *Matsushita Tech. J.*, 2008, **53**, 104.
17. J. Bloem, A. Bouwknegt and G. A. Wesselink, *J. Illum. Eng. Soc.*, 1977, **6**, 141.
18. H. Hein and F. Raiser, *PCT Pat. Appl.*, WO 2008/017654 A1, 2006.
19. W. Hellebreker and L. C. I. Kaldenhoven, *US Pat.*, 7 977 858, 2011.
20. D. Hofmann and E. Rasch, *Technisch-wissenschaftliche Abhandlungen der Osram-Gesellschaft*, 1973, **12**, 394.
21. G. Evans and C. Morehead, *US Pat.*, 3 526 804, 1970.
22. T. Yorifuji and S. Komoda, *US Pat.*, 4 972 118, 1990.
23. G. A. Wesselink and P. Hokkeling, *US Pat.*, 4 157 485, 1979.
24. J. Bloem and A. Bouwknegt, *US Pat.*, 4 093 889, 1978.
25. W. Vrieze, *US Pat.*, 4 636 686, 1987.
26. L. C. I. Kaldenhoven, *US Pat.*, 4 924 142, 1990.
27. J. F. Sarver, D. A. Stafford, S. C. Hansen and T. R. Brumleve, *US Pat.*, 5 882 237, 1999.
28. L. A. Zabdyr and C. Guminski, *J. Phase Equil.*, 1995, **16**, 353.
29. M. Pušelj, Z. Ban and A. Drašner, *Z. Naturforsch. B*, 1982, **37**, 557.
30. T. R. Brumleve, S. C. Hansen and L. Kaczorowski, presented at HTLC-III, Electrochemical Society, New Orleans, 1993.
31. L. F. Kozin and R. Sh. Nigmetova, *Russ. J. Inorg. Chem.*, 1963, **8**, 1338.
32. Y.-W. Yen, J. Gröbner, S. C. Hansen and R. Schmid-Fetzer, *J. Phase Equil.*, 2003, **24**, 151.
33. G. Ptaschek and C. Di Vincenzo, *US Pat. Appl.*, 2010/0130092 A1, 2010.
34. P. della Porta and M. Rebaudo, *US Pat.*, 3 657 589, 1972.
35. P. della Porta and M. Rebaudo, *US Pat.*, 3 733 194, 1973.
36. P. Pietrokowsky, *Trans. AIME*, 1954, **200**, 219.
37. T. R. Brumleve, D. A. Stafford, S. C. Hansen and K. Fukutome, *US Pat.*, 6 910 932, 2005.
38. M. V. Nosek, N. M. Semibratova and G. V. Yan-Sho-Syan, *Russ. Metall.*, 1970, **1**, 117.
39. E. S. Demchenko, A. A. Lange and S. P. Bukhman, *Izv. Akad. Nauk Kazakh. SSR, Ser. Khim.*, 1973, **23**, 76.
40. M. V. Nosek, G. V. Yan Sho-Syan and N. M. Semibratova, *Izv. Akad. Nauk Kazakh. SSR, Ser. Khim.*, 1969, **4**, 11.

41. L. F. Kozin, A. M. Dairova and R. Sh. Nigmetova, *Izv. Akad. Nauk Kazakh. SSR, Ser. Khim.*, 1977, **5**, 71.
42. Z.-C. Wang, X.-H. Zhang, T.-Z. He and Y. H. Bao, *J. Chem. Thermodyn.*, 1989, **21**, 653.
43. T. Ikeda and M. Nagai, *US Pat.*, 5 204 584, 1993.
44. R. Schmid-Fetzer and J. Gröbner, personal communication.
45. S. A. Rushworth, L. M. Smith, A. J. Kingsley, R. Odedra, R. Nickson and P. Hughes, *Microelectron. Reliabil.*, 2005, **45**, 1000.
46. P. Morávek, M. Fulem, K. Růžička, J. Pangrác, E. Hulicius, T. Símeček, V. Růžička and S. A. Rushworth, *J. Chem. Eng. Data*, 2010, **55**, 4095.
47. K. Růžička, M. Fulem and V. Růžička, *J. Chem. Eng. Data*, 2005, **50**, 1956.
48. J. D. Robinson, personal communication, 2005.
49. K. Hilpert and I. Ali-Khan, *J. Nucl. Mater.*, 1978, **78**, 265.
50. K. Hilpert, *Ber. Bunsenges Phys. Chem.*, 1980, **84**, 494.
51. K. Hilpert, *Ber. Bunsen. Phys. Chem.*, 1983, **87**, 818.
52. K. Hilpert, *Ber. Bunsenges. Phys. Chem.*, 1984, **88**, 37.
53. K. Hilpert, *Z. Metallkd.*, 1984, **75**, 70.
54. E. Fischer and H. Hoerster, *US Pat.*, 5 109 181, 1992.
55. L. F. Kozin, *Physicochemical Properties of High-Purity Mercury and Its Alloys*, Naukova Dumka, Kiev, 1992.
56. G. N. Rokhlin, *Gas Discharge Sources of Light*, Energia, Moscow, 1966, p. 560.
57. S. Anderson, *US Pat.*, 4 216 178, 1980.
58. S. Anderson, *US Pat.*, 4 419 303, 1983.
59. I. M. Veselnitsky and G. N. Rokhlin (eds), *High–Pressure Mercury Lamps*, Energia, Moscow, 1971.
60. E. V. Shpolsky, *Atomic Physics*, Nauka, Moscow, 1974, vol. 2, p. 240.
61. K. Schmidt, in *Proceedings of the 6th International Conference on Ionization Phenomena in Gases*, Paris, 1963, vol. 3, pp. 323.
62. J. Maya, *Appl. Phys. Lett.*, 1982, **40**, 933.
63. A. N. Zaidel, V. K. Prokofiev, S. M. Raisky and E. Ya. Shreider, *Tables of Spectral Lines*, Nauka, Moscow, 1969.
64. G. Hertzberg, *Electronic spectra and electronic structure of polyatomic molecules*, Van Nostrand, Princeton, NJ, 1966.
65. V. P. Glushko (ed.), *Thermodynamic Constants of Substances*, VINITI, Moscow, 1972, Releases 5, 6, 10.
66. V. S. Litvinov and A. N. Grigoryan, *Trends in High-Pressure Sodium Lamps with Improved Light Emission*, Informelectro, Moscow, 1984, p. 31.
67. M. B. Dergacheva, Yu. P. Petrenko, N. P. Petrenko and V. S. Litvinov, in *Physicochemical Grounds for Production of Metallic Alloys*, Report Highlights, Republican Conference, Almaty, 12–14 June 1990, Nauka, Almaty, 1990, p. 73.
68. L. F. Kozin, R. Sh. Nigmetova and M. B. Dergacheva, *Thermodynamics of Binary Amalgam Systems*, Nauka, Almaty, 1977, p. 344.
69. M. B. Dergacheva, L. F. Kozin and O. Ya. Smirnova, in *Thermodynamics of Metallic Systems: Papers of the 4th National Meeting on*

Thermodynamics of Metallic Alloys (Melts), Nauka, Almaty, 1979, Part 2, pp. 36–41.
70. M. B. Dergacheva, N. L. Panova, G. R. Hobdanergenova and L. F. Kozin, *Zh. Fiz. Khim.*, 1984, **58**, 305.
71. C. Hirayama, K. F. Andrew and R. J. Kleinowsky, *Thermochim. Acta*, 1981, **45**, 23.
72. R. Hultgren, P. D. Desai, D. T. Hawkins, M. Gleiser and K. K. Kelley, *Selected Values of the Thermodynamic Properties of Binary Alloys*, American Society for Metals, Metals Park, OH, 1973.
73. G. R. Hobdanergenova, *Studies of Thermodynamic Properties of Liquid Alloys Based on Alkali Metals, Indium, Thallium and Mercury*, Author's Summary of a Dissertation for PhD in Chemistry, Almaty, IOCE, Academy of Sciences of the Republic of Kazakhstan, 1984, p. 29.
74. M. B. Dergacheva, Yu. P. Petrenko and G. R. Hobdanergenova, *et al*, *Izv. Akad. Nauk Kazakh. SSR, Ser. Khim.*, 1976, **1**, 54.
75. Y. Watarai, H. Yamazaki, N. Saito, M. Yamaguchi, T. Okamoto and H. Akutsu, *US Pat.*, 4 109 175, 1978.
76. H. Mizuno, *Jpn. Pat.*, 49-8548, 1974.
77. L. F. Kozin, M. B. Dergacheva and G. R. Hobdanergenova, *Electronic Processes in Aqueous Solutions*, Nauka, Almaty, 1981, pp. 36.
78. L. F. Kozin, Yu. A. Meshcheryakov and R. Sh. Nigmetova, *et al.*, *Svetotekhnika*, 1974, **8**, 10.
79. L. F. Kozin, Yu. A. Meshcheryakov and R. Sh. Nigmetova, *et al.*, *Svetotekhnika*, 1977, **15**, 8–9.
80. G. R. Hobdanergenova, M. B. Dergacheva and L. F. Kozin, *Izv. Akad. Nauk Kazakh. SSR, Ser. Khim.*, 1983, **3**, 8–12.
81. G. R. Hobdanergenova, M. B. Dergacheva and L. F. Kozin, *Zh. Fiz. Khim.*, 1983, **57**, 2059.
82. A. J. Heethling, *J. Chem. Thermodyn.*, 1975, **7**, 73.
83. W. Gottschling, K. Guenther, H.-G. Kloss, R. Radtke and F. Serick, *US Pat.*, 4 963 796, 1990.
84. A. Genz, *US Pat. Appl.*, 2011/0204776, 2011.
85. R. Gibson, T. Steere and J. Tu, *US Pat. Appl.*, 2011/0266955, 2011.

CHAPTER 8

Synthesis of Semiconducting Compounds

8.1 Synthesis of Semiconducting Mercury Compounds

Chalcogenides, halides, chalcohalides and other compounds of mercury exhibit semiconducting properties. The chalcogenides HgS, HgSe, HgTe are n-type semiconductors. Figure 8.1 shows the phase diagrams for mercury chalcogenides. For all three systems it is typical to generate only one equiatomic compound. The melting point of the equiatomic compounds are as follows:

HgS	825 ± 2 °C[1]
	820 °C[2]
HgSe	799 °C[3]
	793 °C[4] at a mercury vapor pressure of 9.12×10^6 Pa[4]
HgTe	686 °C[5]
	670 °C[1]

There are three modifications of mercury sulfide (HgS) in the Hg–S system. The red α-modification of cinnabar (cinnabarite) is stable from room temperature up to 315–345 °C. It forms a hexagonal lattice. At intermediate temperatures, from 316–346 to 470–481 °C, the black β-modification (metacinnabarite) is stable. It forms a cubic lattice with $a = 0.5852$ nm. At high temperatures, from 470–481 to 788–804 °C, the bright red γ-modification (hypercinnabarite) is stable. It forms a prismatic lattice with $a = 0.686$ and $c = 1.407$ nm.[1–3] The solubility range of solid mercury sulfide is narrow and it melts congruently at 820 °C (1093 K). It is also apparent, from Figure 8.1, that two monotectic reactions occur in the Hg–S system. These reactions:

$$L_2 \rightleftharpoons L_1 + \gamma\text{–HgS}$$

Mercury Handbook: Chemistry, Applications and Environmental Impact
By Leonid F Kozin and Steve Hansen

Published by the Royal Society of Chemistry, www.rsc.org

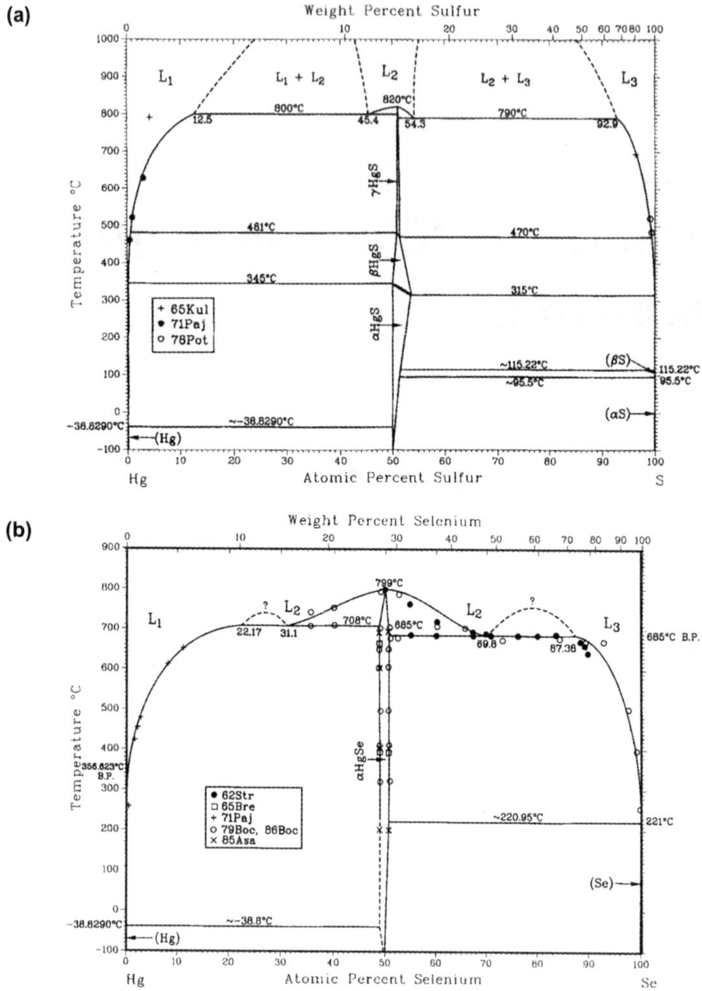

and

$$L_2 \rightleftharpoons \gamma\text{–HgS} + L_3$$

occur at 800 and 790 °C, respectively. Mercury sulfide evaporates congruently through a dissociation mechanism. When evaporating, HgS dissociates with a degree of dissociation close to 1 and decomposes into gaseous atoms of mercury and S_2 molecules.[2,6] Novoselova and Pashinkin[6] recommend the following equations for the characterization of the total pressure of HgS:

$$\log\,(p_{\text{tot}},\,\text{Pa}) = -\,6200/T + 12.3909 \quad (8.1)$$

$$\log\,(p_{\text{tot}},\,\text{Pa}) = -\,5814/T + 11.7879 \quad (8.2)$$

(c)

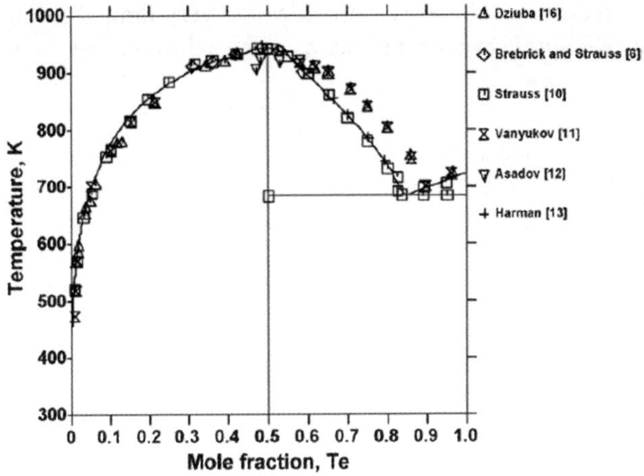

Figure 8.1 Phase diagrams of the systems (top) Hg–S [9a], Hg–Se [9b] and Hg–Te [9c]. a) R. C. Sharma, Y. A. Chang and C. Guminski, *J. Phase Equilibria*, 1993, **14**, 100. b) R. C. Sharma, Y. A. Chang and C. Guminski, *J. Phase Equilibria*, 1992, **13**, 663. c) R. C. Sharma, Y. A. Chang and C. Guminski, *J. Phase Equilibria*, 1995, **16**, 338.

The heat of evaporation of HgS is 183.4 kJ mol^{-1} and its thermodynamic parameters are[1,7] $\Delta G^{\circ}_{298.15} = -52.42\,kJ\,mol^{-1}$, $\Delta H^{\circ}_{298.15} = -58.99\,kJ\,mol^{-1}$ and $\Delta S^{\circ}_{298.15} = 82.42\,J\,mol^{-1}\,K^{-1}$.

Cinnabar is the main mineral in mercury deposits.[8] Cinnabar (α-HgS) exhibits photoconductivity and is highly sensitive to electromagnetic and X-radiation.

In the Hg–Se system, mercury selenide (HgSe) forms a face-centered cubic sphalerite structure with $a = 0.608$ nm, which is stable from room temperature up to 799 °C (872 K).[3] Sharma, Chang and Guminski[9] reported pure HgSe is stable up to 799 °C (1072 K). According to Brebrick,[10] the vapor pressure above HgSe is equal to the sum of the pressures of mercury and selenium vapors. No HgSe molecules were found in studies employing the method for the measurement of the optical density of vapors above solid HgSe at 450–816 °C (723–1089 K).[10] The pressure of the saturated vapor of mercury selenides in the temperature range 340–770 °C (613–1043 K) is described by the equation

$$\log\,(p, Pa) = -6445/T + 11.7349 \qquad (8.3)$$

Thermodynamic parameters of HgSe are[7] $\Delta G^{\circ}_{298.15} = -53.748\,kJ\,mol^{-1}$, $\Delta H^{\circ}_{298.15} = -59.413\,kJ\,mol^{-1}$ and $\Delta S^{\circ}_{298.15} = 99.035 \pm 0.837\,J\,mol^{-1}\,K^{-1}$.

Mercury selenide can be obtained with both n- and p-type conductivity; it falls into the category of narrow bandgap semiconductors where the bandgap is $\Delta E = -0.07$ eV.[1]

HgTe is also a narrow bandgap semiconductor ($\Delta E = -0.02$ eV[1]). Under standard conditions, HgTe has a face-centered cubic lattice with $a = 0.646$ nm.[3] At a pressure of $\sim 1.4 \times 10^3$ MPa, the cubic lattice changes into a hexagonal cinnabar structure.[3] Solid HgTe provides for very limited solubility of Hg and Te. Mercury telluride evaporates incongruently through the following reaction:[11]

$$HgTe_{solid} \rightarrow Te_{solid} + Hg_{gas} \tag{8.4}$$

The degree of dissociation is close to 1 while the mercury partial pressure is greater than the tellurium partial pressure almost by two orders of magnitude.[2,12] Partial pressures of mercury and tellurium in the Hg–Te system were found through the determination of optical density.[13] The relationship between mercury telluride vapor pressure and temperature in coordinates log$p$–1/T is linear and, according to various authors, is described by the equations

$$\log\,(\mathrm{p,\ kPa}) = \frac{-5251.3}{\mathrm{T}} + 9.1549 \text{ at } 486 - 600\,\mathrm{k}^{1} \tag{8.5}$$

and

$$\log\,(\mathrm{p,\ kPa}) = \frac{-5700 \pm 200}{\mathrm{T}} + (8.30 \pm 0.40) \text{ at } 480-730\,\mathrm{k}.^{11} \tag{8.6}$$

Analysis of experimental data[11] for the temperature range 435–950 K resulted in the following vapor pressure equation:

$$\log\,(\mathrm{p,\ kPa}) = \frac{-5700}{\mathrm{T}} + 8.18 \tag{8.7}$$

The standard enthalpy change for HgTe upon evaporation is $\Delta H^{\circ}_{298.15} = 107 \pm 4\,\mathrm{kJ\,mol^{-1}}$, while the entropies of vaporization at 500, 550 and 600 K are $\Delta S = 110.5$, 110.2 and $109.9\,\mathrm{J\,mol^{-1}\,K^{-1}}$, respectively.[11] Thermodynamic parameters of HgTe are[7] $\Delta G^{\circ}_{298.15} = -28.033\,\mathrm{kJ\,mol^{-1}}$, $\Delta H^{\circ}_{298.15} = 32.175\,\mathrm{kJ\,mol^{-1}}$ and $\Delta S^{\circ}_{298.15} = 111.50 \pm 0.628\,\mathrm{J\,mol^{-1}\,K^{-1}}$.

Mercury sulfides, selenides and tellurides form solid solutions with sphalerite structures[14–16] between themselves and also with chalcogenides of subgroups IIB–VIB of the periodic table. These solutions allow one to obtain materials used in optoelectronics and microelectronics.[1,4–6,8,12–20] For instance, single crystals of solid solutions of HgS–HgSe, HgSe–HgTe and others are used for the manufacture of photoconductive infrared detectors in optoelectronics (light sources and receivers, Raman lasers, light flux control devices), in ionizing radiation detectors, in generators and amplifiers of acoustic and microwave-band oscillations, *etc.* Materials based on solid solutions of $Cd_xHg_{1-x}S$, $Cd_xHg_{1-x}Se$, $Cd_xHg_{1-x}Te$ and $Hg_xMn_{1-x}Te$ have been extensively used in optoelectronics, microelectronics and other fields of advanced technology. These n-type semiconductors are used for the manufacture of photoconductive infrared detectors with a pronounced maximum in the 0.3–12 μm band. Figure 8.2 shows quasi-binary phase diagrams of HgS–CdS, HgTe–CdTe and HgTe–MnTe based on data of Mizetskaya *et al.*[15] It is apparent from Figure 8.2 that the mutual solubility of the components in the solid state depends on the temperature. In the HgTe–CdTe system (Figure 8.2b), a broad field of solid solutions can be seen. The large temperature range between the liquidus and solidus curves in the middle part, greater than 100 °C, makes it difficult to obtain homogeneous crystals.

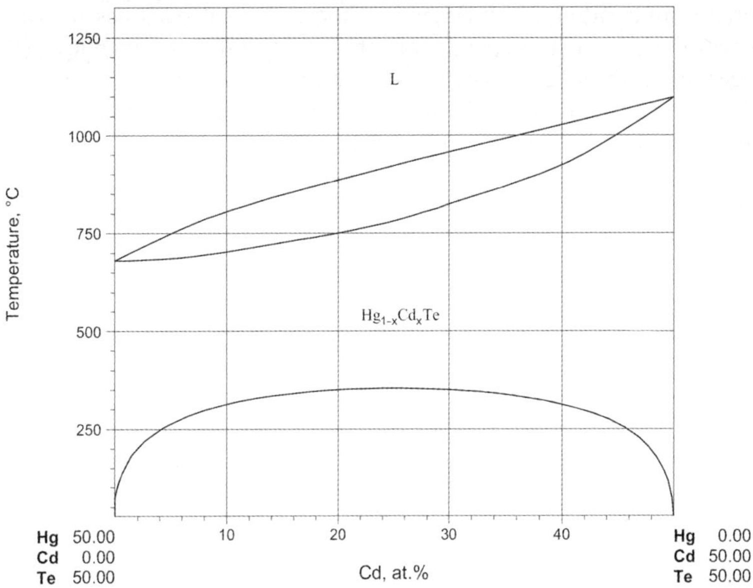

Figure 8.2 Phase diagrams of the quasi-binary systems, HgTe–CdTe. Reproduced with kind permission from Springer Science + Business Media. Landolt-Börnstein – Group IV Physical Chemistry Selected Semiconductor Systems Volume 11C1 2006 Non-Ferrous Metal Systems Part 1, Ed. G. Effenberg, S. Ilyenko, Springer, 2006, 267. Ref. 80.

A number of studies have been aimed at crystallization patterns of solid solutions of $Hg_xCd_{1-x}Te$, owing to the great importance of tellurides of mercury for advanced infrared technology and outer space optoelectronics. That is why there are stringent requirements on homogeneity, which researchers are trying to meet in both ground-based laboratories[1,15–21] and in the weightlessness conditions in space.[22–25]

Solid solutions of HgTe–CdTe and HgTe–MnTe systems (manganese–mercury–tellurium α-phase) have identical band structures and close electrophysical properties. The structure of the HgTe–MnTe equilibrium diagram has not been fully established.[21] From Ref. 21, it is clear that the α-field of solid solutions is stable up to 35 at.% MnTe. At higher concentrations, the α-solid solution breaks down into $MnTe_2$ and a mercury-enriched α-phase. Manganese–mercury–tellurium α-phase crystals feature high homogeneity and n-type conductivity. At 77 K, the average concentration and charge carrier mobility in Mn–Hg–Te single crystals are $2.6\times10^{15}\,cm^{-3}$ and $2.6\times10^{4}\,cm^2\,V^{-1}\,s^{-1}$, respectively.[20]

Close parameters are demonstrated by crystals of $Hg_{1-x}Mn_xTe$ ($x = 0.125$), synthesized from high-purity Hg (7N), Mn (9N) and Te (6N) at around 800 °C for 24–48 h and subjected to solid-phase recrystallization in mercury vapor at around 750 °C for 320 h. The grown crystals had n-type conductivity at room temperature (average concentration of charge carriers $m = 2\times10^{16}\,cm^{-3}$) and p-type conductivity at 77 K ($m = 10^{16}\,cm^{-3}$).[19]

Mercury and cadmium and mercury and zinc chalcogenides display high photosensitivity, which depends on the structural and chemical (purity) homogeneity of the single crystals. The impurity content in the input materials Hg, Te, Cd, Mn and Zn used for the synthesis of chalcogenide-based semiconductors should not exceed 3×10^{-9} mass%.[26]

Chalcogenides can be used as a base for the synthesis of both high- and low-resistance electrooptic crystals, which demonstrate photoconductivity in crystals with electrical resistivity ranging from 1 to $10^{18}\,\Omega\,cm$, with the life of free carriers, which determines photosensitivity, ranging from tens of minutes to 10^{-12} min according to Petrovsky *et al.*[25]). Mercury chalcogenides $Hg_3X_2Y_2$ (where X = S, Se, Te; Y = Cl, Br, I), which demonstrate high optical activity, electrooptical effects and photoconductivity,[27] are valuable for opto- and microelectronics.

Semiconducting mercury compounds decompose when heated.[28] The vapor pressures of mercury chalcogenides and chalcohalogenides become considerable during melting and crystallization. The pressures of mercury vapor at the melting points of HgS, HgSe and HgTe are as follows:

HgS	5.82×10^6 Pa
HgSe	5.28×10^5 Pa
HgTe	$\sim 3 \times 10^6$ Pa

Very high vapor pressures of mercury sulfide, selenide and telluride often cause ampoules to burst. Therefore, during the synthesis of mercury chalcogenides, protective counter-pressure containers are often used. These containers are also used for cadmium–mercury–tellurium recrystallization in zero gravity.[22,23,25]

Therefore, to avoid losing a component during the synthesis of semiconducting mercury compounds, the operation should take place in an enclosed volume with ampoules heated to temperatures that exceed the vapor condensation temperatures of the volatile components. To prevent ampoules from breaking at high temperatures, the ampoule with the synthesis components is placed inside a sealed container filled with an estimated amount of input components to ensure an equivalent counter-pressure.[15,22,28] This is the so-called ampoule method. The method completely eliminates any loss of the components (oxidation, carry-over). The ampoules are made of high-purity fused-silica glass, *i.e.* quartz tubes. Methods for the synthesis of mercury chalcogenides HgX (X = S, Se, Te) and solid solutions of the type $Hg_xMe_{1-x}X$ (Me = Cd, Mn, Zn, *etc.*) have been described.[1,4–6,8,14,15,20,22,23,29]

Semiconducting mercury compounds are synthesized using direct or indirect methods. Direct methods include synthesis from pure components in elemental form, from the gas phase and melt according to the following equations:

$$Hg + Te \rightarrow HgTe \tag{8.8}$$

$$Hg + Se \rightarrow HgSe \tag{8.9}$$

$$Hg + S \rightarrow HgS \tag{8.10}$$

$$xHg + (1 - x)Cd + Te \rightarrow Hg_xCd_{1-x}Te \tag{8.11}$$

$$xHg + (1 - x)Mn + Te \rightarrow Hg_xMn_{1-x}Te \tag{8.12}$$

Chalcogenide synthesis is performed in high vacuum at background pressures of 10^{-3}–10^{-5} mmHg [$(13.3–0.133)\times10^{-2}$ Pa] or in an inert atmosphere in a three-section oven, at temperatures T_1, T_2 and T_3, ensuring vaporization of mercury (section one) and tellurium (section three) and their interaction inside the reaction chamber (section two), where the vapors of mercury and chalcogens mix, react and form homogeneous nuclei, which crystallize on a solid surface layer by layer, creating a densely packed surface.

Direct synthesis methods also include the method of cultivating chalcogenide (cadmium–mercury–tellurium, manganese–mercury–tellurium) crystals in the liquid phase with one of the components (tellurium, cadmium, mercury) present in excess. Guminski[30] shows a plot of the solubility of CdTe in mercury, according to Guminski.[30] It can be seen that the function $\log S_{CdTe}$ *versus* $1/T$ is linear and, as shown by Guminski,[30] is described by

$$\log S_{CdTe} = \frac{-3070}{T} + 3.68 \tag{8.13}$$

in the temperature interval 500–1250 K. At 298 K, the solubility of CdTe in Hg is 2.4×10^{-7} at.%.

Special attention has been devoted to studies of the equilibrium between solid CdTe and its saturated amalgam, which is characterized by the solubility product:

$$L_p = [Cd][Te] \tag{8.14}$$

A reaction occurs between equimolar amounts of cadmium and tellurium in mercury according to the equation

$$Cd(Hg) + Te(Hg) \rightleftharpoons CdTe\downarrow + \infty Hg \tag{8.15}$$

However, if tellurium is in excess with respect to cadmium in Cd–Te–Hg solutions, some amount of mercury becomes engaged (jointly deposited) with the residuum, according to the stoichiometry of the reaction

$$xCd(Hg) + Te(Hg) \rightleftharpoons Cd_xHg_{1-x}Te\downarrow + \infty Hg \tag{8.16}$$

The affinity of cadmium and mercury that are part of Cd–Hg–Te system is very high; therefore, if tellurium is in excess with respect to cadmium, $Cd_xHg_{1-x}Te$ will be in equilibrium with the liquid phase of the solution. According to Vydyanath,[31] the activity coefficients of mercury and cadmium in epitaxial layers of $Hg_{0.65}Cd_{0.35}Te$, grown on CdTe substrates at 550 °C, are $\gamma_{Hg} = 0.143$ and $\gamma_{Cd} < 6.8\times10^{-5}$, respectively.

The activity coefficients of cadmium are close to γ_{Cd} for the tellurium-enriched binary alloy $CdTe_{solid}$. Therefore, in several papers,[15,18] reaction (8.16) is used for the synthesis of solid solutions of cadmium–mercury tellurides. This reaction is used to perform recrystallization of $Cd_xHg_{1-x}Te$ and grow single crystals.[15,18,20,22–26,29] For industrial-scale production, chalcogenide single crystals are grown in autoclaves[32] with programmed temperature ramps. Cruceanu and Nistor[33] grew $Hg_{1-x}Zn_xTe$ single crystals from a mercury solution. They dissolved 5 at.% $Hg_{1-x}Zn_xTe$ in mercury and grew the crystals in vacuum-sealed quartz ampoules at <923 K at a cooling rate of 6 K min^{-1}. However, because of the high cooling rate, the resulting crystals featured high dislocation densities.

8.1.1 Sublimation and Resublimation Methods

To crystallize solid $Cd_xHg_{1-x}Te$ solutions, resublimation of input binary chalcogenides CdTe and HgTe in an inert atmosphere is used. The crystals are grown in two-section electric furnaces; one of the designs is shown in ref. 15. Section one of the oven, used for trays with the input material, has the maximum temperature (T_2), and the end product section the minimum temperature (T_1). In section one at T_2 the original chalcogenide evaporates and, with the help of a carrier gas, is carried over to section two with a low temperature T_1 where conditions are suitable for vapor condensation due to supersaturation ($T_1 \ll T_2$) and crystallization. The structure of the crystals depends on the degree of supersaturation, cooling rate, flow rate and other factors.

Therefore, to obtain complex semiconducting compounds and their single crystals, a two-phase process is used. Phase one consists in preparing binary compounds of mercury telluride and cadmium telluride and melting these together to obtain cadmium–mercury telluride ($Cd_xHg_{1-x}Te$) of the required proportions. In phase two, the compounds obtained are used to grow single crystals of solid solutions of $Cd_xHg_{1-x}Te$ using sublimation,[15] zone melting *via* solution in tellurium melt[20] and recrystallization from melt.[15,22–23] Studies of $Hg_{1-x}Cd_xTe$ recrystallization in zero gravity conditions demonstrated that the quality of crystals depended, as in normal gravity conditions,[20] on growth rate.[22,23] At growth rates below 3 mm h^{-1}, the experiments produced a homogeneous $Hg_{1-x}Cd_xTe$ specimen with usable output of 50% of input material volume.[22] The resulting homogeneous specimen was different from that obtained on Earth under the same thermal conditions[22] and demonstrated p-type conductivity. At $T = 50$ K, the concentration and charge carrier mobility were 10^{17} cm^3 and 100 cm^2 s^{-1}. According to Galonzka *et al.*,[22] low carrier mobility was due to a high dispersion at defects.

The annealing of $Hg_{1-x}Cd_xTe$ specimens in mercury vapor at 265 °C for 6 weeks produced n-type specimens with concentration and charge carrier mobility of $(2.5–4.0)\times10^{16}$ cm^3 and $(6–9)\times10^4$ cm^2 s^{-1}, respectively (theoretical mobility 1.9×10^5 cm^2 s^{-1}).[22] Earth-grown $Hg_{1-x}Cd_xTe$ single crystals produced under the same conditions had cellular substructures, consisted of grains of high dislocation density (around 10^6 cm^{-2}), and contained second-phase impurities.

It should be mentioned that the crystallization conditions of solid solutions of $Hg_{1-x}Cd_xTe$, $Hg_{1-x}Mn_xTe$ and $Hg_{1-x}Cd_{x-x}Mn_yTe$ have been the subject of many studies, as these materials have physicochemical and electrical properties of great value for fiber-optic communications, infrared equipment and optoelectronics.[34–42]

8.1.2 Methods Used to Grow Single Crystals

The Bridgman method and its modifications are commonly used to grow single crystals of solid solutions of $Hg_{1-x}Cd_xTe$. In order to obtain a planar crystallization front, Bagai and Borle[43] placed the ampoule with its flat bottom slanting along the horizontal axis in a position offset with respect to the vertical axis of the oven, which created a specific temperature gradient along the crystal bar. The input components – high-purity mercury, cadmium and tellurium – were loaded into the ampoule, which was evacuated to a pressure of 10^{-5} mm Hg (1.33×10^{-3} Pa) and sealed. The synthesis of single crystals or polycrystals of HgCdTe followed at a growth rate of 1.0–4.5 mm h^{-1} and a vertical temperature gradient of >50 K cm^{-1}. The described method helped to improve the homogeneity of HgCdTe crystals. A modified Bridgman method was used to grow single crystals of more complex systems – quaternary solid solutions of $Cd_xZn_yHg_{1-x-y}Te$.[44]

The solid solutions of this system outperform the solid solutions of the quasi-binary system HgTe–CdTe in terms of their physicochemical and electrical properties, because they have higher melting points, greater mechanical strength, better deformation resistance and better time-independent photosensitivity. In this case, the crystals were synthesized in thick-walled quartz ampoules from the input components (in at.%) 50% Te, 36% Hg, 12.5% Cd and 1.4% Zn. These were held for 50 h at a temperature 30 °C above the liquidus temperature while being continuously stirred. Single crystals of solid solutions of the Hg–Cd–Zn–Te system were grown in an ampoule traveling at a rate of ~1 mm h^{-1} inside the oven with a temperature gradient of ~40 K cm^{-1}. For homogenization, the crystal was annealed at 550 °C for 250 h.[44] Specimens cut out of the crystal demonstrated n-type conductivity with an electron concentration of 3×10^{15}–1×10^{17} cm^3 and a mobility of ~10^6 cm^2 V^{-1} s^{-1}. Property mapping of $Cd_xZn_yHg_{1-x-y}Te$ and $Cd_xZn_yHg_{1-x}Te$ crystals showed that zinc-containing crystals have greater stability and strength. According to Virt *et al.*,[45] the high plasticity of $Cd_xHg_{1-x}Te$ contributes to the formation of structural disturbances: spots of integrity failure surrounded by areas of increased dislocation density, and ultimately redistribution of components. It has been established that mechanical disturbances in the $Cd_xHg_{1-x}Te$ crystal structure result in local breakdown of the solid solution either at the time of mechanical attack or during subsequent improper storage of the crystals and crystal-based products under natural conditions.[45] Interestingly, according to Vasiliev *et al.*,[46] the stability of solid solutions $(CdTe)_x(HgTe)_{1-x}$ in the pseudo-binary system CdTe–HgTe decreases with decrease in temperature. Vasiliev *et al.*[46] gave the integral Gibbs free energies of the formation of tellurium-saturated solid solutions of this system shown in Table 8.1.

8.2 Indirect Synthesis of Mercury Chalcogenides

Indirect synthesis methods for mercury chalcogenides also merit attention. These include methods in which at least one of the input components is used in the form of a chemical compound. In this case, during synthesis, the

Table 8.1 Integral Gibbs free energy of formation of tellurium-saturated solid solutions in pseudo-binary CdTe–HgTe.

	$-\Delta G^f$ *(kJ mol⁻¹)*	
x_{CdTe}	*600 K*	*700 K*
0.01	0.42	0.50
0.2	1.04	1.18
1.3	1.50	1.66
0.4	1.80	1.97
0.5	1.92	2.10
0.6	1.94	2.15
0.7	1.89	2.14
0.8	1.69	1.96
1.0	0	0

components interact *via* exchange or oxidation–reduction reactions according to the following equations:

$$HgCl_2 + Na_2S \rightarrow HgS + 2NaCl \tag{8.17}$$

$$HgBr_2 + H_2Te \rightarrow HgTe + 2HBr \tag{8.18}$$

$$Hg + Na_2S_6 \rightarrow HgS + Na_2S_4 \tag{8.19}$$

$$HgCl_2 + H_2Te \rightarrow HgTe + 2HCl \tag{8.20}$$

$$(1-x)HgCl_2 + xCdCl_2 + H_2Te \rightarrow Hg_{1-x}Cd_xTe + 2HCl \tag{8.21}$$

$$(1-x)HgBr_2 + xCdBr_2 + H_2Te \rightarrow Hg_{1-x}Cd_xTe + 2HBr \tag{8.22}$$

These reactions also demonstrate the high affinity of chalcogens towards cadmium and mercury and also zinc, which permits the formation of binary chalcogenides of the type $A^{II}B^{VI}$ (A = Zn, Cd, Hg; B = S, Se, Te) and pseudo-binary systems $A^{II}B^{IV}$–$A^{II}B^{VI}$.[1,15]

8.2.1 Transport Reactions Method

Semiconducting compounds of mercury and chalcogenides and solid solutions based on them are produced using transport reactions.[16,17,47–58] Transport reactions are also used for fine purification and cultivation of single crystals, films and epitaxial layers. The chemical transport of Hg–Te and Hg–Cd–Te system components occurs *via* reversible heterogeneous solid–gas transport reactions at high temperature T_2 and precipitation gas–solid reactions at a lower temperature. Thus, the chemical transport of Hg–Cd–Te system components is accomplished with the help of NH_4Cl,[47] NH_4Br,[59,60] NH_4I,[60,61] HgI_2,[62] and I_2,[63] while HgTe is produced using diethyltellurium $Te(C_2H_5)_2$–H_2[64] and other organometallic compounds as carriers.[16,17,38,47–58]

For transport reactions, the transport and precipitation conditions in each temperature zone (T_i, Δp_i, C_i, ...) are adjusted so as to permit reactions in equilibrium conditions. The number of such chemical reactions even in relatively simple systems, *e.g.* $Hg_{1-x}Cd_xTe$–NH_4Cl, $Hg_{1-x}Cd_xTe$–NH_4Br and

$Hg_{1-x}Cd_xTe$–NH_4I, may reach 21–25. Thus, the technology used to produce high-purity $Hg_{1-x}Cd_xTe$ by precipitation *via* transport reactions in the system $Hg_{1-x}Cd_xTe$–NH_4X (X = Cl, Br, I) produces the following gaseous components:

1 Hg	6 CdX_2	11 Te	16 H_2Te	21 NH_2
2 H_2	7 HgX_2	12 H	17 CdX	22 TeX_2
3 HX	8 NH_3	13 N	18 HgX	23 Cd
4 N_2	9 X_2	14 CdH	19 N_2H_4	24 NH
5 Te_2	10 X	15 HgH	20 N_3	25 N_2H_2

To analyze and select optimum conditions for an experiment to grow layers of solid solutions of $Hg_{1-x}Cd_xTe$ using the system $Hg_{1-x}Cd_xTe$–NH_4X ($X = Cl^-$, Br^-, I^-), Akhromenko and co-workers[47,59–61] calculated the equilibrium composition of the gas phase and partial pressures of the components of the above systems. The total pressure in gas-phase composition calculations was taken as the sum of the pressures of saturated vapors of mercury or tellurium (p_{Te_2}), plus the pressure generated by the products of reaction between ammonium iodide and the solid matter. Temperature dependences of the partial pressures (in Pa) of mercury (p_{Hg}) and tellurium (p_{Te_2}) at the edge of the homogeneity area in a system with NH_4I on the side of Hg and Te for $Hg_{1-x}Cd_xTe$ with $x = 0.4$ are determined according to the following equations:

$$\log p_{Hg} = 0.193062 \times 10^{-4} + 0.143345/T - 0.991008T \times 10^{-7} - 8.951198 \times 10^5/T^2 + 0.34860485 \log T \tag{8.23}$$

$$\log p_{Te_2} = -4.00002 + 0.121644/T - 0.95965T \times 10^{-7} - 2.214170/T^2 + 0.678135 \log T \tag{8.24}$$

whereas in iodide and bromide systems the transport of cadmium, mercury and tellurium is effected by dihalides of cadmium, mercury and ditellurium (Te_2).[59–61] At high temperatures, T_2, the partial pressures of these components are higher than at lower temperatures, which causes the mass transfer of cadmium, mercury and tellurium into the zone with low temperature T_1 producing the solid solution $Hg_{1-x}Cd_xTe$.

In chloride systems, metallic cadmium also takes part in mass transfer along with $CdCl_2$, Hg and Te_2.[47] At high temperatures in such systems, the partial pressures of mercury-, cadmium- and tellurium-containing gaseous substances are greater that at lower temperatures, which also determines mass transfer towards the low-temperature zone T_1.[47] However, in ammonium halide systems containing solid solutions $Hg_{1-x}Cd_xTe$, the partial pressure of Te_2 increases by about four orders of magnitude, the pressures of CdI_2 and $CdCl_2$ decrease by about 1.4- and 7.2-fold, respectively, and the Hg pressure decreases 86-fold.[47,61]

Transport reactions have also been used to obtain epitaxial layers of mercury-containing and other semiconducting compounds.[16,17,42,48–58] Epitaxy is controlled deposition of single-crystal layers of semiconductors and their simultaneous alloying. Three methods are used to deposit epitaxial layers of solid solutions over various substrates: (1) crystallization from solutions,[42,65–67] (2) crystallization from the gas phase[16,17,38,48–58] and (3) molecular beam epitaxy.[68]

8.2.2 Epitaxial Layer Growth

To grow epitaxial semiconducting layers of mercury-containing compounds $A^{II}B^{IV}$ and their solid solutions of $A^{II}B^{IV}$–A_iB^{IV} type $A_{1-x}A_{ix}B$ (*e.g.* $Hg_{1-x}Cd_xTe$, $Hg_{1-x}Zn_xTe$), the substrate (CdTe, CdZnTe, CdTeSe, GaAs, *etc.*) is introduced into a melt supersaturated relative to the epitaxial layer components at temperature T_1.[42,65–67] Nevsky *et al.*[67] described an experiment used to produce layers with $x = 0.35$ out of a solution of $Hg_{0.203}Cd_{0.022}Te_{0.755}$ at $T = 820$–719 K.

Interesting results were obtained *via* low-temperature liquid-phase epitaxy of $Hg_{1-x}Cd_xTe$ from tellurium-enriched melts of mercury.[66] In this case, epitaxial layers of solid solutions of $Hg_{1-x}Cd_xTe$ with $x \approx 0.2$ were grown from solutions of $(Hg_{1-x}Cd_z)_{1-x}Te_y$ (where $z = 0.056$; $y = 0.83$–0.92 depending on temperature). The melt used to grow epitaxial layers was obtained by melting Te, HgTe and CdTe in a graphite tray immediately prior to the experiment. As substrates, $12 \times 12 \times 1\ mm^3$ (111) CdTe crystals were used after chemical–mechanical polishing and etching. Epitaxial layers were grown from supercooled melts of mercury. The temperatures of the growing zone and mercury vessel were controlled separately during the growth process. Epitaxial layers were grown at temperatures between 400 and 500 °C. For a growth temperature of 450 °C, the saturation and homogenization of the solution were performed at 475 °C, *i.e.*, 25 °C above the temperature of epitaxial layer growth. Layer growth begins with a 5 °C supercooling followed by cooling at a constant rate of 0.15 °C min^{-1}. The process produces epitaxial layers of solid solutions of $Hg_{1-x}Cd_xTe$ with low defect concentrations.[42,66] Epitaxial layers of $Hg_{1-x}Cd_xTe$ can also be grown using the thermal zinc 'shift' method of epitaxy.[42]

The process of growing epitaxial layers of mercury-containing semiconductors suggested by Chiang and Wu [66] is very labor intensive and requires high-precision control. Therefore, the method of growth *via* the reaction of thermal dissociation of compounds in the gas phase appears to be more practical.[16,17,38,48–58] Epitaxial layers of solid solutions of $Hg_{1-x}Cd_xTe$ are grown from the gas phase with the help of organometallic compounds of tellurium (diethyl, dialkyl, dimethylallyl, diisopropyl telluride) and cadmium (dimethyl-, diethylcadmium) in a ratio of 1:1 and at a concentration of 1×10^{-4} mol L^{-1} and metallic mercury.[48–53,55–58] To grow epitaxial layers of HgTe and CdTe, Brebrick[13] used dimethylmercury, dimethylcadmium and methylallyl telluride. The preferred use of high-purity metallic mercury rather than organomercuric compounds is dictated by its greater thermal stability compared with other components at growing temperatures.[48,69]

CdTe with (100) and (111)[48,50–51,54–55] orientations, GaAs,[38,51,58] InSb,[54] HgCdTe,[47] CdTeSe[53] and Al_2O_3 (0001) – sapphire[70] have been used as substrates. Epitaxial layers were grown in horizontal and vertical reactors at temperatures below 180 °C,[49] 225 °C,[48] 230–410 °C,[55] 250–350 °C,[51,54] ~370 °C,[53] 395 °C[50] or 415 °C.[38]

Epitaxial films of $Hg_{1-x}Cd_xTe$ (at $x=0.3$) with n-type conductivity have a Hall conductivity of $2.3\times10^4\,cm^2\,V^{-1}\,s^{-1}$ at 40 K and a carrier concentration $3.5\times10^{15}\,cm^{-3}$.[58] Alloying with arsenic (AsH_3) produced p-type conductivity with a charge carrier concentration of 3.5×10^{15}–$1.1\times10^{16}\,cm^{-3}$.[58]

Molecular beam epitaxy is not usually used for depositing epitaxial layers of decomposing semiconducting compounds. However, Faurie *et al.*[68] used this method to obtain epitaxial layers of $Hg_{1-x}Cd_xTe$ with $x\approx0.34$ over a CdTe substrate. Layers with a thickness of 12 μm were deposited at 195 °C. The epitaxial layers demonstrated p-type conductivity, which was due to the presence of mercury vacancies, as the layers were grown in excess tellurium. Charge carrier mobility and concentration in the epitaxial layer at 77 K were $8.0\times10^2\,cm^2\,V^{-1}\,s^{-1}$ and $3.6\times10^{15}\,cm^{-3}$, respectively, which were similar to those of the best specimens of solid solutions of $Hg_{1-x}Cd_xTe$ at $x=0.34$.

Ion implantation of boron into the surface layers of p-type substrates $Hg_{1-x}Cd_xTe$ ($x\approx0.15$) produced n-type semiconductors due to defects induced by implanting. Bubulac[71] used ion implantation (B or Be) to produce a series of p–n junctions of different nature and electrical profiles that were used to manufacture high-quality instruments based on $Hg_{1-x}Cd_xTe$ ($x\approx0.2$), with n- to p- or p- to n-type junctions with As implanted into the p-region and In as background into the n-region. Ion implantation (Hg, B, In) followed by annealing of epitaxial layers of $Hg_{1-x}Cd_xTe$ ($x=0.22$–0.23) over CdTe and CdZnTe substrates in mercury vapor allowed Destefanis[72] to achieve high-quality p–n junctions and produce diodes for matrix optical detectors.

Over the years, intensive research efforts have been dedicated to growing semiconducting compounds, especially ternary compounds $Hg_{1-x}Cd_xTe$, binary HgTe–HgI_2 and ternary systems HgCdTe–HgI_2, in microgravity conditions and in space.[102–104] Special attention is being devoted to studies of the laws of mass transfer during seedless growth of bulk crystals, deposition of epitaxial layers and the effects of microgravity on the homogeneity, structure and electrical properties of the above semiconducting compounds, compared with their Earth-made counterparts. It has been demonstrated that samples of ternary compounds $Hg_{1-x}Cd_xTe$ grown in microgravity have uniform composition and low segregation and are much more homogeneous. The experiments have demonstrated the advantages of microgravity processing of $Hg_{1-x}Cd_xTe$ crystals for various technologies, including annealing in mercury vapor to produce crystals of uniform composition.[75]

In conclusion, it will be observed that the materials of type $A^{II}B^{VI}$ semiconductors and derivative solid solutions are based on high-purity metallic mercury. The production of solid-state semiconductors and epitaxial films based on solid solutions of $Hg_{1-x}Cd_xTe$, $Hg_{1-x}Mn_xTe$, *etc.*, relies on the processes of alloying metallic mercury, cadmium and tellurium or mercury, manganese and tellurium, and in the case of films, on reactions of thermal dissociation of high-purity organic compounds of mercury, cadmium and tellurium or with a combination of thermal dissociation of dimethyl- or diethylcadmium and the corresponding organic compounds of tellurium and the interaction with high-purity metallic mercury vapor.

Among the important process operations is the annealing of $Hg_{1-x}Cd_xTe$, $Hg_{1-x}Zn_xTe$, $Hg_{1-x-y}Cd_xMn_yTe$ in saturated vapors at 400–200 °C for 24–170 h.[58,73–77] Here the fundamental factor is the production culture, which affects not only the quality of technologically critical semiconducting materials with tailored properties, but also the ecology of working areas and whole communities.[78,79]

References

1. N. Kh. Abrikosov, V. F. Bankina, L. V. Poretskaya, E. V. Skudnova and S. N. Chizhevskaya, *Semiconducting Chalcogenides and Their Alloys*, Nauka, Moscow, 1975, p. 219.
2. Yu. V. Levinsky, *p–T–x State Diagrams of Binary Metal Systems: Reference Book*, Metallurgia, Moscow, 1990, vol. 2, p. 360.
3. F. A. Shunk, *Constitution of Binary Alloys*, Second Supplement, McGraw-Hill, New York, 1969.
4. P. V. Usachev, A. B. Golubkov and N. S. Volosatova, *Zh. Prikl. Khim*, 1960, **33**, 2771.
5. S. D. Gromakov, I. V. Zaritskaya and Z. M. Latipov, *Zh. Neorg. Khim.*, 1964, **9**, 2485.
6. A. V. Novoselova and A. S. Pashinkin, *Vapor Pressure of Volatile Chalcogenides of Metals*, Nauka, Moscow, 1978, p. 111.
7. V. P. Glushko (ed.), *Thermodynamic Constants of Substances: Reference Book*, VINITI, Moscow, 1972, Releases 5, 6, 10.
8. B. F. Bilenkay and A. K. Filatova, *Mercury Sulfide: Production and Application*, Vysshaya Shkola, Lvov, 1988, p. 192.
9. (a) R. C. Sharma, Y. A. Chang and C. Guminski, *J. Phase Equilibria*, 1993, **14**, 100; (b) R. C. Sharma, Y. A. Chang and C. Guminski, *J. Phase Equilibria*, 1992, **13**, 663; (c) R. C. Sharma, Y. A. Chang and C. Guminski, *J. Phase Equilibria*, 1995, **16**, 338.
10. R. F. Brebrick, *J. Chem. Phys.*, 1965, **43**, 3846.
11. D. Ferro, P. Imperatory and G. Chirulli, *Thermochim. Acta*, 1984, **77**, 87.
12. V. P. Zlomanov and A. V. Novoselova, *p–T–x Metal–Chalcogen Phase Diagrams*, Nauka, Moscow, 1987, p. 208.
13. R. F. Brebrick, *J. Phys. Chem. Solids*, 1965, **26**, 989.
14. I. B. Mizetskaya, L. D. Budyonnaya and Ya. D. Oleinik, *Physicochemical Grounds for Synthesis of Semiconducting Monocrystals*, Naukova Dumka, Kiev, 1975, p. 76.
15. I. V. Mizetskaya, G. S. Oleinik, L. D. Budyonnaya, *et al.*, *Physicochemical Grounds for Synthesis of Single Crystals of Solid Semiconducting Solutions of $A^{II}B^{VI}$ Compounds*, Naukova Dumka, Kiev, 1986, p. 158.

16. S. K. Ghandhi, in *Epitaxy of Semiconductor Layered Structures Symposium*, Boston, MA, 30 November–4 December 1987, pp. 65–75.
17. C. G. Bethea, B. F. Levine, P. Y. Lu, L. M. Williams and M. H. Ross, *J. Appl. Phys. Lett.*, 1988, **53**, 1629.
18. E. I. Petrik and V. A. Presnov, Production and properties of $A^{II}B^{VI}$ and $A^{IV}B^{VI}$-type semiconducting compounds and solid solutions based on them, Report Highlights, Moscow Institute of Steel and Alloys, Moscow, 1977, vol. 1, p. 224.
19. R. G. Mani, T. McNair, C. R. Lu and R. Grober, *J. Cryst. Growth*, 1989, **97**, 617.
20. N. G. Garbuz, S. V. Kondrakov and S. A. Popov, *Izv. Akad. Nauk SSSR, Neorg. Mater.*, 1990, **26**, 536.
21. (a) V. N. Tomashik and V. I. Grytsiv, *Phase Diagrams of Systems Based on Semiconductor $A^{II}B^{VI}$ Compounds*, Naukova Dumka, Kiev, 1982, p. 168; (b) V. Tomashik, L. Shcherbak and P. Perrot, *Cadmium – Mercury – Tellurium*, New Series IV/11C1, ed. Landolt-Börnstein, MSIT, p. 255.
22. R. R. Galonzka, Yu. Aulyaytner, T. Warminski, *et al.*, 'Salyut-6' – 'Soyuz', *Materials Science and Technology*, Nauka, Moscow, 1985, pp. 164–175.
23. L. N. Kurbatov, I. V. Barmin, B. I. Golovin, V. S. Zinov'ev, A. V. Lz'yurov, A. P. Cherkasov, N. V. Komaro, V. I. Stafeev, E. N. Kholina, A. V. Frolov and É. P. Bochkarev, *Soviet Physics Doklady*, 1983, **28**, 986.
24. R. R. Gałazka, T. Warmiński, J. Bak, J. Auleytner, T. Dietl, A. S. Okhotin, R. P. Borovikova and I. A. Zubritskij, *J. Cryst. Growth*, 1981, **53**, 397.
25. G. T. Petrovsky, Yu. V. Popov, A. A. Berezhnoy and I. V. Semeshkin, *Materials and Processes in Space Technology*, Nauka, Moscow, 1980, pp. 21–27.
26. H. E. Hirsch, S. C. Liang and A. G. White, in *Semiconductors and Semimetals, Vol. 18, Mercury–Cadmium–Telluride*, ed. P. K. Willardson and A. C. Beer, Academic Press, New York, 1981, pp. 21–45.
27. D. M. Bercha, Yu. V. Voroshilov, V. Yu. Slivka, I. D. Turyanitsa, and D. V. Chepur, *Complex Chalcogenides and Chalcohalogenides*, Vysshaya Shkola, Lvov, 1983, p. 182.
28. A. Ya. Nashelsky, *Technology of Semiconductor Materials*, Metallurgia, Moscow, 1972, p. 432.
29. H. E. Hirsch, S. C. Liang and A. G. White, in *Semiconductors and Semimetals, Vol. 18, Mercury–Cadmium–Telluride*, ed. P. K. Willardson and A. C. Beer, Academic Press, New York, 1981, p. 388.
30. C. Guminski, *J. Less-Common Metals*, 1986, **116**, 16.
31. H. R. Vydyanath, *J. Appl. Phys.*, 1986, **59**, 958.
32. L. A. Sysoev and E. K. Raiskin, *U.S. Patent 3414387*, 1968.
33. E. Cruceanu and N. Nistor, *J. Electrochem. Soc.*, 1966, **113**, 955.
34. A. V. Vanyukov, A. M. Sokolov, A. A. Ofitserov and S. N. Repenko, *Izv. Akad. Nauk SSSR, Neorg. Mater.*, 1971, **7**(4), 141.

35. F. P. Volkova, N. S. Averyanov and A. P. Cherkasov, *Izv. Akad. Nauk SSSR, Neorg. Mater.*, 1971, **7**(10), 1853.
36. A. M. Sokolov, A. A. Ofitserov and A. V. Vanyukov, *Izv. Akad. Nauk SSSR, Neorg. Mater.*, 1975, **11**, 1617.
37. A. M. Sokolov and A. Ofitserov, *Izv. Akad. Nauk SSSR, Neorg. Mater.*, 1975, **11**, 1884.
38. V. Natarajan, N. R. Taskar, I. B. Bhat and S. K. Ghandhi, *J. Electron. Mater.*, 1988, **17**, 479.
39. K.-J. Ma, J. Shen, X.-Y. Song and S.-L. Wen, *Acta Phys. Sin.*, 1985, **34**, 1641.
40. J.-R. Yang, Z.-Z. Yu and D.-Y. Tang, *J. Cryst. Growth*, 1985, **72**, 275.
41. J. Schilz and T. N. Guyen Duy, *Semicond. Science Technol.*, 1988, **3**, 992.
42. B. Pelliciari, State of the art of LPE HgCdTe at LIR, *J. Cryst. Growth*, 1990, **86**, 146–160.
43. R. K. Bagai and W. N. Borle, *J. Cryst. Growth*, 1989, **94**, 561.
44. I. S. Virt, D. I. Tsyutsyura and D. D. Shuptar, *Izv. Akad. Nauk SSSR, Neorg. Mater.*, 1990, **26**, 2219.
45. I. S. Virt, V. I. Kempnik, D. I. Tsyutsyura and D. D. Shuptar, *Izv. Akad. Nauk SSSR, Neorg. Mater.*, 1988, **24**, 1394.
46. V. P. Vasil'ev, E. N. Holina and M. N. Mamontov, *Izv. Akad. Nauk SSSR, Neorg. Mater.*, 1990, **26**, 1632.
47. Yu. G. Akhromenko, G. A. Ilchuk and S. P. Pavlishin, *et al.*, *Izv. Akad. Nauk SSSR, Neorg. Mater.*, 1990, **26**, 739.
48. P. Y. Lu, L. M. Williams, S. N. C. Chu and H. Ros, *Appl. Phys. Lett.*, 1989, **54**, 2021.
49. D. V. Shenai-Khatkhate, P. Webb, D. J. Cole-Hamilton, D. W. Blackmore and J. B. Mullin, *J. Cryst. Growth*, 1988, **93**, 744.
50. J. T. Mullins, P. A. Clifton, A. W. Brinkman and J. Woods, *J. Cryst. Growth*, 1988, **93**, 755.
51. S. R. Glanvill, C. J. Rossouw, M. S. Kwietniak, G. N. Pain, T. Warminski and L. S. Wielunski, Structure and polarity of {111} CdTe on {100} GaAs. *J. Appl. Phys.*, 1989, **66**(2), 619–624.
52. T. McAllister, *J. Cryst. Growth*, 1988, **91**, 218.
53. S. K. Ghandhi, I. B. Bhat and H. Fardi, *Appl. Phys. Lett.*, 1988, **52**, 392.
54. J. D. Parsons and L. S. Lichtmann, *J. Cryst. Growth*, 1990, **86**, 222.
55. P. Capper, C. D. Maxey, P. A. C. Whiffin and B. C. Easton, *J. Cryst. Growth*, 1989, **97**, 833.
56. S. P. Kobeleva, in *Thermodynamics and Material Science of Semiconductors: Report of Highlights of the Third All-Union Conference, Moscow, May 1986*, Nauka, Moscow, 1986, vol. 2 , pp. 34–35.
57. V. N. Martynov and M. B. Slavin, in *Thermodynamics and Material Science of Semiconductors: Report of Highlights of the Third All-Union Conference, Moscow, May 1986*, Nauka, Moscow, 1986, vol. 2, pp. 262–263.
58. S. K. Grandy, N. R. Taskar, K. K. Parat, D. Terry and I. B. Bhat, *J. Appl. Phys. Lett.*, 1988, **53**, 1641.

59. V. I. Ivanov-Omskiy, Yu. G. Akhromenko and V. I. Vibliy, *et al*, *Electron. Technol. Mater. Ser.*, 1984, **11/196**, 38–42.
60. V. V. Ratnikov, V. N. Sorokin, V. I. Ivanov-Omskiy, K. E. Mironov, I. A. Gerko, V. K. Erganov and V. M. Merinov, *Pisíma v JTF*, 1988, **14**, 1410.
61. Yu. G. Akhromenko, G. A. Ilchuk, S. P. Pavlishin, S. I. Petrenko and O. I. Gorbova, *Inorg. Materials*, 1987, **23**, 762.
62. H. Wiedemeier and D. Chandra, *Z. Anorg Allg. Chem.*, 1987, **545**, 109.
63. D. Chandra and H. Wiedemeier, *Z. Anorg Allg. Chem.*, 1987, **545**, 98.
64. L. I. Dyakonov, V. N. Ivlev and Y. I. Lipatova, *Izv. Akad. Nauk SSSR, Neorg. Mater.*, 1990, **26**, 69.
65. N. N. Berchenko, V. E. Krevs and V. G. Sredin, *Semiconducting Solid Solutions and Their Application*, Voenizdat, Moscow, 1982, p. 208.
66. C. D. Chiang and T. B. Wu, *J. Cryst. Growth*, 1989, **94**, 499.
67. O. B. Nevsky, Yu. I. Rastegin and V. A. Fedorov, *Izv. Akad. Nauk SSSR, Neorg. Mater.*, 1988, **24**, 1963.
68. J. P. Faurie, S. Sivananthan, M. Lange, R. E. Dewames, A. M. B. Vandewyck, G. M. Williams, D. Yamini and E. Yao, *Appl. Phys. Letters*, 1988, **52**, 2151.
69. J. M. Centeno, C. Gonzalez, J. Sanz-Maudes and T. Rodriguez, *Mater. Lett.*, 1990, **9**, 60.
70. E. R. Gertner, W. E. Tennant, J. D. Blackwell, J. P. Rode, *J. Cryst. Growth*, 1985, **72** , 462.
71. L. O. Bubulac, *J. Cryst. Growth*, 1990, **86**, 723.
72. G. L. Destefanis, *J. Cryst. Growth*, 1990, **86**, 700.
73. L. L. Regel, *Sci. Technol. Rev. Space Explor. Ser.*, VINITI, Moscow, 1984, **21**, 224.
74. L. L. Regel, *Sci. Technol. Rev. Space Explor. Ser.*, VINITI, Moscow, 1987, **29**, 296.
75. L. L. Regel, *Sci. Technol. Rev. Space Explor. Ser.*, VINITI, Moscow, 1990, **34**, 335.
76. J. J. Kennedy, P. M. Amirtharaj, R. P. Boyd, S. B. Qadri, R. C. Dobbyn and G. G. Long, *J. Crystal Growth*, 1990, **86**, 93.
77. T. Piotrowski, *J. Cryst. Growth*, 1985, **71**, 453.
78. F. C. Lu, P. E. Berteau and D. J. Clegg, *Mercury Contamination in Man and His Environment*, Technical Reports Series No. 134, IAEA, Vienna, 1972, pp. 67–85.
79. J. Vostal, *Mercury in the Environment*, CRC Press, Cleveland, OH, 1982, pp. 15–27.
80. Landolt-Börnstein – Group IV Physical Chemistry Selected Semiconductor Systems Volume 11C1 2006 Non-Ferrous Metal Systems Part 1, Ed. G. Effenberg, S. Ilyenko, Springer, 2006, 267.

CHAPTER 9

Chlor-Alkali Process

9.1 Introduction

Early in the nineteenth century, Berzelius and Davy discovered that mercury cathode electrolysis of saline solutions of alkali metals, *i.e.* NaCl or KCl, produced liquid alkali metal amalgams. Upon contact with water, the amalgams decomposed and produced alkali metal hydroxides (Me_iOH, where Me_i is the one of the cations K^+, Na^+, Li^+), hydrogen (H_2) and the original mercury (Hg). This experimental result was taken the basis for the industrial production of chlorine and alkalis, such as NaOH. The first patent for a mercury cathode electrolyzer designed to produce chlorine and alkali was granted to Nolf in 1882 and the first Kastner bath-based industrial facility to produce these chemicals was put into service in 1894 in Oldbury, UK.[1–4]

Industrial production of chlorine totaled 2.0 metric tons in 1950, 10 metric tons in 1975, 12 metric tons in 2000 and over 15 metric tons in 2010. Out of that quantity, over 7 metric tons of chlorine were produced *via* mercury cathode electrolysis of sodium chloride. The future of the chlorine and alkali production method will depend on the future developments in the 'chlor-alkali' process and in competitive methods, the so-called diaphragm and membrane processes.[1,3] This chapter discusses electrochemical aspects of mercury cathode electrolysis, the metallurgy of the sodium–mercury system and the design of electrolysis units.

9.2 Electrochemistry of the Mercury Cathode Process

When dissolved in water, a salt, *e.g.* sodium chloride, dissociates into sodium and chloride ions according to the equation

$$NaCl \rightarrow Na^+ + Cl^- \quad (9.1)$$

Mercury Handbook: Chemistry, Applications and Environmental Impact
By Leonid F Kozin and Steve Hansen

Published by the Royal Society of Chemistry, www.rsc.org

In pure water, a much smaller number of molecules also dissociate into ions:

$$H_2O \rightarrow H^+ + OH^- \tag{9.2}$$

and produce positively charged protons H^+ and negatively charged hydroxyl ions OH^-. The two ions carry the same charge but have different signs. Mercury(I) and -(II) reactions have the following standard electrode potentials:[5]

$$Hg_2^{2+} + 2e \rightarrow 2Hg \quad E^\circ = 0.788\,V \tag{9.3}$$

$$Hg^{2+} + 2e \rightarrow Hg \quad E^\circ = 0.854\,V \tag{9.4}$$

In aqueous solutions, the zero charge potential of mercury is[5]

$$E_{\text{zero charge Hg}} = -0.193V \tag{9.5}$$

versus the normal hydrogen electrode (NHE) and the standard electrode potential of the sodium half-reaction:

$$Na^+ + e \rightleftharpoons Na \tag{9.6}$$

is $E^\circ = -2.728$ V.[6] In 0.1 and 1.0 M aqueous NaCl solutions, the zero charge potential of mercury is $E_{\text{zero charge Hg}} = -0.559$ V and -0.557 V (*versus* NHE), respectively.[5] Therefore, the surface of the mercury electrode in sodium chloride solutions has a negative charge and, in concentrated solutions designed for the cathode process, is described by the equation

$$E_i = E^\circ_{Na} + \frac{2.303RT}{zF} \ln C_{Na^+} \times f_{Na^+} = E^\circ_{Na} + \frac{0.05912}{zF} \log C_{Na^+} \times f_{Na^+} \tag{9.7}$$

where E_i = reversible electrode potential at ion concentration C, E°_{Na} = normal electrode potential of sodium, R = gas constant (J K^{-1} mol^{-1}), T = temperature (K), z = charge on the ion (for sodium $z = 1$), C_{Na^+} = concentration of sodium ions (Na^+) (mol L^{-1} solution), F = Faraday constant (K mol^{-1}) and f_{Na^+} = Na^+ ionic activity coefficient.

9.3 Sodium–Mercury Phase Diagram

Sodium demonstrates a very strong affinity to mercury. The sodium–mercury equilibrium phase diagram is given in Figure 9.1.[7] The Na–Hg system is characterized by one congruently melting compound, $NaHg_2$, melting at ~340 °C, and five incongruently melting compounds which produce peritectic reactions. The incongruently melting compounds decompose *via* peritectic reactions: $Na_{11}Hg_{52}$ at ~156 °C, NaHg at 215 °C, Na_3Hg_2 at 121 °C, Na_8Hg_3 at 66 °C and Na_3Hg at 60 °C.[7] Na_3Hg has α and β polymorphs and both NaHg and Na_8Hg_3 have α, β and γ polymorphs. Appendix I considers these phases in more detail.

Owing to the high affinity of sodium to mercury, the sodium amalgam potential becomes more positive compared with the sodium potential by around 0.7 V. The difference between the standard potentials of pure mercury

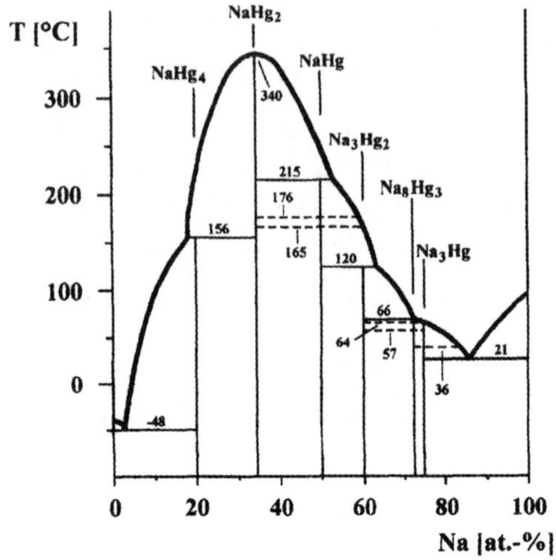

Figure 9.1 Structure of the Na–Hg equilibrium diagram.[7] See Appendix I for more detailed information about the Na–Hg phase diagram. HgNa has recently been shown to be $Na_{11}Hg_{52}$.[7b]
Reproduced with permission from Pergamon (Ref. 7b).

and sodium [electromotive force (EMF)] is $\Delta E = E^{\circ}_{Hg} - E^{\circ}_{Na} = 0.7973 - (-2.728) = 3.5253$ V.[5,6] As mercury interacts with sodium, the activity of sodium is greatly reduced, its amalgam electrode potential shifts towards the electropositive side and the real value of the EMF of the concentration cell is greatly reduced. The potentials, in volts, of electrodes in actual sodium amalgam systems may be calculated with the help of activity coefficients using the following equations:

$$E = E_{Na(Hg)^0} - \frac{2.303RT}{F} \ln\left(\frac{C_{Na^+} f_{Na^+}}{C_{Na} f_{Na}}\right) \tag{9.8}$$

$$E = -1.8490 - 0.05915 \log\left(\frac{C_{Na^+} f_{Na^+}}{C_{Na} f_{Na}}\right) \tag{9.9}$$

The relationship of sodium and mercury activities in the Na–Hg system at 375 °C, according to the literature,[8–12] is given in Figure 9.2.

The activity curves of sodium and mercury demonstrate strong negative deviations from ideal solution behavior. Formation of intermetallic compounds occurs even in liquid amalgams.[11,12] In the dilute sodium region, activity is a linear function of concentration; at higher sodium concentrations, the sodium activity is approximately proportional to the square of concentration, C^2_{Na-Hg}.

Studies of the physicochemical properties of amalgams prompted the development of the EMF analysis method, which consists in measuring the EMFs of amalgam concentration circuits of the type

$$Me(Hg)|Me^+, MeX, H_2O|Me(Hg)_x \tag{9.10}$$

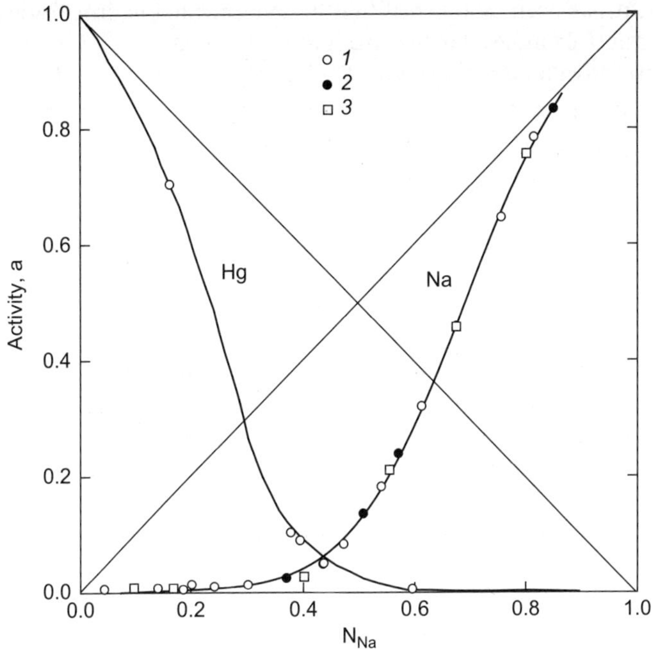

Figure 9.2 Activities of the Na–Hg system components at 375 °C according to different authors: (1) Ref. 8, (2) Ref. 9; (3) Refs 10–12.

In this case, the role of the reference electrode is played by diluted amalgam (on the left-hand side) with a constant concentration of the studied metal throughout the experiment. The potentials of each electrode of the cell with respect to the NHE can be found from the following equations:

$$E_1 = \text{constant} + \frac{2.303RT}{zF} \log a_{\mathrm{Me}_i(\mathrm{Hg})} \tag{9.11}$$

$$E_2 = \text{constant} + \frac{2.303RT}{zF} \log a_{\mathrm{Me}_i(\mathrm{Hg})} \tag{9.12}$$

where constant $= E^\circ + E^s$, and $E^s = -\Delta\overline{G}^{\mathrm{exc}}/zF$ or $E^s = (RT/zF)\ln\gamma_1$, where $\Delta\overline{G}^{\mathrm{exc}}$ is the partial excess Gibbs free energy of Na in the amalgam, $z = 1$ and γ_1 is the activity coefficient of Me = sodium.[11] E^s values are a function of the change in chemical potentials and EMFs of the concentration cells. Values of E^s are closely related to the structure of the Na–Hg equilibrium diagram. This problem was addressed in detail by Kozin *et al.*[11]

Because the zero charge potential of mercury is –0.193 V (*versus* NHE), *i.e.* more negative when compared with the equilibrium potential of the hydrogen electrode, mercury may dissolve slightly in an acidic solution, *e.g.* hydrochloric acid solution:[11,12]

$$2Hg + 2HCl \rightarrow Hg_2Cl_2 + H_2 \tag{9.13}$$

Moreover, it has been established that in concentrated hydrochloric acid both adsorbed and free molecular hydrogen are generated.[11]

The thermodynamic properties of the Na–Hg system at 375 °C, *i.e.* activities of sodium and mercury, a_{Na} and a_{Hg}, partial molar Gibbs free energies, $\Delta\overline{G}_{Na}$ and $\Delta\overline{G}_{Hg}$, partial molar enthalpies, $\Delta\overline{H}_{Na}$ and $\Delta\overline{H}_{Hg}$, partial molar entropies, $\Delta\overline{S}_{Na}$ and $\Delta\overline{S}_{Hg}$, and integral thermodynamic quantities ΔG, ΔH and ΔS, are presented in Figure 9.3.[11–14]

Sodium–mercury amalgam forms extremely strong and weakly dissociated intermetallic compounds. The structure of the Na–Hg equilibrium phase diagram and the activities of the components (sodium and mercury) are interrelated. Formation of the intermetallic compounds in the Na–Hg system leads to a negative departure of sodium and mercury activities from Raoult's law.[11] Comparing the negative Gibbs partial and integral free energies with the structure of Na–Hg equilibrium diagram also indicates that the system has deviated from the ideal solution law. Furthermore, the partial and integral enthalpies of mixing and partial and integral entropies of mixing also indicate that Na–Hg system has deviated from the ideal solution law.[11]

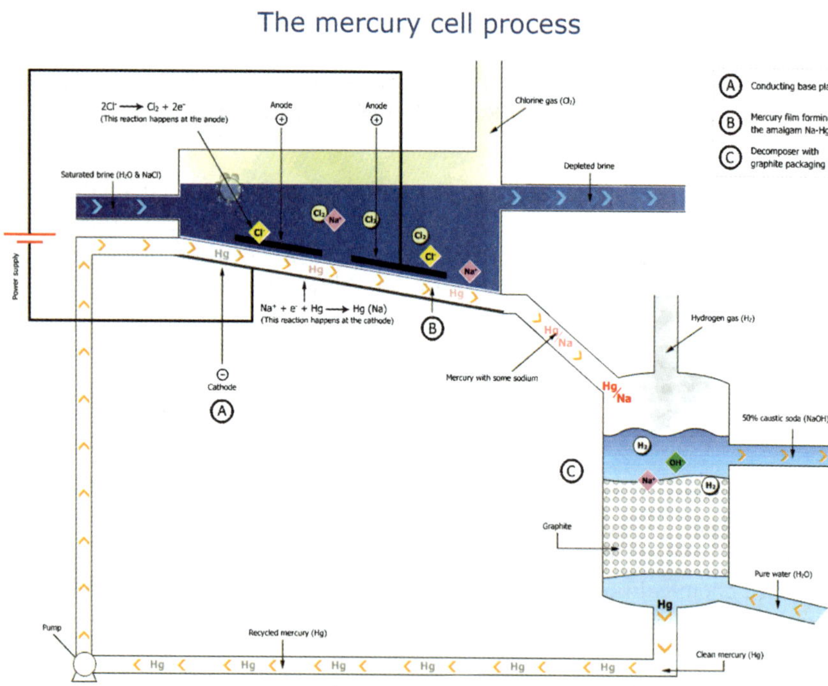

Figure 9.3 The mercury cell process. In the mercury cell process, sodium forms an amalgam (a 'mixture' of two metals) with the mercury at the cathode. The amalgam reacts with the water in a separate reactor called a decomposer where hydrogen gas and caustic soda solution at 50% are produced. Reproduced with permission from Ref. 3b.

Table 9.1 Thermodynamic characteristics of intermetallic compounds of sodium and mercury at 18, 25 and 375 °C.

Intermetallic compound	*Volume ratio Na,* N_1	$\Delta\overline{G}_{Na}$[14] *[kJ (mol atoms)*$^{-1}$*]*	ΔG[14] *[kJ (mol atoms)*$^{-1}$*]*	ΔH *(18 °C)*[15] *[kJ (mol atoms)*$^{-1}$*]*	ΔH *(25 °C)*[16] *[kJ (mol atoms)*$^{-1}$*]*	ΔH *(375 °C)*[8] *[kJ (mol atoms)*$^{-1}$*]*	ΔS *(25 °C)*[14] *(J mol*$^{-1}$ *K*$^{-1}$*)*
Na_3Hg	0.750	–5.9	–11.5	–8.9	–	–8.5	–92.0
Na_8Hg_3[a]	0.727						
Na_3Hg_2	0.600	–5.4	–17.2	–43.5	–19.2	–13.8	–87.0
NaHg	0.500	–7.1	–20.7	–47.3	–21.3	–16.1	–90.0
$NaHg_2$	0.333	–23.3	–23.0	–39.3	–25.5	–17.0	–54.8
$Na_{11}Hg_{52}$[b]	0.175	–20.2	–15.1	–23.4	–16.7	–12.7	–27.2

[a] Stoichiometry was originally assigned as Na_5Hg_2.
[b] Stoichiometry was originally assigned as $NaHg_4$.

Bent and Forziati[13] analyzed the EMF of dilute sodium amalgams and found that an amalgam containing $NaHg_4$ (recently found[7] to be $Na_{11}Hg_{52}$). $Na_{11}Hg_{52}$ has a free energy of formation $\Delta G = -20.2$ kJ (mol atoms)$^{-1}$. Deiseroth[7] performed multiple studies on the intermetallic compounds present in the Na–Hg system. His results and others are tabulated in Appendix I. Bent and Swift[14] also calculated the free energies of formation of the intermetallic compounds using the EMF method.

Recommended values for the excess Gibbs free energy and excess entropy of mixing at infinite dilution are $\Delta\overline{G}^{exc}_{Na} = -74.1$ kJ mol^{-1} and $\Delta\overline{S}^{exc}_{Na} = -29.3$ J mol^{-1} K^{-1}.[11,12] The thermodynamic characteristics of sodium–mercury intermetallic compounds at 18, 25 and 375 °C are given in Table 9.1. The highest negative partial free energy values are for $NaHg_2$ [$\Delta\overline{G}_{Na} = -23.3$ kJ (mol atoms)$^{-1}$] and $Na_{11}Hg_{52}$ [$\Delta\overline{G}_{Na} = -20.2$ kJ (mol atoms)$^{-1}$]. Table 9.1 also shows that as the temperature increases from 18 to 375 °C, the integral enthalpy of formation of $NaHg_2$ and $Na_{11}Hg_{52}$ decreases from $\Delta H = -39.3$ and –23.4 to $\Delta H = -17.0$ and –12.7 kJ mol^{-1}, respectively.

9.4 Production of Chlorine

The production of chlorine, caustic soda (or potassium hydroxide) and hydrogen *via* mercury cathode electrolysis implies the use of graphite anodes. Anodes are normally treated with flax-seed oil in autoclaves at temperatures ≤ 400 °C[4] to reduce porosity. Over the past 25 years, a new anode technology has been introduced that uses sheets of titanium with oxidized surfaces covered with special microlayers of ruthenium, which after heat treatment turn the surfaces into layers of ruthenium and titanium oxides, which constitute the so-called oxides of ruthenium and titanium anodes (ORTA) that feature a small chlorine overvoltage ($\eta_{Cl_2} = 0.05$ V) and a higher oxygen overvoltage ($\eta_{O_2} = 0.6$ V).[3]

The cathode, preferably composed of high-purity (99.9999%) metallic mercury, should be used during the electrolysis of chloride solutions. The technology used to obtain it is described in a Soviet Patent SU 401, 747[17] and Chapter 4. The patterns related to mercury cathode polarization during electrodeposition of sodium ions have been addressed.[1–4] Several groups[18–26] found that the sodium overvoltage is determined by the concentration polarization. De Nora,[22] inventor of mercury electrode electrolyzers, found the sodium overvoltage to be 80 mV on an amalgam containing 0.1 wt% Na at a temperature of 70 °C and a current density of 5250 A m^{-2}. Schmidt and Holzinger[23] suggested an empirical equation for the relationship between the cathode potential, E, of sodium amalgam and the current density, i, in A m^{-2}:

$$E = 1.81 + 0.000085i \tag{9.14}$$

However, from a theoretical point of view, E should have a logarithmic relationship to i. Therefore, various workers[18–26] undertook measurements of the amalgam cathode potentials of a horizontal electrolyzer with a mercury cathode. The electrolyzer had a bottom slope of 5 mm m^{-1} and was 30 mm wide and 340 mm long. The supply of brine of concentration 310 g L^{-1} NaCl and temperature 75 °C was adjusted so that its concentration at the output of the electrolyzer was 280–290 g L^{-1}. A 1 L bottle[11] was provided in a mercury recirculation system upstream of the decomposer so that the amalgam concentration did not fluctuate abruptly during the variable current density test. Amalgam was supplied at 300 mL min^{-1} or 100 mL min^{-1} cm^{-1} of electrolyzer width.

Polarized cathode potentials were measured by a salt bridge introduced *via* the lid from the amalgam output side. The bent end of the salt bridge should touch the amalgam surface. Potentials were measured with reference to a saturated calomel electrode. The results demonstrate the relationship between the sodium overvoltage (η) and the logarithm current density, i. The Tafel equation is then proposed:

$$\eta = a + b\log i \tag{9.15}$$

with b close to RT/nF at $n = 1$. During the deposition of sodium, polarization decreases with growing amalgam concentration. The potential of the polarized cathode, E, may be represented as the difference between the equilibrium potential of the amalgam, E_p, and overvoltage, η:

$$E = E_p - \eta \tag{9.16}$$

The relationship between polarization at a current density of 1 A cm^{-2} and amalgam concentration is shown as a straight line 1 in Figure 9.6 and may be expressed by the equation

$$\eta = 0.2 - 0.33\sqrt{c_{Na}} + 0.068\log i \tag{9.17}$$

where c_{Na} is the sodium concentration (wt%) in the Na–Hg amalgam.

There is also a relationship between the equilibrium potential of the amalgam and its concentration.[18] In the amalgam concentration range 0.05–0.5 wt% Na this relationship may be expressed as

$$E_p = -1.68 - 0.23\sqrt{c_{Na}} \tag{9.18}$$

By inserting values of η and E_p from eqns (9.17) and (9.18) into eqn (9.2), we find

$$E_p = -1.88 - 0.068\log i \tag{9.19}$$

Thus, according to Volkov and Klitsa,[18] the polarized cathode potential does not depend on amalgam concentration. Table 9.2 illustrates the relationship between cathode potential and amalgam concentration at a current density of 10 000 A m^{-2}.

According to Volkov and Klitsa,[18] the independence of the polarized cathode potential and amalgam concentration means that the sodium activity at 75 °C is independent of the composition. This is stipulated by the exchange equilibrium of the Hg–Na binary system and the invariability of the sodium activity, a_{Na} = constant. Hence independence of the hydrogen discharge rate at the amalgam cathode from the sodium activity in the amalgam was proved on a laboratory electrolyzer model similar to that described by Bent and Forziati.[13] The hydrogen discharge rate was measured *via* its chlorine content. The results of experiments performed by Bent and Forziati[13] are in Table 9.3.

Table 9.2 Relationship between cathode potential and amalgam concentration at current density 10 000 A m^{-2} and 75 °C.

Amalgam concentration (wt%)	*Polarized cathode potential (V)*
0.040	1.898
0.070	1.900
0.130	1.898
0.180	1.899
0.220	1.899
0.300	1.900
0.340	1.900

Table 9.3 Hydrogen content in chlorine at different amalgam concentrations. Current density, 0.5 A cm^{-2}; temperature, 60 °C; specific mercury supply rate, 150 mm min^{-1} cm^{-1}; bottom slope, 10 mm.

Amalgam concentration (%)	*Hydrogen content in chlorine (%)*	*Amalgam concentration (%)*	*Hydrogen content in chlorine (%)*
0.02	0.29	0.34	0.58
0.03	0.29	0.45	0.49
0.05	0.38	0.50	0.33
0.10	0.50	0.60	0.40
0.18	0.56	0.65	0.50
0.20	0.46	0.70	0.34

Source: Ref. 13.

The independence of cathode potential and its hydrogen discharge rate from amalgam activity shows that the discharge rate of hydrogen ions is determined only by the cathode potential, *i.e.* only by the activity of sodium in the surface layer of the amalgam. According to Volkov and Klitsa,[18] the sodium transfer rate i (if we ignore the effect of the cathode layer thickness, since transfer occurs faster inside the layer than on the surface) can be described by

$$i = kv^{\frac{3}{2}}c_{\mathrm{Na}}\,a \tag{9.20}$$

where v = average linear flow rate, c_{Na} = sodium concentration in the surface layer of the amalgam and k = kinetic constant. Taking into account that the average linear flow rate v is related to specific inflow Q and layer thickness h by

$$v = \frac{Q}{h} \tag{9.21}$$

and layer thickness h is related to specific supply Q and cathode inclination p by

$$h = k_2 Q^{\frac{1}{2}} p^{-\frac{1}{3}} \tag{9.22}$$

and by inserting eqns (9.21) and (9.22) into eqn (9.20), Volkov and Klitsa[18] arrived at the equation

$$i = k_3 c_{\mathrm{Na}} Q^{\frac{3}{4}} p^{\frac{1}{2}} \tag{9.23}$$

where c_{Na} is a function of cathode potential:

$$c_{\mathrm{Na}} = \exp\left[-\frac{nF(E - E^{\circ})}{RT}\right] \tag{9.24}$$

The hydrogen discharge current density is then another similar function of cathode potential:

$$i_{\mathrm{H_2}} = a[\mathrm{H^+}]\exp\left[-\frac{nF(E - E^{\circ})}{2RT}\right] \tag{9.25}$$

Therefore, eqn (9.23) may appear as follows:

$$i = k_4 \left(i_{\mathrm{H_2}}\right)^2 Q^{\frac{3}{4}} p^{\frac{1}{2}} \tag{9.26}$$

where, according to Volkov and Klitsa,[18]

$$i_{\mathrm{H_2}} = k_5 i^{\frac{1}{2}} Q p^{-\frac{1}{4}} \tag{9.27}$$

The current efficiency, Ψ, is determined by the following relationship (if we ignore current losses on reduction of active chlorine):

$$\Psi = -\frac{i - i_{\mathrm{H_2}}}{i} = 1 - \frac{k_5}{Q^{\frac{3}{8}} p^{\frac{1}{4}} i^{\frac{1}{2}}} \tag{9.28}$$

Given Q and p as constants, eqn (9.28) becomes

$$\Psi = 1 - \frac{k_6}{i^{\frac{1}{2}}} \tag{9.29}$$

according to which the share of current spent for hydrogen discharge decreases with increase in current density, which is in agreement with experimental data. Equations (9.27) and (9.28) also agree with experience. For example, when the current density in the baths was increased from 0.25 to 0.5 A cm^{-2}, the specific mercury supply rate of 140 mL min^{-1} cm^{-1} was unchanged and the slope changed from 2 to 10 mm min^{-1}. In accordance with eqn (9.28), the hydrogen content in chlorine decreased from 1.2 to 0.5%. The calculated value was 0.56%.

It should be remembered that it is difficult to perform an accurate verification of eqns (9.27) and (9.28) owing to the high sensitivity of the hydrogen release rate at the cathode and to impurities in the electrolyte and amalgam phase. It should also be mentioned that the derived relationships are only true when there is concentration polarization. The system will break down if the process takes place at current densities that are lower than the exchange current density for the amalgam at a given concentration.

The process of mercury cathode electrolysis of sodium chloride has been examined in detail.[1–4,17–23] During electrolysis, mercury was used in a closed cycle, which included cathode reduction of adsorbed sodium ions on the surface of a negatively charged ($E_{\text{zero charge Hg}} = -0.557$ V *versus* NHE) mercury cathode, according to the reaction

$$Na^+ + Hg^{2-} + e \rightarrow HgNa^+_{ads} + e \rightarrow HgNa \quad (9.30)$$

at current densities of 5000–7500 A m^{-2}. The concentration of sodium chloride in a carefully purified solution was $310 \pm 5\,g\,L^{-1}$.[1] The cathode output of sodium during the formation of sodium amalgam $Na_{11}Hg_{52}$ was 96%. For ORTA, the primary anodic output product (chlorine) reaches 99–99.9% (depending on the production 'culture'). Therefore, the following processes occur at the anode:

$$\text{Primary process}: 2Cl^- - 2e \rightarrow Cl_2\uparrow \qquad E^\circ_{Cl^-/Cl_2} = 1.359V \quad (9.31)$$

$$\text{Secondary process}: 2H_2O - 4e \rightarrow O_2\uparrow + 4H^+ \quad E^\circ_{H_2O/O_2,H^+} = 1.229V \quad (9.32)$$

The oxygen current efficiency is 0.1–1.0%. An electrolyzer made for an aqueous solution of sodium chloride with pump-driven circulation of the mercury cathode within a closed cycle and with contact to the inserted graphite electrodes is illustrated in Figure 9.3. Upon contact of sodium amalgam with the graphite electrode inserts, a galvanic element $C_{graphite}|NaHg_n$ is formed, in which the functions of the anode are performed by the sodium amalgam and those of the cathode by graphite.

Decomposition of sodium amalgam occurs as a result of operation of the galvanic element, with release of hydrogen according to the reaction

$$\frac{1}{11}Na_{11}Hg_{52} + H_2O \rightarrow NaOH + \frac{1}{2}H_2 + \frac{52}{11}Hg \quad (9.33)$$

The resulting sodium hydroxide solution with concentration 600–700 g dm^{-3} flows out from the decomposer into a collection tank. The decomposer also outputs the released hydrogen, which is then distributed to consumers or stored into cylinders and gas tanks. The metallic mercury obtained according to eqn (9.33) is transferred into the electrolyzer for the next chlorine production cycle and to become sodium amalgam.

To date there are high-performance industrial mercury cathode electrolyzers.[3,4] Key technical data on medium-capacity industrial electrolyzers are provided in Table 9.4.

All mercury cathode electrolyzers consist of the following parts: electrolysis bath, sodium amalgam decomposer and mercury pump (mechanical or electromagnetic). Industry uses monopolar horizontal electrolyzers. All electrolyzers are designed similarly and consist of the following three elements: one-piece flat steel bottom, rubber-coated steel frame and mercury electrolyzer lid.[3,4]

The P-101 mercury electrolyzer, shown in Figure 9.3, is a widely used model. The P-101 commonly uses a one-piece steel flat bottom that is 13.2 m long and 1.2 m wide. The frame and lid consist of two parts connected *via* a flange. The bottom supports a rubber-coated steel frame. The frame has two pockets – alkali and acid. All these elements together form an electrolyzer housing.

The alkali input chamber is located on the mercury pump side and the acid chamber is located on the opposite side. The pockets are equipped with mercury and liquid valves which seal the electrolyzer from gaseous hydrogen, chlorine and liquid. The electrolyzer is topped with a rubber-coated lid, which accommodates 284 sealed graphite anodes or ORTA. The anodes have gland

Table 9.4 Technical specifications of industrial mercury cathode electrolyzers[1–4,18]

	Ukrainian			*Others*		
Parameter	*P-101*	*P-20M*	*P-300*	*De Nora*	*Krebs*	*Hoechst–Uhde*
Load (kA)	100	150	300	400	300	300
Anode material	Graphite	Graphite	ORTA	ORTA	ORTA	ORTA
Current density (kA m^{-2})	5.3	7.85	10.4	12.5	12.0	12.5
Voltage (V)	4.4	4.6	4.3	4.0	4.25	4.25
Current efficiency (%)	95	96	96.5	96–97	96	95
Cathode dimensions (m)	14.5×2.4	19.4×1.95	19.7×2.3	14.8×2.3	14.4×1.61	12.5×2.4
Chlorine output (t day^{-1})	3.0	4.5	9.0	12.0	9.0	9.0
Mass of mercury in cathode (t)	2.2	3.1	3.9	4.8	3.75	3.5
Power consumption (kWh t^{-1} of chlorine)	3660	3780	3530	3020	3350	3300

seals. The electrolyzer lid connects to the positive bus and its bottom connects to the negative bus. Therefore, the lid and the bottom must be electrically isolated.

The electrolyzer housing is insulated and mounted at a slope of 10 mm per meter of length. Mercury and electrolyte are supplied into the input pocket with a mercury pump. The anode reaction releases chlorine, while the cathode reaction forms dilute sodium amalgam, which flows into the amalgam decomposer. The horizontal-type decomposer is located parallel to the main housing. The decomposer is a welded $13\,\text{m long} \times 0.5\,\text{m}$ wide box sloping 18 mm per meter of length (Figure 9.8) towards the side opposite that of the electrolyzer.

The sodium amalgam decomposer outputs the resulting alkali, which contains NaOH 42–50 wt%, NaCl 0.01–0.05 wt%, Na_2CO_3 0.2 wt% and Hg up to $3\,\text{g}\,\text{m}^3$.

A fairly high content of mercury in the alkali is due to the specifics of the behavior of mercury during the mercury cathode electrolysis process. Different methods are used to reduce the mercury content in the end products. The behavior of mercury during amalgam electrolysis has been analyzed in detail.[23–26] The cited studies offer different methods of reducing the mercury content in the end products.

Uhde (Germany) produces electrolyzers of various capacities, which depend on the cathode area.[4] Available areas are 4, 15 and 35 m^2. The design uses batch suspension of anodes, which are either graphitized or metal oxide coated (ORTA). ORTA deliver chlorine gas, which contains Cl_2 99.5 vol.%, CO_2 0.2 vol.%, H_2 0.1 vol.% and air 0.2 vol.%. For graphitized anodes the figures are Cl_2 99.0 vol.%, CO_2 0.6 vol.%, H_2 0.2 vol.% and air 0.2 vol.%. The resulting caustic soda (NaOH) solution has a concentration of 50% and contains 0.03% NaCl, 0.2% Na_2CO_3, 0.002% Fe and 0.05–3 mg L^{-1} of Hg.[4]

References

1. L. N. Sheludyakov, L. A. Saltovskaya and V. V. Stender, *Appl. Chem. Mag.*, 1953, **26**, 160.
2. J. Billiter, *Die Technische Elektrolyse der Nichtmetalle*, Springer, Vienna, 1954.
3. (a) A. K. Gorbachov, *Technical Electrochemistry*, Prapor, Kharkov, 2002, p. 254; (b) Euro Chlor Institute, *The mercury cell process*. Available online at http://www.eurochlor.org/the-chlorine-universe/how-is-chlorine-produced/the-mercury-cell-process.aspx, Last accessed 21 June 2013.
4. G. I. Volkov, *Production of Chlorine and Caustic Soda via Mercury–Cathode Electrolysis*, Khimiya, Moscow 1968.
5. A. M. Sukhotin (ed.), *Electrochemistry Handbook*, Khimiya, Leningrad, 1981.
6. G. Milaazzo and S. Caroli, *Tables of Standard Electrode Potentials*, Wiley, Chichester, 1978.

7. (a) C. Hoch and A. Simon, *Angew. Chem. Int. Ed.*, 2012, **51**, 3029; (b) H. J. Deiseroth, *Prog. Solid State Chem.*, 1997, **25**, 73.
8. M. A. Bykova and A. G. Morachevsky, *Appl. Chem. Mag.*, 1973, **46**, 312.
9. K. Hauffe, *Z. Elektrochem.*, 1940, **46**, 348.
10. M. L. Iverson and H. L. Recht, *J. Chem. Eng. Data*, 1967, **12**, 262.
11. L. F. Kozin, R. Sh. Nigmetova and M. B. Dergacheva, *Thermodynamics of Binary Amalgam Systems*, Nauka, Alma-Ata, 1977, p. 343.
12. R. Hultgren, P. Desai, D. Hawkins, M. Gleiser and K. Kelley, *Selected Values of Thermodynamic Properties of Binary Alloys*, ASM, New York, 1973.
13. H. E. Bent and A. F. Forziati, *J. Am. Chem. Soc.*, 1936, **58**, 2216.
14. H. E. Bent and E. Swift, *J. Am. Chem. Soc.*, 1936, **58**, 2220.
15. W. Biltz and F. Meyer, *Z. Anorg. Chem.*, 1928, **176**, 23.
16. O. Kubaschewski, E. L. Evans and C. B. Alcock, *Metallurgical Thermochemistry*, Pergamon Press, Oxford, 1967.
17. L. F. Kozin, G. M. Cherniy and A. A. Nikotin, Method of Amalgam refining of mercury in a four-section electrolyzer, *USSR Pat.*, 401 747, *Bulletin*, 1973, (41), 109.
18. G. I. Volkov and Z. L. Klitsa, *Elektrokhimiya*, 1968, **4**, 1347.
19. L. S. Genin, *Electrolysis of Sodium Chloride Solutions*, Gostekhizdat, Moscow, 1960, p. 208.
20. Tekhnika, *Development of a Chlorine Electrochemical Process*, http:www.industring.ru/chemical/chemical12.html, 2009 (in Russian) (last accessed 9 March 2013).
21. H. S. Schmidt and F. Holzinger, *Chem.-Ing.-Tech.*, 1963, **35**, 37–44.
22. V. De Nora, *J. Electrochem. Soc.*, 1950, **97**, 346–351.
23. L. F. Kozin and S. V. Volkov, Chemistry and Technology of High-Purity Metals and Metalloids, Naukova Dumka, Kiev, 2002, vol. 1, p. 540 and 2003, vol. 2, p. 351.
24. L. F. Kozin and Y. P. Sushkov, *Ukr. Chem. Mag.*, 1991, **57**, 611.
25. L. F. Kozin, *Ukr. Chem. Mag.*, 1991, **57**, 733.
26. V. De Nora, presented at the XXII Congres de Chimique Industrielle, Barcelona, October 1949.

CHAPTER 10

Use of Mercury in Small-scale Gold Mining

CEZARY GUMINSKI

Department of Chemistry, University of Warsaw, Warsaw, Poland

10.1 Introduction

Gold and mercury are metals that have been known since ancient times. Gold has been known since antiquity and mercury was used as early as the fifth century BC. The two metals mix fairly easily and the first civilizations probably used their ability to alloy with each other as both a gilding process and a process for extraction of gold from ore. Gilding was abandoned after the very unfortunate incident that occurred in 1858 during the amalgamation gilding of the cathedral cupola in St Petersburg, when about 60 workers died from mercury vapor inhalation.[1]

Gold mining via mercury amalgamation was probably one of the first metallurgical extraction processes developed by humankind. Unfortunately, this dangerous practice persists to this day, largely unchanged since ancient times. Small-scale and artisanal gold mining continues in many sites around the world and nothing suggests that this method, hazardous to people and the environment, will be abandoned soon. Although the use of mercury to extract gold is illegal in several countries, it still continues.

When mercury processing (distillation, reclamation, synthesis) is performed in industrialized countries, there are generally stringent restrictions concerning mercury release to the atmosphere. However, in regions where primitive, small-scale methods are performed in the open air, there is the potential to do considerable harm to the miners and the surroundings.

Mercury Handbook: Chemistry, Applications and Environmental Impact
By Leonid F Kozin and Steve Hansen

Published by the Royal Society of Chemistry, www.rsc.org

10.1.1 Reasons for Artisanal Gold Mining

Mercury is a poisonous element without color or odor. Its toxic features are not immediately manifested in the human body until some months or years after exposure to mercury vapor. It is probable that the main reason why people in developing countries continually apply such 'technology' is because the effects of poisoning with mercury are not obvious until some time after the exposure. Chapter 14 discusses the medical effects of metallic mercury poisoning.

The second reason is that performing this risky method is relatively cheap and the product (by contrast) is very expensive. Therefore, it is difficult to imagine that the procedure will soon be given up by artisanal gold miners, even if they were properly informed about the irreversible damage that mercury and its compounds do to humans, animals and ecosystems.

10.1.2 Mercury Pollution

Small-scale gold mining is one of the largest sources of anthropogenic (human-made) mercury entering the atmosphere every year.[2] The largest contribution to elemental mercury is from burning of coal in power plants. Detailed studies of air, water and ground contamination in recent decades established that about one-third of the world pollution with Hg in various forms comes from small-scale gold mining. Owing to the strict regulations applied to industry in developed countries, one may predict that the amount of Hg emitted by official sources will decrease every year.

Mercury pollution from artisanal gold mining is likely to increase owing to the high price of gold. On the other hand, secondary mercury emissions from other sources, such as fossil fuels, dental amalgams, fluorescent lighting and incineration of medical and municipal waste, have been decreasing owing to elimination of mercury and environmental regulations.

10.2 Method of Artisanal Gold Mining

The extraction of gold particles by liquid mercury is relatively simple. Generally, the procedure is characterized by the following sequential stages:[3]

1. powdering of the ore containing gold
2. enrichment of the crumbled ore for gold content
3. mixing of the ore with mercury for gold extraction
4. separation of excess mercury from the Au amalgam
5. distillation of mercury from the concentrated Au–Hg amalgam
6. purification of the gold alloy by remelting
7. purification of mercury for its reuse.

The method may be applied not only to Au but also, with some modifications (with the use of Zn amalgam), to the extraction of Pt, Pd, Rh and Ir from ores.[4] Gold is especially tractable for the amalgamation because it is easily wetted by mercury.[5,6] If a drop of Hg is in contact with an Au surface, the Hg loses its high surface tension and is almost immediately (within seconds) propagated over all Au surfaces. The process seems to be controlled by surface diffusion and at this stage it does not reflect a tendency of Au to form Au–Hg intermetallics or a tendency towards dissolution and saturation of liquid Hg with Au (which is only 0.14 at.% Au at 298 K) or solubility of Hg in solid Au (which may be estimated at about 13 at.% Hg at room temperature).[7] Figure 10.1 shows the binary Au–Hg phase diagram. The structures present in the diagram are discussed in Appendix I.

In practice, when drops of Hg contact small grains of Au, the Au particles are very quickly soaked up inside the Hg drops. However, the real dissolution process of Au and the formation of Au–Hg intermetallic compounds is comparatively slow. If Au forms relatively large lumps in an ore, a miner is very fortunate, since no further treatment, except Au remelting or refining, is not needed. Unfortunately, such situations seldom arise.

The very traditional method of gold extraction from river sands starts with panning, which is based on the significant density difference of Au ($d = 19.3\ g\,cm^{-3}$) and sand ($d \approx 2–3\ g\,cm^{-3}$). This step concentrates Au particles. The next step is the amalgamation of the particles with Hg. Gold ore in Au-containing rocks must start as a fine powdery consistency. The particles of Au must be liberated from the rock sheath to allow efficient contact with Hg, otherwise Hg would

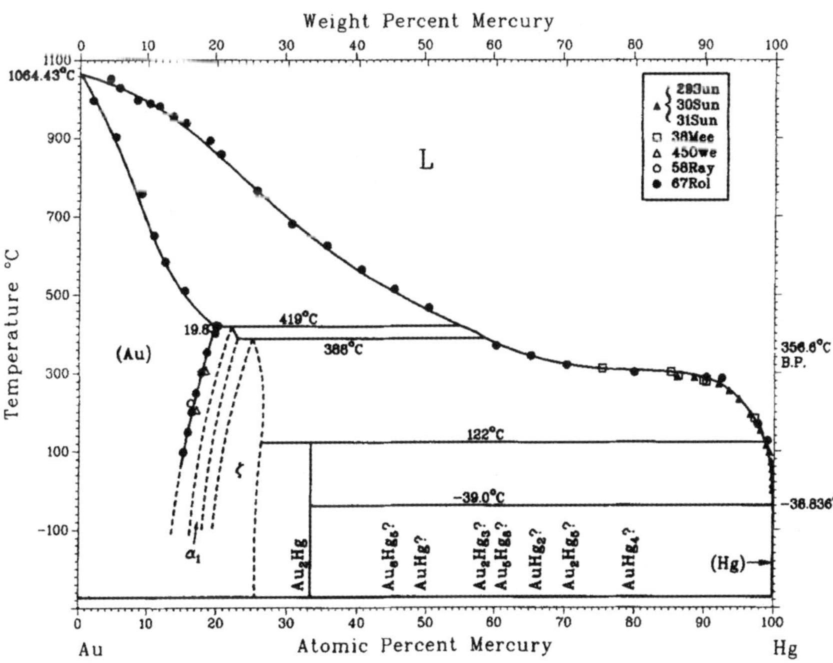

Figure 10.1 The Au–Hg binary phase diagram.
Reproduced with kind permission from ASM International.[7]

not be able to wash out the gold grains. Miners who are able to automate the grinding process use grinding mills, whereas those who are unable to do so use hand-operated mortars and pestles to crush the gangue into powder.

The next step is effective mixing of the river sands or the crushed Au ore with Hg for about 1 h to form the Au amalgam. A discontinuous process is more effective than a continuous process. It may be performed in grinding mills, on copper plates covered with the Hg or simply in pots. The amalgamation process is not effective when the Hg is contaminated with other metals or minerals because small drops of Hg are covered with impurity particles on its surface and the dirt may hamper its proper contact with Au grains. This situation frequently occurs after reusing Hg. The wetting properties of Hg may be improved by addition of an active metal (such as Zn) or by polarization of the Hg in water with a battery to restore its bright surface. The wetting properties of Hg can also be significantly improved by a single distillation, although an additional apparatus is needed.

The concentrate of Au amalgam thus formed may be further separated from ore by manual or mechanical panning or elutriation, since the sand and the amalgam have very different densities (2–3 and >13.5 g cm^{-3}, respectively). Excess Hg is then removed from the Au amalgam by squeezing it through leather, chamois or a fabric, and by centrifuging. The resulting Hg is, or should be, reused and eventually purged of the impurities.

The concentrated Au amalgam obtained in this way is further heated and Hg is distilled off; this step should be performed at 700–800 K (423–523 °C). When the distillation is not complete, the remaining Hg is alloyed with Au. If the container used for the distillation was made from a metal which readily combines with Au (*e.g.* Zn, Fe, Sn), then the solid Au left after the distillation may be contaminated by these metals and thus be less valuable. The distilled Hg is, or should be, carefully collected and reused.

In the last stage, the solid Au-rich alloy is remelted at about 1400 K (1123 °C). Mercury and other volatile metals are released during this final distillation. This step is frequently carried out in a middleman's shop in towns or villages. The last two steps (steps 5 and 6) should be performed in closed apparatus with coolers dipped in water to condense Hg. With proper distillation equipment, it is possible to reclaim and recycle 50–90% of the Hg in the amalgam.[8]

10.3 Environmental Degradation Caused by Small-scale Gold Mining

Inexperienced miners typically introduce three types of environmental degradation while mining:

1. They discard tailings of the ore after the amalgamation step. Tailings contain highly disintegrated Hg that is later washed into the ecosystem and can be converted into methylmercury.

2. They distill Hg from the Au amalgam in the open air on a bonfire. Mercury is released directly to the atmosphere and travels long distances.
3. They remelt solid Au containing Hg in the open air.

Sometimes miners use nitric acid to dissolve Hg when Au is not dissolved. However, if the Hg is not subsequently reduced to its metallic state with the use of metallic Al, Zn or Fe, then this very toxic and corrosive solution produces mercury nitrate and creates very serious degradation of the surrounding ecosystem.

Elemental Hg left on the Earth's surface continuously evaporates. However, more harmful is the runoff into inland waters from the ore after the amalgamation step. Metallic Hg is a noble metal with a low solubility in water (5.2×10^{-7} mol% at 298 K). Its solubility is about 10% lower in sea water and generally decreases with increasing inorganic salt concentration.[9] The majority of Hg salts are also sparingly soluble in water. Surprisingly, Hg in both the metallic combined forms is readily absorbed by some bacteria living in waters and the most toxic organometallic compounds of Hg (methyl-, ethyl-, phenylmercury) are produced. These compounds further accumulate in fish and finally enter the ecosystem.[10,11] Therefore, every portion of the tailings, with even small amounts of Hg discarded into waters, adds more Hg to the environment. Residues of Hg in the tailings left in the open air evaporate continuously and likewise contribute to pollution of the environment; however, such processes occur slowly.

Perhaps the worst procedure in the gold extraction process, unfortunately sometimes applied by miners, is the use of cyanides for the extraction of traces of Au left in ore that also contains Hg. The cyanide reacts not only with Au but also with Hg, forming mercury cyanide compounds that are extremely toxic, especially when they run off into inland waters or are involuntarily acidified.

10.4 Remedies or Improvements to Small-scale Gold Mining

Appealing to miners to keep Hg-containing ores in closed containers or carefully to collect Hg and reuse it seems to be the only practical method of protecting wastes containing Hg from evaporation or from being discarded into waters. The remainder of the Hg may also be trapped by contact of the ore with Ag or amalgamated Cu plates, but such effective procedures are expensive.

It is much easier to convince miners to improve the Hg distillation process from the liquid and the solid Au amalgam (which the present author experienced during educational lectures in Peru). If the distillation is performed in a closed system (a retort made of steel, cast iron or enameled material) with an effectively cooled condenser made of metal that does not amalgamate with Hg, then the Hg obtained in this way is fairly pure and can be reused many times for Au extraction.

A very detailed report on many aspects of small-scale of Au mining with the use of amalgam techniques is available[12] and also in a more condensed form.[8]

References

1. S. I. Venetskii, Mercury, in *Tales about Metals*, Mir, Moscow, 1981, p. 186.
2. L. D de Lacerda, *Environ. Geol.*, 2003, **43**, 308.
3. M. M. Veiga, S. M. Metcalf, R. F. Baker, B. Klein, G. Davis, A. Bamber, S. Siegel and P. Singo, *Global Mercury Project Manual for Training Artisanal and Small-Scale Gold Miners*, UNIDO, Vienna, 2006.
4. V. S. Shemyakin, K. A. Brik, S. P. Bukhman and K. A. Karasev, *Izv. Akad. Nauk Kaz. SSR, Ser. Khim.*, 1979, **3**, 73.
5. M. N. Gavze, *Korroziya i Smachivaemost Metallov Rtutyu*, Nauka, Moscow, 1969, p. 170.
6. H. Winterhager and W. Schlosser, *Metall*, 1960, **14**, 1.
7. H. Okamoto and T. B. Massalski, *Phase Diagrams of Binary Gold Alloys*, ASM International, Materials Park, OH, 1987, 132.
8. K. H. Telmer and M. M. Veiga, in *Mercury Fate and Transport in the Global Atmosphere: Missions, Measurements and Models*, ed. N. Pirrone and R. Mason, Springer, Dordrecht, 2009, p. 131.
9. L. Clever, *Mercury in Liquids, Compressed Gases, Molten Salts and Other Elements*, IUPAC Solubility Data Series, vol. 29, Pergamon Press, Oxford, 1987, p. 7.
10. R. Von Burg and M. R. Greenwood, Mercury, in *Metals and Their Compounds in the Environment*, ed. E. Merian, VCH, Weinheim, 1991, p. 1052.
11. I. Drabaek and A. Iverfeldt, Mercury, in *Hazardous Metals in the Environment*, ed. M. Stoeppler, Elsevier, Amsterdam, 1992, p. 257.
12. M. M. Veiga, P. A. Maxson and L. Hylander, *J. Cleaner Prod.*, 2006, **14**, 436.

General References

P. A. Maxson, in *Dynamics of Mercury Pollution on Regional and Global Scales*, ed. N. Pirrone and K. R. Mahaffey, Springer, New York, 2005, p. 25.

J. O. Nriagu and J. M. Pacyna, *Nature*, 1988, **333**, 134.

E. G. Pacyna and J. M. Pacyna, *Water Air Soil Pollut.*, 2002, **137**, 149.

J. M. Pacyna, J. Munthe, K. Larjava and E. G. Pacyna, in *Dynamics of Mercury Pollution on Regional and Global Scales*, ed. N. Pirrone and K. R. Mahaffey, Springer, New York, 2005, p. 51.

R. P. Mason, W. F. Fitzgerald and F. M. M. Morel, *Geochim. Cosmochim. Acta.*, 1994, **58**, 3191.

S. J. Spiegel and M. M. Veiga, *J. Cleaner Prod.*, 2010, **18**, 375.

CHAPTER 11

Mercury Legislation in the United States

11.1 Introduction

The use of mercury is heavily regulated owing to its toxicity. Activities that release mercury to the environment include chlor-alkali plants, steel and metal refiners, coal-based power plants, artisanal and small-scale gold mining, cement producers, municipal waste incinerators, dental amalgam, fluorescent and metal halide lamps and disposal of mercury switches, barometers, thermometers, etc. Legislation on mercury occurs to some extent in most countries. A detailed look at all of the major directives regarding mercury from industrial countries is beyond the scope of this book. European directives concerning mercury have been reviewed elsewhere.[1] This chapter summarizes the regulation of mercury in the United States.

11.2 Mercury Legislation[2]

The following Acts of Congress have been promulgated for the control and reduction of mercury pollution in the United States. Specific to mercury are the Acts:

- Mercury Export Ban Act (2008)
- Mercury-containing and Rechargeable Battery Management Act (1996).

Other broader legislation controlling mercury release includes:

- Federal Food, Drug and Cosmetic Act (1938)
- Federal Insecticide, Fungicide and Rodenticide Act (1947)
- Clean Air Act (1970) and Amendments

Mercury Handbook: Chemistry, Applications and Environmental Impact
By Leonid F Kozin and Steve Hansen

Published by the Royal Society of Chemistry, www.rsc.org

- Safe Drinking Water Act (1974)
- Toxic Substances Control Act (1976)
- Clean Water Act (1977)
- Comprehensive Environmental Response, Compensation and Liability Act (1980)
- Superfund Amendments and Reauthorization Act (1986)
- Emergency Planning and Community Right-to-Know Act (1986)
- Resource Conservation and Recovery Act (1984)
- Food and Drug Administration Modernization Act (1997).

11.2.1 Mercury Export Ban Act

The goal of the US Mercury Export Ban Act is to remove mercury from the world market. In October 2009, the US Environmental Protection Agency (EPA) released its Report to Congress on Mercury Compounds.[3] The report, required by Congress under Section 4 of the Mercury Export Ban Act of 2008 (MEBA), identifies sources of mercury compounds in the USA and reports quantities in imports, exports and uses of these compounds in products and processes. The report also assesses the potential for key mercury compounds to be exported for reduction into elemental mercury. Table 11.1 gives the mercury compounds included in the Report.

1. Specifically required for this report by MEBA.
2. More than 25 000 pounds (11 340 kilograms) were produced at single site in any single reporting year since 1986.
3. Manufactured or imported as a specialty chemical.
4. Technologically feasible to export and convert to elemental mercury abroad.
5. Produced in potentially significant quantities, including as a waste or byproduct.

Table 11.1 Mercury compounds by criteria for inclusion in the Report to Congress on Mercury Compounds.[3]

Compound	*CAS No.*	*Criteria for inclusion*[a]
Mercury(I) chloride	10112-91-1	1, 2, 3, 4, 5
Mercury(II) acetate	600-27-7	2, 3
Mercury(II) chloride	7487-94-7	1, 2, 3
Mercury(II) iodide	7774-29-0	3
Mercury(II) nitrate	10045-94-0	3, 4
Mercury(II) oxide	21908-53-2	1, 3, 4
Mercury(II) selenide	20601-83-6	5
Mercury(II) sulfate	7783-35-9	3, 4, 5
Mercury(II) sulfide	1344-48-5	3, 5
Mercury(II) thiocyanate	592-85-8	3
Phenylmercury(II) acetate	62-38-4	2
Thimerosal	54-64-8	3

[a]Criteria for inclusion:

The Act's three main provisions are as follows:

1. Federal agencies are prohibited from conveying, selling or distributing elemental mercury that is under their control or jurisdiction. This includes stockpiles held by the Departments of Energy and Defense.
2. Export of elemental mercury from the USA is prohibited beginning 1 January 2013.
3. The Department of Energy (DOE) shall designate one or more DOE facilities for long-term management and storage of elemental mercury generated within the USA. This designation must occur no later than 1 January 2010.

11.2.2 Mercury-containing and Rechargeable Battery Management Act

The Mercury-containing and Rechargeable Battery Management Act of 1996 (Battery Act) phases out the use of mercury in batteries and provides for the efficient and cost-effective collection and recycling, or proper disposal, of used nickel–cadmium batteries, small sealed lead–acid batteries and certain other regulated batteries. The statute applies to battery and product manufacturers, battery waste handlers and certain battery and product importers and retailers. It phased out the use of mercury in batteries and established labeling, collection and recycling and disposal requirements for certain regulated batteries.

11.2.3 Legislation Controlling Mercury Release

The *Clean Air Act* was introduced in 1970 and was amended in 1977 and 1990. Mercury was added as a hazardous air pollutant in 1990. The Clean Air Act regulates 188 air toxics, also known as 'hazardous air pollutants.' The Act directs the US Environmental Protection Agency (EPA) to establish standards for certain sources that emit mercury. Those sources are also required to obtain Clean Air Act operating permits and to comply with all applicable emission standards. The Act also establishes emission limits for sources of mercury emissions, such as medical waste and solid waste incinerators, hazardous waste combustors and chlor-alkali plants. The law includes special provisions for dealing with air toxics emitted from utilities, giving the EPA the authority to regulate power plant mercury emissions by establishing 'performance standards' or 'maximum achievable control technology' (MACT), whichever the Agency deems most appropriate. MACT standards apply to both new and *existing* sources of pollution.

Mercury generated by coal-burning power plants is regulated through the use of MACT standards.[4] The Clean Air Act Amendments of 1977 and 1990 contain an exemption that permits older coal-burning power plants to release between four and ten times the amount of mercury that new plants may release. The stricter rules apply only to new or modified power plants.[5]

Table 11.2 Clean Water Act effluent guidelines for one day.[5]

Type of wastewater	*BAT*[a]	*NSPS*[b]
Smelter wet air pollution control	0.325 mg troy oz^{-1}	0.195 mg troy oz^{-1}
Silver chloride reduction, spent solution	0.010 mg troy oz^{-1}	0.060 mg troy oz^{-1}
Electrolytic cells, wet air pollution control	49.50 mg troy oz^{-1}	2.97 mg troy oz^{-1}
Electrolytic preparation wet air pollution control	0.013 mg troy oz^{-1}	0.008 mg troy oz^{-1}
Calciner wet air pollution control	46.55 mg kg^{-1}	3.30 mg kg^{-1}
Calciner quench water	4.40 mg kg^{-1}	2.640 mg kg^{-1}
Calciner stack gas contact cooling water	1.038 mg kg^{-1}	0.623 mg kg^{-1}
Condenser blowdown	3.45 mg kg^{-1}	2.07 mg kg^{-1}
Mercury cleaning bath water	0.35 mg kg^{-1}	0.21 mg kg^{-1}

[a]BAT = best available technology.
[b]NSPS = new source performance standards.

The *Clean Water Act* of 1977 regulates the discharge of mercury into surface waters by using a permit system to regulate industrial discharges. Permits may assign a facility a specific mercury discharge limit or require them to monitor and report on mercury discharges. The Clean Water Act authorizes the EPA to enact pollution control programs and set water quality standards for all contaminants in surface waters.[5] There is a bifurcation of standards under Title III of the Clean Water Act.[5] Existing point sources of hazardous waste are subject to the 'best available technology' (BAT). New point sources must abide by stricter 'new source performance standards' (NSPS). The permissible mercury levels in 2004 for wastewater are given in Table 11.2.

The *Comprehensive Environmental Response, Compensation and Liability Act* (CERCLA), commonly known as Superfund, was enacted by Congress on 11 December 1980. CERCLA requires that mercury spills of ≥1 lb be reported to the National Response Center. This law created a tax on the chemical and petroleum industries and provided broad Federal authority to respond directly to releases or threatened releases of hazardous substances that may endanger public health or the environment. CERCLA:

- established prohibitions and requirements concerning closed and abandoned hazardous waste sites
- provided for liability of persons responsible for releases of hazardous waste at these sites and
- established a trust fund to provide for cleanup when no responsible party could be identified.

CERCLA was amended by the Superfund Amendments and Reauthorization Act (SARA) on 17 October 1986.

The *Superfund Amendments and Reauthorization Act* (SARA) was implemented to provide funding for the clean-up of heavily polluted sites. Under the authority of the Superfund Amendments and Reauthorization Act of 1986, the Agency for Toxic Substances and Disease Registry (ATSDR) works to prevent or mitigate adverse human health effects and diminished quality of life resulting from exposure to hazardous substances in the environment. To do this, ATSDR provides expert support to Federal, State and local health officials.

The *Resource Conservation and Recovery Act* (RCRA) authorizes the EPA to control all stages of the life cycle of hazardous waste. This includes hazardous waste generation, transportation, storage and disposal. It also contains two separate regulatory pathways, one for new facilities and the other for old facilities. Older facilities are not required to meet all of the standards of new facilities. The Resource Conservation and Recovery Act establishes disposal requirements for wastes that contain mercury (*e.g.* thermometers, medical and dental wastes and mercury switches). It also allows States to adopt less stringent 'Universal Waste Rules' if certain often-used, mercury-containing wastes are recycled (*e.g.* thermostats, fluorescent and high-intensity discharge lamps and batteries).

Under the RCRA, the EPA has specifically listed many chemical wastes as hazardous. Mercury is listed as a hazardous waste under the RCRA and has been assigned EPA Hazardous Waste No. U151. This substance has been banned from land disposal until treated by retorting or roasting.

The RCRA requires that the EPA manage hazardous wastes, including mercury wastes, from the time they are generated, through storage and transportation, to their ultimate treatment and disposal. The EPA has established treatment and recycling standards that must be met before these wastes can be disposed of. Certain mercury wastes – mercury-containing household hazardous waste and waste generated in very small quantities – are exempt from some RCRA hazardous waste requirements. The RCRA also sets emission limits for mercury-containing hazardous waste that is combusted. US States are largely responsible for implementing the RCRA program. Individual States may have stricter requirements than Federal requirements.

Under the *Safe Drinking Water Act*, the EPA sets standards for drinking water that apply to public water systems. These standards protect people by limiting levels of mercury and other contaminants in drinking water. Mercury contamination in drinking water can come from erosion of natural deposits of mercury, discharges into water from refineries and factories and runoff from landfills and cropland. US States have the primary responsibility for enforcing drinking water standards. The maximum contaminant level for mercury in drinking water, established under the Safe Drinking Water Act, is 0.002 mg L^{-1}.

The *Toxic Substances Control Act* (TSCA) of 1976 provides the EPA with the authority to require reporting, record-keeping and testing requirements, and also to set restrictions relating to chemical substances and/or mixtures. Certain substances are generally excluded from TSCA, including food, drugs, cosmetics and pesticides. The objective of the TSCA is to allow the EPA to regulate new commercial chemicals before they enter the market, to regulate existing chemicals when they pose an unreasonable risk to health or to the environment and to regulate their distribution and use.

The *Emergency Planning and Community Right-to-Know Act* forces employers who own or operate facilities in SIC codes 20–39 that employ 10 or more workers and that manufacture 25 000 lb or more of mercury per calendar year, or otherwise use 10 000 lb or more of mercury per calendar year, are required by the EPA to submit a Toxic Chemical Release Inventory form (Form R) to the EPA reporting the amount of mercury emitted or released from their facility annually. In 1999, the limit for the amount of mercury consumed was lowered from 10 000 lb to 10 lbs. The Emergency Planning and Community Right-to-Know Act established three new regulations. It:

1. established reporting requirements for accidental and intentional releases
2. established requirements to report inventory information to state and local authorities
3. required facilities to submit a report to the Toxics Release Inventory when they manufacture, process or otherwise use 10 lb or more of mercury.

11.2.4 Food and Drug Administration

The US Food and Drug Administration (FDA) regulates the mercury content in food, drugs and cosmetics. It limits the use of mercury as an antimicrobial or preservative in cosmetics and regulates the use of mercury in dental amalgams. The *Federal Food, Drug and Cosmetic Act* (FFDCA) establishes an FDA action level for methylmercury in fish at 1 ppm. The *Food and Drug Administration Modernization Act* of 1997 (FDAMA), Amended 21 USC 301, required the FDA to compile a list of drugs and foods that contain intentionally introduced mercury compounds and to provide a quantitative and qualitative analysis of the mercury compounds in the list.

In 1999, the FDA undertook what they considered to be a comprehensive review of the use of thimerosal in childhood vaccines. Although they found no evidence of harm, they did find that some infants could be exposed to cumulative levels of mercury that exceeded the EPA's guidelines for safe intake of methylmercury.

Limitations on mercury-added products have been specified in regulations given in the *Federal Insecticide, Fungicide and Rodenticide Act* (FIFRA). The objective of FIFRA is to provide Federal control of pesticide distribution, sale and use. All pesticides used in the USA must be registered (licensed) by the EPA. Registration assures that pesticides will be properly labeled and that, if used in accordance with specifications, they will not cause unreasonable harm to the environment. The use of each registered pesticide must be consistent with use directions contained on the label or labeling.

11.3 Mercury Regulations and Standards

In October 2007, the EPA issued a *Significant New Use Rule* (SNUR) to require notification to the EPA 90 days prior to US manufacture, import or processing of elemental mercury for use in convenience light switches, anti-lock brake system switches and active ride control system switches in certain motor vehicles. In July 2010, the EPA issued a final SNUR for elemental mercury used in flow meters, natural gas manometers and pyrometers. The Agency requires

90 days' notice prior to US manufacture, import or processing of elemental mercury for use in flow meters, natural gas manometers and pyrometers. The Rule is promulgated under Section 5(a)(2) of the Toxic Substances Control Act for elemental mercury.

11.3.1 Measurement of Mercury in Water

Method 1631 allows for the determination of mercury at a minimum level of 0.5 ppt and supports measurements for mercury published in the National Toxics Rule and in the Final Water Quality Guidance for the Great Lakes System. Revision E of the directive replaces the currently approved version of Method 1631 and Method 1631 Revision C. This revision clarifies quality control and sample handling requirements and allows flexibility to incorporate additional available technologies. This rule also amends the requirements regarding preservation, storage and holding time for very low-concentration mercury samples.

The *Total Maximum Daily Load* (TMDL) Regulations and Guidance gave the EPA's regulations and guidance for the Total Maximum Daily Load, *i.e.* the maximum amount of a pollutant (including mercury) that a waterbody can receive and still meet water quality standards.

11.3.2 Land Disposal Restrictions

The primary goal of the Resource Conservation and Recovery Act (RCRA) Subtitle C program is to protect human health and the environment from the dangers associated with generation, transportation, treatment, storage and disposal of hazardous waste. Land disposal restrictions (LDRs) provide a second measure of protection from threats posed by hazardous waste disposal. The LDR program ensures that hazardous waste cannot be placed on the land until the waste meets specific treatment standards to reduce the mobility or toxicity of the hazardous constituents in the waste.

The LDR program found in 40 CFR Part 268 requires waste handlers to treat hazardous waste or meet specified levels for hazardous constituents before disposing of the waste on the land. This is called the disposal prohibition. To ensure proper treatment, the EPA establishes a treatment standard for each type of hazardous waste. The EPA lists these treatment standards in Part 268, Subpart D.

For non-wastewaters, the waste handler prepares an extract that reflects the leaching potential of hazardous constituents in the waste. The waste meets the treatment standard if the concentration of regulated constituents in the liquid extract falls below the regulatory levels given for the waste code.

11.3.3 Mercury in Air

Reduction of *Toxic Air Emissions from Industrial, Commercial and Institutional Boilers and Process Heaters Final Rule* reduces toxic air pollutants, including mercury, from industrial, commercial and institutional boilers and process heaters. This rule limits the amount of air toxics that may be released from exhaust stacks of all new (built after 13 January 2003) and existing large and limited-use solid-fuel boilers and process heaters located at facilities that are considered to be major sources of air toxics.

On 19 December 2003, the EPA introduced the *Reduction of Toxic Air Pollutants from Mercury Cell Chlor-Alkali Plants Final Rule*. This Final Rule reduces mercury emissions from mercury cell chlor-alkali plants that are considered 'major sources' of hazardous air pollutants and also facilities considered to be 'area sources.' Mercury cell chlor-alkali plants produce chlorine and caustic using mercury cells. A detailed discussion of chlor-alkali plant operation is given in Chapter 9.

In April 2004, the EPA issued a regulation to control emissions from iron and steel foundries. The rule included emission limits for manufacturing processes and pollution prevention-based requirements to reduce air toxics from furnace charge materials and coating/binder formulations. The rule also included a work practice requirement to ensure removal of mercury switches from automobile scrap.

On 28 December 2007, the EPA issued a final *National Emission Standards for Hazardous Air Pollutants* (NESHAP) rule for electric arc furnace steel-making facilities. This Final Rule established requirements for the control of mercury emissions that are based on the maximum achievable control technology and requirements for the control of other hazardous air pollutants that are based on generally available control technology or management practices. The final amendments to the NESHAP add or revise, as applicable, emission limits for mercury, total hydrocarbons (THCs) and particulate matter (PM) from new and existing kilns located at major and area sources and for hydrochloric acid from new and existing kilns located at major sources. The standards for new kilns apply to facilities that commenced construction, modification or reconstruction after 6 May 2009.

In August 2010, the EPA established the first regulations for mercury emissions from cement factories. The production of Portland cement is believed to account for 7% of US mercury emissions. Mercury is emitted when cement components such as clay, limestone and shale are heated in a kiln.

In December 2010, the EPA issued final regulations and added gold mine ore processing and production area to the list of source categories to be regulated under Section 112(c)(6) of the Clean Air Act. This source category was added because of its significant mercury emissions. Gold ore processing was believed to be the seventh largest source of mercury air emission in the USA.

In February 2011, the EPA established practical and protective Clean Air Act emissions standards for large and small boilers and incinerators that burn solid waste and sewage sludge. These standards cover more than 200 000 boilers and incinerators that emit harmful air pollutants, including mercury, cadmium and particle pollution. The EPA also announced that it will reconsider certain aspects of the boiler and commercial/industrial solid waste incinerator (CISWI) rules.

In July 2011, the EPA finalized the *Cross-State Air Pollution Rule* (CSAPR). The CSAPR requires States to improve air quality significantly by reducing power plant emissions that contribute to ozone and/or fine particle pollution in other States. This rule replaces EPA's 2005 Clean Air Interstate Rule (CAIR). A December 2008 court decision kept the requirements of CAIR in place temporarily but directed the EPA to issue a new rule to implement Clean Air Act requirements concerning the transport of air pollution across State boundaries. This action responds to the court's concerns. The CSAPR aims to improve air quality throughout the eastern half of the USA, helping States achieve national clean air standards. The emission reductions expected from the EPA's recently finalized Mercury and Air Toxics Standards (MATS) are not included in the estimated emission reductions from the CSAPR.

In December 2011, the EPA, under the authority of the Clean Air Act, signed the *Mercury and Air Toxics Standards* (MATS). The purpose of MATS was to reduce emissions of toxic air pollutants from power plants and set performance standards for fossil fuel-fired electric utility, industrial–commercial–institutional and small industrial–commercial–institutional steam-generating units. Specifically, these mercury and air toxics standards for power plants will reduce emissions from new and existing coal- and oil-fired electric utility steam-generating units. MATS will reduce emissions of heavy metals, including mercury, arsenic, chromium and nickel. The new law will also reduce emissions of acid gases, including hydrochloric acid and hydrofluoric acid.

11.4 Occupational Safety and Health Administration

In 1974, the Occupational Safety and Health Administration (OSHA) set a permissible exposure limit for mercury in workplace settings at 0.1 mg m^{-3} as an upper limit. OSHA's revised Respiratory Protection Standard (29 CFR 1910.134 and 29 CFR 1926.103) came into effect on 8 April 1998. The final standard replaces the respiratory protection standards adopted by OSHA in 1971.

Respirators are needed by people working with mercury. To comply with the OSHA Respiratory Protection Standard, employers should institute a complete respiratory protection program that, at a minimum, complies with the requirements of OSHA's Respiratory Protection Standard. Such a program must include respirator selection, an evaluation of the worker's ability to perform the work while wearing a respirator, the regular training of personnel, respirator fit testing, periodic workplace monitoring and regular respirator maintenance, inspection and cleaning. A medical surveillance program must also instituted. A workplace respirator protection program is discussed in Chapter 14.

11.5 Department of Transportation and International Air Transport Association

The transportation of mercury and mercury devices is covered by hazardous materials regulations under the Department of Transportation (DOT) and International Air Transport Association (IATA). These agencies require a hazardous material warning label for all air shipments regardless of the amount of mercury and for land freight in amounts of 1 lb or more.

References

1. N. Emmott and M. Slayne, in *Dynamics of Mercury Pollution on Regional and Global Scales*, ed. N. Pirrone and K. R. Mahaffey, Springer, New York, 2005.
2. EPA, *Mercury: Laws and Regulations*, 2012, http://www.epa.gov/hg/regs.htm (last accessed 9 March 2013).
3. EPA, *Report to Congress: Potential Export of Mercury Compounds from the United States for Conversion to Elemental Mercury*, 2009, http://www.epa.gov/hg/pdfs/mercury-rpt-to-congress.pdf (last accessed 9 March 2013).
4. M. Sweeney, MS Thesis, Massachusetts Institute of Technology, 2006.
5. W. Thomas, *Columbia J. Environ. Law*, 2004, **29**, 156.

CHAPTER 12

Environmental Aspects of the Industrial Application of Mercury

LEONID F. KOZIN,[a] STEVE C. HANSEN,[b] NIKOLAI F. ZAKHARCHENKO[a] AND JASON GRAY[c]

[a] V. I. Vernadsky Institute of General and Inorganic Chemistry, National Academy of Sciences of Ukraine, Kiev, Ukraine; [b] Consultant; [c] Nippon Instruments North America, College Station, TX, USA

12.1 History and Uses of Mercury

Mercury has been used and is still used commercially despite its high toxicity. Mercury has valuable physicochemical properties and has been known for about 2000 years.[1] The properties of mercury have been described by Aristotle, Pliny the Elder, Paracelsus, Theophrastus, Vitruvius and Dioscorides. Ancient people made extensive use of the medicinal properties of some inorganic mercury compounds. For example, yellow mercury(II) oxide was used then as a component of eye and skin ointments. Mercury(II) chloride ($HgCl_2$) has been used as a strong disinfectant in medicine and as a fungicide in agriculture. Calomel (Hg_2Cl_2) has been used in medicine, in pyrotechnics and as a catalyst. Thiomersal is mainly used as an antiseptic and antifungal agent in medicines and vaccines.

In the sixteenth century, Paracelsus (1493–1541), a Swiss doctor and natural scientist, created pharmaceutical chemistry.[2] Paracelsus struggled to obtain the purest possible compounds of mercury, arsenic, copper, lead and silver and used small doses of them as medicines for various diseases. Before Paracelsus,

Mercury Handbook: Chemistry, Applications and Environmental Impact
By Leonid F Kozin and Steve Hansen

Published by the Royal Society of Chemistry, www.rsc.org

mercury formulas were mostly used as extremely strong poisons. The lethal outcome depended on the concentration of mercury administered to the human body.

Mercury has been used to produce ship fouling-prevention paints for seawater[1,3] and other paints. In agriculture, mercury is a long-established component of pesticides.[1,3,4] Mercury is used in micro- and optoelectronics,[5] the electrical engineering industry,[6,7] the instrument-making industry,[8] discharge lighting (see Chapter 7), technical electrochemistry in the production of chlorine and alkali (see Chapter 9), electrosynthesis of organomercury compounds,[9] amalgam metallurgy (production of highly refined metals),[10–13] aluminum[14] and gold production,[15–19] stripping voltammetry,[20,21] medicine for dental amalgams based on mercury, tin and silver, polarography with mercury electrodes,[22–25] inorganic synthesis of sodium sulfide and hydrosulfite,[26] organometallic synthesis and other products.[1,27,28]

Solvent extraction of mercury from chloride media[29] is in common use. Mercury oxide has recently gained prominence as a key component of high-T_c superconductors.[30] Among the most widely used technologies is the growth of epitaxial films $Hg_{1-x}Cd_xTe$[31] (see Chapter 8), colloidal nanocrystals of HgTe[32] and the production of complex epitaxial heterostructures. These structures are based on cadmium–mercury–tellurium solid solutions *via* metallorganic chemical vapor deposition over carriers of gallium arsenide,[33] indium phosphide, *etc.*

This chapter discusses the occurrence of mercury in Nature and its physicochemical and thermodynamic properties in the mercury–water system. It also demonstrates that mercury belongs to hazard class I and is toxic for humans and warm-blooded animals. The discussion addresses the maximum allowable concentrations (MACs) of inorganic and organic mercury compounds in water and air – including potable water and effluents – the mechanisms of transforming mercury from inorganic to organic compounds by the action of anaerobic and aerobic bacteria and microorganisms and mercury circulation cycles within the Earth's ecosystem. A method to measure atmospheric mercury is briefly discussed.

12.2 Occurrence of Mercury in Nature

Mercury is a trace element, its average content in the Earth's crust being 7×10^{-6} wt% (0.5 g t^{-1}).[1,34] Mercury's geochemical Clarke value is 7×10^{-7} wt% (0.007 g t^{-1}) and its industrial Clarke value is 4.2×10^{-4} wt%,[35] *i.e.* 600 times higher, which is an indication of the extensive use of mercury in industry, technology and science.[1]

Global reserves of mercury total about 600 000 t. It is believed that only 0.02% of mercury reserves are concentrated in deposits of hydrothermal origin. There is a registered total of 324 000 t of mercury in deposits in Spain (which accounts for 26% of the total), Russia (14%), Kirghizia (Kirghizstan) (13%) and Ukraine – Nikitovka (8%). Among the other countries that hold significant portions of the total reserves are the USA, Mexico, Turkey, China and Slovenia.

Table 12.1 Mercury-bearing ores.[36]

Name	*Ore*	*Hg (at.%)*	*Hg (wt%)*
Velikite	Cu_2HgSnS_4	12.5	34.9
Galkhaite	$Tl(Cu,Hg,Zn)_{12}As_8S_{24}$	0–26.6	0–60.5
Temagamite	Pd_3HgTe_3	14.2	22.2
Atenaite	$(Pd,Hg)_3As$	0–75	88.9
Khakite	$(Cu,Hg)_3SbSe_3$	0–42.3	0–62.9
Balkanite	$Ag_5Cu_9HgS_8$	4.3	12.8
Saukovite	$(Hg,Zn)S$	0–50.0	0–86.2
Timanite	$HgSe$	50.0	71.8
Lorodaite	$HgTe$	50.0	61.1
Korderoite	α-$Hg_3S_2Cl_2$	42.3	81.7
Petrovicite	$Cu_3PbHgBiSe_5$	9.1	16.7
Gruzdevite	$Cu_6Hg_3Sb_4S_{12}$	11.5	32.4
Aktashite	$Cu_6Hg_3As_5S_{12}$	11.5	34.5
Livingstonite	$HgSb_4S_8$	7.7	21.3
Tvalchrelidzeite	$Hg_{12}(Sb,As)_8S_{15}$	34.2	69.0
Christite	$TlHgAsS_3$	16.6	34.8
Laffittite	$AgHgAsS_3$	16.6	41.8

While the mineral reserve base of the global mercury industry is adequate to satisfy the global demand quantity wise, its quality characteristics are not completely satisfactory. The highest quality ores with an average metal content above 1.5% are found only in Spain and Algeria. In all other countries the mercury ore is at least three times less rich, which makes it very difficult to achieve cost-effective production given the current prices. Moreover, most mercury deposits contain relatively moderate reserves of the metal. Out of around 2000 of the world's mercury deposits (1000 of which were mined at different periods), only seven – Almaden, El Entredicho and Las Nueva Concepcion (Spain), Fendek (Algeria), Wanshan and Danchjai (China) and Khaidarkan (Kirghizia) – contain large reserves, together accounting for 80% of world reserves.

Cinnabar (HgS) is the most widely distributed mercury mineral and the most stable natural mercury compound. Mercury is also found in a wide range of complex minerals and complementary polymetallic ores.[36] There are 35 known mercury-containing minerals, some of which are listed in Table 12.1. Mercury complements copper-, arsenic-, antimony-, lead-, thallium- and selenium–tellurium-bearing ores. Mercury ores are categorized as rich (containing ~1 wt% or more Hg), common (0.2–0.3 wt% Hg) and lean (0.06–0.12 wt% Hg).

12.3 Recovery of Mercury from HgS

Pyrometallurgical and hydrometallurgical processes are used to produce mercury. In the pyrometallurgical process, ore or mercury concentrates are roasted at 673–1173 K in fluidized-bed furnaces. Elemental mercury is produced through the following reactions:

$$HgS \rightarrow Hg + {}^{1}\!/_{2}S_2 \tag{12.1}$$

$$Me_i(HgMe_j)S_n \rightarrow Hg + Me_iMe_jS_{n-{}^{1}\!/_{2}} + {}^{1}\!/_{2}S_2 \tag{12.2}$$

When mercury undergoes one of these reactions, it evaporates and condenses inside special chambers where it is collected and purified by either physico-chemical or electrochemical methods[1] (see Chapter 4). Reactions 12.3 and 12.4 show the oxidation of the non-mercury metals present in the mercury-bearing ore:

$$Me_iMe_jS_{n-{}^{1}\!/_{2}} + (n+m)O_2 \rightarrow Me_iO_n + Me_jO_m \tag{12.3}$$

$${}^{1}\!/_{2}S_2 + O_2 \rightarrow SO_2 \tag{12.4}$$

In hydrometallurgical processes, cinnabar contained in concentrates and ores is leached with the help of ligands (Cl^-, Br^-, *etc.*) and the resulting solutions are subjected to electrolysis, electrolytic precipitation, hydrolytic reprecipitation, *etc.* Hydrometallurgical processes recover only ~90–95% of mercury from ores and concentrates. Open sludge dumps that build up around mercury production facilities are permanent sources of mercury vapor.

Polymetallic sulfide ores contain from 10^{-4}–10^{-2} up to 1.0–2.5 wt% of mercury. In the course of polymetallic ore processing, mercury, being a metallic impurity, is normally not extracted but instead is distributed between the end products and often concentrates into one of them. Being highly volatile (the metallic mercury content at 0, 20 and 100 °C is 2.0, 14.0 and 242 g dm^{-3} of air, respectively) and easily restorable in the course of pyrometallurgical processing of ore or concentrate, mercury mostly disperses in the environment, which, as shown below, creates a lethal hazard for humans and warm-blooded animals[4,37–39]

As noted above, mercury and mercury-based compounds have been known since ancient times. In the five centuries before 1925, the world produced around 10^6 t of metallic mercury, which was mostly used for the production of gold and silver.[40] It should be noted that 10^6 t of metallic mercury has, over the years, already been dispersed in the environment. Metallic mercury and its ions (Hg_2^{2+}, Hg^{2+}) and hydroxo and other compounds of varying complexity are formed from ore deposits in which mercury is contained in the form of the above-mentioned minerals, by atmospheric and hydroerosive processes and redox reactions.

12.4 Amounts of Mercury Used in Industry

As noted above, the world's geological reserves of mercury are currently estimated at 600 000 t and global reserves of mercury make up 209 200 t.[41] Estimates of mercury use throughout the 1980s are given in Table 12.2.[42] Therefore, it is assumed that global reserves will last at least 85–90 years.

Table 12.2 Mercury use during the 1980s.

Year	*Hg usage (t)*[42a]	*Hg usage (t)*[42b]
1980	9244	6818
1984	6909	6036
1987	7255	5906
1988	7866	–

However, world production of mercury has begun to fall rapidly. According to the US Geological Survey, mercury extraction in 1997 was 3447 t and in 2006–2007 was only 1500 t.[43] The main reason for the sharp reduction in mercury extraction and the reduced demand for it at the end of the twentieth century was the realization that mercury and its compounds are highly toxic, causing increasing damage to ecological areas of human habitation.

In recent years, China has been supplying ~70–75% of world mercury extraction. Kirghizia, the world's second largest producer, extracted 200–300 t of mercury annually. Total mercury extraction in other countries is 100–150 t per year, although not in the form of a commercial product. For many countries the main source of mercury can be recycled mercury and unclaimed mercury reserves in mine waste piles and slurry pits. Mercury from *secondary sources*, such as dental amalgam, fluorescent lamps, thermostats, fossil-fueled electric power stations, car switches, batteries, emissions from cement plants and metallurgical refineries, PVC production and artisanal and small scale gold mining,[43–45] account for much of the mercury used in commerce.

For example, in the USA, recycled waste furnishes 29% of the mercury consumed.[41] In 1970, the US population had more than 150 million mercury–silver amalgam dental fillings installed. The noted reduction in mercury consumption over time is associated with decreased production of mercury batteries, paints, chlor-alkali plants and dental amalgams and reduced amounts of mercury in fluorescent lamps, pesticides, insecticides, *etc.* From 1950 to 1980, the Nikitovsky Mercury Plant in Ukraine produced about 800 tons of mercury for the USSR gold industry; from 1981 to 1991 the extraction of mercury was reduced to 400 tons per year, but the production of high-purity mercury 99.99999–99.999999 (7N–8N) wt% for the semiconductor industry was 5–8 tons per year. After about 1992, mining of mercury in the Nikitovsky mercury plant in Ukraine was reduced to 50 t per year or less.

The USA is an importer of mercury (importing between 131 and 696 t per year). Domestic production in the USA, which amounted to 577.1 tin 1985 and 416.5 t in 1986, did not satisfy the country's demand for this metal.[6] The production of secondary (recycled or reclaimed) mercury rose steadily up to 1988, as follows:

1985	185 t
1986	219 t
1987	265 t
1988	278 t

The current task s to demercurize emissions sufficiently from the following: sulfur production, Waelz kiln production of zinc,[1,46] aqueous solutions of acids,[1,47–53] chlor-alkali production,[1,54–59] aqueous solutions and solutions of alkalis and hydrosulfides with sulfur.[1,47–53,60–63] The following processes are currently under development: sorption methods to extract mercury ions from solution,[49,64–67] methods of mercury extraction from solutions using liquid membranes,[68] using anion-exchange resins,[69–73] extraction of mercury by electrolysis[1,74–77] and electrolytic precipitation,[78,79] radiochemical[80] and biological purification[81] and mercury sulfide precipitation.[82,83]

12.5 The Role of Industry in Environmental Mercury Pollution

Metallic mercury has a relatively high vapor pressure at ambient temperature and low heats of fusion and vaporization. Therefore, metallic mercury easily evaporates into the atmosphere and is dispersed around the planet with air masses and precipitation. Metallic mercury also partly dissolves in environmental waters. Figure 12.1 illustrates the exchange equilibrium between air and water, which is in a complex functional relationship with many factors (temperature, vapor and gas pressure, concentration of salts in water, *etc.*). Figure 12.2 summarizes data[1,6] on the solubility of metallic mercury water in relation to temperature (273–773 K).[1,84]

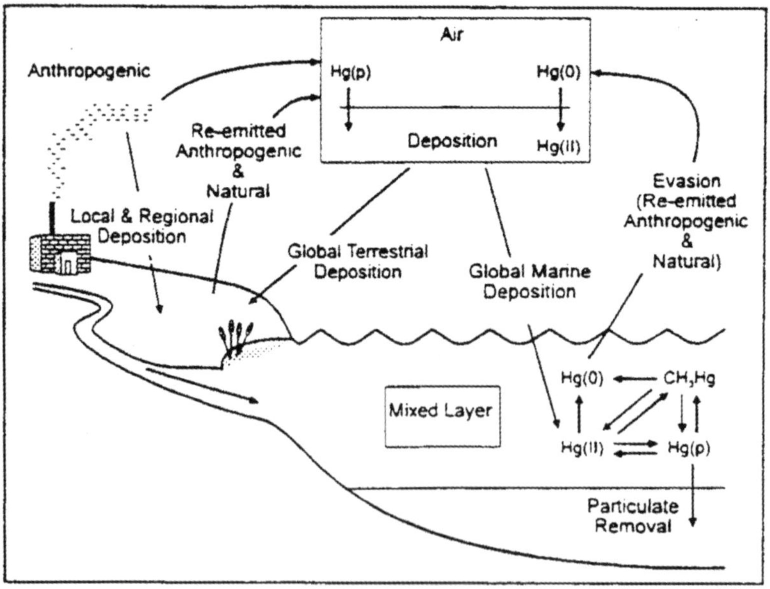

Figure 12.1 Mercury circulation within the Earth's ecosystem. *Source*: US EPA.[114]

As can be seen in Figure 5.1, in the temperature range 273–500 K, the solubility of metallic mercury is directly proportional to temperature. At temperatures >500 K (marked with an arrow) the curve deviates from the linear ln*m*–*T* relation. The solubility of metallic mercury in water, expressed in mole fractions (x_{Hg}), is described by the equation[84]

$$\log x_{Hg} = -147.56 + 5581.3/T + 48.7231 \log T \quad (12.5)$$

where T is the temperature (K). It has been reported that the solubility of metallic mercury in water is $(3.0 \pm 0.1) \times 10^{-7}$ mol dm^{-3} (0.602 mg L^{-1}) at 298 K.[85] It has been determined that the solubility of metallic mercury in water decreases in the presence of salts of inorganic compounds.[1] A systematic analysis of the solubility of metallic mercury in aqueous solutions of electrolytes (NaCl, NaOH) has been performed.[86–89]

Figure 2.1 also illustrates that mercury enters water as a result of industrial activities and ablation from the Earth's crust. Mercury minerals demonstrate a certain, albeit low, solubility in water. The solubility product of mercury sulfide is small and equal to 2×10^{-49} at 283 K,[3] while the solubility at 291 K is 1.25×10^{-6}%.[90] The solubility products of HgS, HgSe and HgTe[91] decrease in that order, with values of 1.4×10^{-45}, 2.4×10^{-61} and 1.0×10^{-64}, respectively. Thus, the solubility of mercury in water decreases with the transition from sulfides to tellurides. Therefore, being a trace element, mercury is omnipresent under natural conditions but its concentrations are normally very low. It has been established that in Nature inorganic forms of mercury chalcogenides are converted into metallic mercury under the action of enzymes of anaerobic bacteria:

$$\mathrm{HgS(Se, Te)} \rightleftharpoons \mathrm{Hg^{2+}} \rightleftharpoons \mathrm{Hg_2^{2+}} \rightleftharpoons \mathrm{Hg^0} + \mathrm{S^{2-}}(\mathrm{Se^{2-}}, \mathrm{Te^{2-}}) \quad (12.6)$$

It is also proven that divalent mercury ions are converted into metallic mercury under the action of *Pseudomonas* bacteria in aerobic conditions.[92] The transition of mercury chalcogenides into metallic mercury may also take place as a result of electrochemical reactions, given a favorable mercury potential. The reduction may occur as follows:

$$\mathrm{HgX} + 2e \rightarrow \mathrm{Hg^0} + \mathrm{X^{2-}} \quad (12.7)$$

where $X^{2-} = S^{2-}$, Se^{2-}, Te^{2-}. The opposite is also true, given the appropriate oxidation potential of mercury, as seen from the equation

$$E = 0.850 + \frac{RT}{2F} \ln\left(\frac{[\mathrm{Hg^{2+}}]}{\alpha}\right) \quad (12.8)$$

According to Jensen and Jernelov,[92] mercury will dissolve if α is greater than 1021:

$$\mathrm{Hg^0} \rightleftharpoons \mathrm{Hg^{2+}} + 2e \quad (12.9)$$

Redox reactions and aerobic/anaerobic bacteria result in the formation of metallic mercury and its inorganic and organic compounds, which circulate in nature. Ultimately, up to 5000 t per year of metallic mercury and its compounds are deposited by natural causes (volcanoes, *etc.*) into the oceans.

12.6 Mercury Pollution

In recent years, another 5000 t per year of mercury have been added to the above amount as a result of anthropogenic (man-made) activities.[3] These mercury 'loss' figures amount to half of the world's industrial production of mercury, which has reached between 5000 and 10 000 t per year.[2,3,6,39,40]

Mercury-based industrial processes involve huge losses of the metal. According to an estimate,[3] the industrial production of chlorine and alkali alone has accounted for 10^6 t of mercury losses. For example, an electrochemical chlorine facility in Ontario lost 15 kg of mercury per day, which added up to 100 t of mercury lost over 20 years.[3] In addition, the world's oceans have accumulated around 50×10^6 t of mercury as a consequence of erosion, underwater volcanic activity and anthropogenic activity. The concentration of mercury in ocean water is 3×10^{-5} mg L^{-1}. Consequently, atmospheric and hydroerosive processes result in the 'spreading' of mercury over the planet. Due to evaporation, mercury becomes airborne at a concentration of 20 ng m^{-3}, which is equivalent to 20 ng cm^{-2} of the Earth's surface.[84] Natural mercury concentrations range from 3 to 9 ng m^{-3}. Near mercury factories, concentrations may reach 50 mg cm^{-2} of the Earth's surface.[93] The airborne concentration of mercury over the Mazatzal Mountain pit in Arizona is 20 μg m^{-3}. This mercury vapor concentration is far below the saturation limit.

In Chapter 1, we summarized data that measure the vapor pressure of mercury. It was seen that vapor pressure increases exponentially with increase in temperature. The airborne concentration of mercury at room temperature at its saturation point lies in the range 10–15 mg m^{-3}. Such high airborne concentrations of mercury occur only inside closed spaces, *e.g.* a laboratory room containing spilled or splashed mercury with a large evaporation area. The maximum allowable airborne concentration of mercury accepted in most countries, including the USA, is 100 μg m^{-3}, and in CIS countries it is 10 μg m^{-3} and for alkylmercury compounds 5 μg m^{-3}.[94,95] The maximum allowable mean daily airborne concentration of mercury in populated areas is 0.3 μg m^{-3}.[95]

The mercury concentration in rain water was stated to be 200 ng L^{-1}[8] in one report[4] and 50–500 ng L^{-1} in another.[96] The mercury concentration in snow is 0.07–0.21 ng L^{-1},[97] but in the vicinity of mercury facilities these snow concentration levels may be exceeded 100–1000-fold.[97] Based on the average mercury content in snow and rain water, it was found that every year atmospheric precipitation enriched with mercury from various sources, including natural sources, deposit around 100 000 t of mercury to the surface of the Earth.[98] It should be noted that burning fossil fuels containing considerable quantities of mercury (brown coals 2.5×10^{-6}%, anthracites 2.7×10^{-4}%, petroleum products $1.9–21.0\times10^{-4}$% contributes substantially to mercury emissions into the atmosphere.[3]

The oceans receive mercury mostly in the form of the Hg^{2+} ion, which then reacts with organic substances and anaerobic microorganisms to become the toxic methylmercury (CH_3Hg^+) and dimethylmercury [$(CH_3)_2Hg$].[3] It is precisely the anaerobic and aerobic microorganisms that transform the inorganic mercury compounds into the extremely harmful organic mercury compounds – methylmercury, dimethylmercury, *etc.* According to Stock and Cucuel,[96] ocean water has a mercury content of about 30 ng L^{-1}. The mercury content in ocean water depends on depth. It is 60–240 ng L^{-1} at a depth of 500 m and 150–270 ng L^{-1} at 3000 m.[3] Methylmercury, just like dimethylmercury, easily dissolves in water and quickly infiltrates aquatic organisms. The mercury concentration in freshwater fish is 76–167 ng g^{-1}, while fish from various sources in Sweden, Finland, Norway and Switzerland contain 60–2500 ng g^{-1} of mercury.[93] Notably, mercury contained in fish exhibits bioaccumulation with an enrichment factor of 3000–9200.[93] Ocean fish contain less mercury. Mercury is also present in the meat of domestic grazing animals (sheep, cows, *etc.*) and fowl. Consequently, mercury enters the human body through the food chain. The actual quantity of mercury entering the human body depends on the person's living environment and the type of food. The migration of mercury and its transport within the environment through water, air and plankton stimulated by industry and plants is illustrated in Figure 12.1. The mercury migration chain closes with humans. It is important to note that in the hydrosphere the pollution effect is increased owing to the ability of the biosphere to concentrate mercury and other microelements to thousands or millions of times the levels of the surrounding water environment.

Ostroumov *et al.*[99] studied the mercury content of clams in the ocean ecosystems. They demonstrated that clams contain mercury at levels of 133–217 ng g^{-1} of the dry weight of the soft tissue. As mussels grow older, the mercury content increases to 2.0–5.8 μg g^{-1}, *i.e.*, 15–26-fold. The microgram levels of mercury in mussels indicate the bioaccumulation effect demonstrated by organisms in the oceans. Owing to bioaccumulation, the natural levels of mercury in some organisms are close to the threshold safety level of 0.5×10^{-4}%.[3] Thus, commercially produced fish coming from near to industrial regions on average contain 0.5×10^{-4}% of mercury, which is exactly the threshold safety level. It has been demonstrated in the early 1980s that the high mercury content in sea fish had hardly changed over the past century: 0.95×10^{-4}% in the meat of tuna caught in 1878 and 1909 and 0.91×10^{-4}% in commercial species caught in 1981.[3] Therefore, the problem of mercury poisoning depends not only on the mercury concentration in fish, but also on the amount of fish consumed. The human body receives 20–50 μg of mercury daily.[100] According to Masters,[6] the temporarily allowable total weekly dose is 0.3 mg for mercury and 0.2 mg for methylmercury.[3,4,101,102]

12.7 Environmental Mercury

The release of mercury and its organic compounds into water basins, rivers and seas may lead to environmental disasters. From 1932 to 1968, the Chisso chemical plant, located on Kyushu island off the coast of Japan, dumped 600 t of mercury in the form of methylmercury and other organomercury compounds into the Shiranui Sea and Minamata Bay, where it was absorbed by shellfish, plankton and microorganisms. These microorganisms were consumed by small fish, which in turn became food for larger fish. Thus the biological mercury transformation chain was transferred to humans, as shown in Figure 12.1. The high concentration of organic mercury compounds in the waters of Minamata Bay and the Shiranui Sea led to high mercury contents in fish and clams used for food by the local population, and the inhabitants of Kyushu Island were stricken by a formerly unknown disease dubbed 'Minamata disease.' The victims suffered from disruption of the central nervous system, which presented itself variously as psychiatric disorders, even insanity, loss of coordination, loss of pain sensitivity, loss of hearing, eyesight and speech and convulsions leading to torpor and coma.[3,4,82,100] Among the approximately 2500 victims, the mortality rate was 32.8%.[100]

The health impact of metallic mercury is based on oxidation reactions producing Hg_2^{2+} and Hg^{2+} ions, which, in the presence of chlorine ions, take part in exchange reactions leading to the production of calomel [mercury(I) chloride] and mercury(II) chloride. The toxicity of inorganic mercury compounds depends on their solubility in water, blood and gastric juice. Owing to the high toxicity of mercury, according to the World Health Organization, the maximum allowable quantity of mercury consumed per person per day must not exceed 0.3 mg, including no more than 0.2 mg of methylmercury.[3]

The total mass content of mercury in the human body is $1\times10^{-6}\%$.[99] Symptoms of mercury poisoning occur at a body burden level of $2–6\times10^{-5}\%$, which corresponds to 14.0–42.0 mg of mercury for an individual weighing 70 kg. Mercury is mostly concentrated in the kidneys and less so in the liver.

The symptoms of Minamata disease correspond to the properties of the mercury compounds. The action of mercury depends on the nature of its compounds. Methylmercury administered into the body quickly enters the bloodstream and brain tissue, destroying the cerebellum and the brain cortex and thereby leading to a loss of spatial orientation and partial loss of eyesight. Inorganic mercury compounds are also highly toxic; however the effects of toxicity depend on the nature of the mercury consumed. Prolonged inhalation of mercury vapor at a concentration of $0.6–2\,mg\,m^{-3}$ exposes the human body to a macro-concentration of mercury with a highly developed (at the molecular level of Hg_2) surface, thereby leading to acute intoxication, which is a forerunner of chronic poisoning. Acute mercury vapor poisoning is marked by a disturbance of calcium metabolism, modification of blood proteins and mercury accumulation in the liver, kidneys, brain and spleen, which in turn suffer actual damage. Chapter 14 discusses the symptoms and effects of acute metallic mercury poisoning in more detail.

Mercury is a material of hazard class I.[82] The threshold concentration of mercury affecting the functional capacity of the central nervous system is

2–5 mg^{-3},[96] while the maximum allowable mean daily airborne concentration of mercury for populated areas is 0.3 mg m^{-3}.[98] The mercury MAC for potable water is 0.001 mg-L^{-1} according to international standards, 0.002 mg-L^{-1} according to US standards and 0.005 mg L^{-1} for utility facilities.[82] Japanese law allows no mercury content in domestic potable water reservoirs or wastewater. European standards limit the mercury concentration in potable water to 0.01 mg L^{-1}.[82]

Nevertheless, despite centuries of accumulated knowledge about the toxic effect of mercury and its compounds on warm-blooded animals and humans, the environmental issue of mercury control and pollution prevention was first addressed only in the 1950s and escalated in 1984–2007.[6,46,54–59,103–108]

The following factors are of extreme importance: promotion of industrial use of recycled mercury, improving the culture of handling mercury-containing devices (medical thermometers, technical thermometers, electrochemical batteries, fluorescent lamps, *etc.*) in households and the development of green technologies for both the production and utilization of mercury in industry, agriculture, science and medicine.[6,46,54–59,103–108]

12.8 Mercury Detection by Atomic Fluorescence Spectrometry

Many methods of mercury detection are now available.[109] Of the many analytical methods available, cold vapor atomic absorption spectrometry (CVAAS), inductively coupled plasma mass spectrometry (ICP-MS), plasma atomic emission spectrometry (plasma AES) and cold vapor atomic fluorescence spectrometry (CVAFS) are in widespread use. They can be used to determine mercury in water at the picogram level. CVAAS has detection limits of 0.01–1 ng g^{-1}, ICP-MS 0.01 ng g^{-1} and CVAFS 0.001–0.01 ng g^{-1}.[109,110] A brief description of atomic fluorescence spectrometry (AFS) is given below.

12.8.1 Atomic Fluorescence Spectrometry

One of the most sensitive methods for measuring environmental mercury is AFS. With this technique, it is possible, with appropriate preparative methods, to measure mercury in water, soil and air. With cold vapor preconcentration of mercury, CVAFS is a well-known analytical method where many improvements have been made to automate and improve the sensitivity and limit of detection. One advantage of AFS over atomic absorption spectrometry (AAS) is the ability to use more intense light sources. AFS has a larger linear dynamic range, higher sensitivity and less interference than AAS.[111] Of these two methods, CVAFS is the more sensitive. The fundamentals of AFS have been reviewed by Morita *et al.*[112]

12.8.2 Application of CVAFS for the Determination of Mercury in Water

CVAFS is a widely used and powerful analytical method capable of measuring mercury in water at levels of $<10^{-9}\,g\,L^{-1}$ ($<1\,ng\,L^{-1}$). To be successful, the method described below requires significant skill, particularly with sample collection and handling. Mercury determination, according to US EPA Method 1631E,[113] focuses on monitoring waste effluents at the lowest EPA Water Quality Certification (WQC) levels. Method 1631E has a prescribed analytical range for mercury of 0.5 ng L^{-1}–100 ng L^{-1}.

Samples are first oxidized using a solution of bromine monochloride (BrCl) in order to liberate and transform the mercury present into its inorganic (Hg^{2+}), water-soluble form. Once the samples have been completely oxidized, they are then treated with hydroxylamine hydrochloride ($NH_2OH{\cdot}HCl$), a mild reducing agent, to destroy free halogens. Samples are then further reduced with stannous chloride solution ($SnCl_2$), which transforms mercury into its elemental and volatile state that can be easily removed from solution. The elemental mercury is transferred to the gaseous phase and is then collected on a gold trap to isolate and concentrate the mercury. Subsequently, the gold trap is heated to liberate elemental mercury and the mercury is transferred through an optical cell for detection *via* AFS.

Figure 12.2 shows a schematic diagram of an older CVAFS system without the manual purging apparatus, taken from EPA Method 1631E.[113] The lamp in older

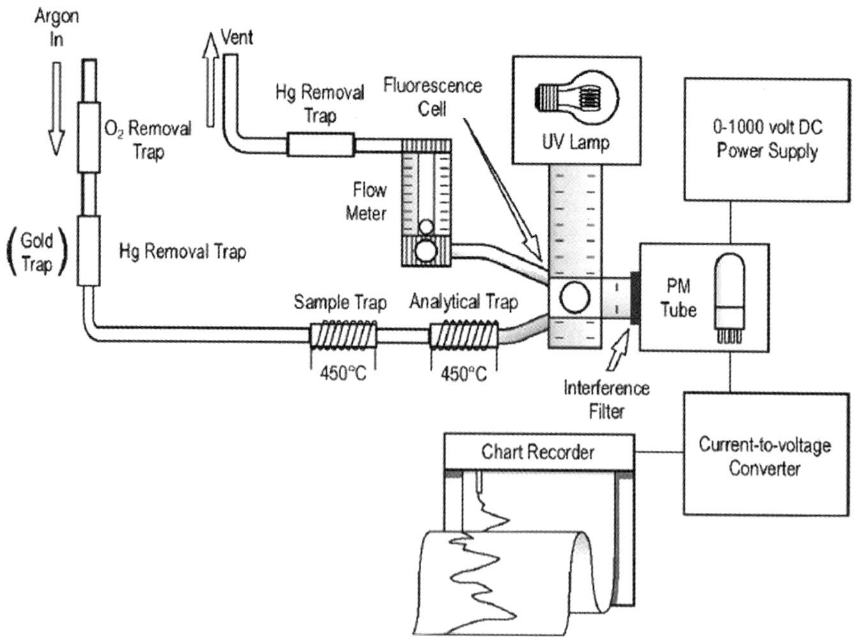

Figure 12.2 Schematic diagram of the manual AFS analyzer without the manual purging apparatus.
Reproduced with permission from Ref. 113.

systems would have emitted a diffuse spectrum and so an interference filter was needed to limit the light source to 254 nm radiation. In more modern systems, a mercury discharge lamp is used since it has a narrow emission spectrum around 254 nm. The chart recorder output gives time on the *x*-axis and absorbance on the *y*-axis. In modern systems, software is used instead of a chart recorder, but the output will still be a similar portrayal of absorbance *versus* time. The part labeled 'Sample Trap' is a gold trap used to collect mercury from the aqueous sample. The sample is purged to get the mercury out of solution and on to the gold trap. This trap is then placed in the analyzer, where it is heated to release the mercury on to the part labeled 'Analytical Trap'. The 'Analytical Trap' is a second gold trap that is then heated to release the mercury for analysis. In this setup, the first gold trap can be one of multiple different gold traps for multiple samples, each collected on a different trap. All samples are desorbed on analytical traps. Multiple gold traps help increase the consistency from sample to sample.

Figure 12.3 shows the peak responses for a typical calibration curve. Table 12.3 shows typical quality control figures for calibration data that were

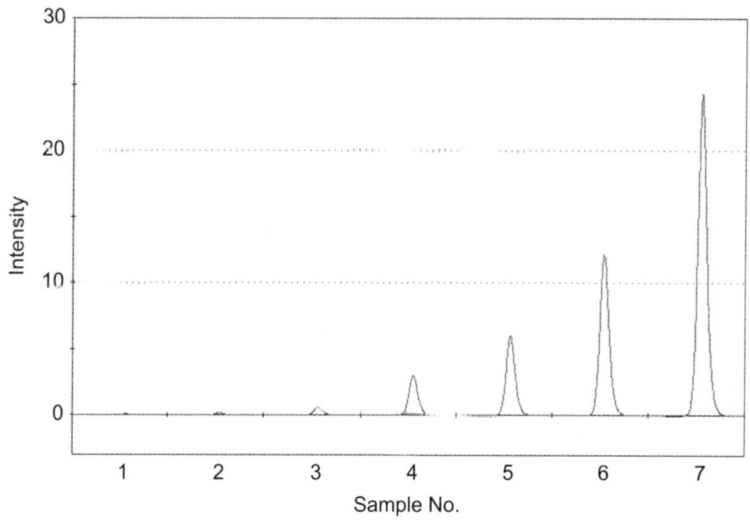

Figure 12.3 Calibration peaks from blank to 50 ppt Hg for mercury determination in water. Courtesy of Nippon Instruments North America.

Table 12.3 Typical control chart from data collected on a commercial AFS instrument.

No.	*Quality control checks*[a]	*Acceptance criteria*	*Result*	*Pass*
1	Bubbler blank average	0.5 ng L^{-1}	0.093 ng L^{-1}	OK
2	Bubbler blank standard deviation	0.1 ng L^{-1}	0.012 ng L^{-1}	OK
3	Calibration factor RSD	15%	2.56%	OK
4	Calibration factor recovery	75–125%	100.49%	OK
6	Method blank	0.5 ng L^{-1}	0.137 ng L^{-1}	OK
7	OPR	77–123%	107.10%	OK

[a]RSD, relative standard deviation; OPR, ongoing precision and recovery.

collected on a commercial instrument. Substances that produce very stable complexes with mercury ions may interfere with the reduction of the ions to the elemental form. Complexes of bromides, iodides, cysteine and sulfide, thiosulfate and Se(IV) have been reported to cause interference[112] in the determination of mercury by CVAFS unless they are decomposed before reduction.

References

1. L. F. Kozin, *Physicochemistry and Metallurgy of High-Purity Mercury and Its Alloys*, Naukova Dumka, Kiev, 1992.
2. V. Shtrube, *Avenues for Development of Chemistry*, Mir, Moscow, 1984, vol. 1.
3. J. O'M Bockris (ed.), *Environmental Chemistry*, Springer US, 1977, pp. 1–18.
4. I. M. Trachtenberg and M. N. Korshun, *Mercury and Its Compounds in the Environment*, Vysha Shkola, Kiev, 1990.
5. G. T. Petrovskiy, Yu. V. Popov, A. A. Berezhnoy and I. V. Semeshkin, *Materials and Processes in Space Technology*, Nauka, Moscow, 1980, pp. 21–27.
6. H. B. Masters, *Metals and Mining Annual Review 1997*, Mercury. Mining J. Ltd., 1997.
7. A. G. Morachevskii, V. B. Busse-Machukas and I. A. Shesterkin, *Natrii: Svoistva, proizvodstvo, primenenie (Sodium: Properties, Manufacture, and Use)*, St. Petersburg, Khimiya, 1992.
8. V.-S. S. Zaretskas and V. L. Ragulskene, *Mercury Switching Components for Devices and Automating Control*, Energiya, Moscow, 1971, p. 105.
9. A. P. Tomilov and L. G. Feokstistov (eds), *Electrochemistry of Organic Compounds*, Mir, Moscow, 1973, p. 733.
10. L. F. Kozin, A. G. Morachevskiy and I. A. Sheka, *Ukr. Khim. Zh.*, 1989, **55**, 495.
11. L. F. Kozin and A. G. Morachevskiy, *Physicochemistry and Metallurgy of High-Purity Lead*, Metallurgiya, Moscow, 1991.
12. L. F. Kozin and S. V. Volkov, *Chemistry and Technology of High-Purity Metals and Metalloids. Vol. 1, Chemical and Electrochemical Methods for Ultrapurification*, Naukova Dumka, Kiev, 2002, p. 542.
13. L. F. Kozin, Kinetics and electrode processes in aqueous solutions, in *Collection of Scientific Papers of the Institute of General and Inorganic Chemistry of Ukrainian Academy of Sciences*, Naukova Dumka, Kiev, 1983, pp. 37–63.
14. Method for refining aluminum alloys, *USSR Pat.*, 199 038.
15. I. N. Maslenitskiy and L. V. Chugayev, *Metallurgy of Precious Metals*, Metallurgiya, Moscow, 1972.
16. M. J. Richardson, *Min. J.*, 1991, **317**, Supplement, 6–7.
17. V. I. Lebedev, N. V. Burdin, V. A. Metkin and T. Kh. Samdanchap, *Gorny Zh.*, 1997, **2**, 41.
18. C. Washburn and E. Hill, *JOM*, 2003, **55**, 45.

19. K. V. Fedotov, A. E. Senchenko and A. A. Romanchenko, Abstracts from presentations at the I. N. Plaksin Jubilee, Moscow, 10–14 October 2000, National Mining Research Center – Institute of Mining, 2000, pp. 253–254.
20. A. A. Bakibaeva and A. G. Stromberg (eds), *Fortieth Anniversary of Tomsk Electrochemical School and Problem-Solving Laboratory of Microimpurities (1962–2002)*, Publishing House of Tomsk Polytechnic University, Tomsk, 2004.
21. A. G. Stromberg and A. A. Kaplin, *Proc. Siber. Branch USSR Acad. Sci., Chem. Sci. Ser.*, 1975, **2**, 58.
22. Ya. Geyrovskiy and Ya. Kuta, *Basic Polarography*, Mir, Moscow, 1965.
23. Ya. P. Stradin' and S. G. Mairanovskiy (eds), *Polarography: Problems and Prospects*, Zinatne Publishing House, Riga, 1977.
24. T. A. Kryukova, S. I. Sinyakova and T. V. Aref'eva, *Polarographic Analysis*, Goskhimizdat, Moscow, 1959.
25. A. M. Bond, *Polarographic Methods in Analytical Chemistry*, Khimiya, Moscow, 1983.
26. *Jpn. Ind. Chem. Soc.*, 1965, **68**, 1248.
27. H. Lund and O. Hammerich (eds), *Organic Electrochemistry*, 4th edn, Marcel Dekker, New York, 2001.
28. A. M. Bond, R. T. Gettar and N. M. McLachiam, *Inorg. Chim. Acta*, 1989, **166**, 279.
29. M. F. Francine and B. M. Marcelo, *Hydrometallurgy*, 2007, **87**, 83.
30. J. Wu and H. Zhao, in *High Temperature Superconductors*, R. Bhattacharya and M. Paranthaman, Wiley-VCH, Weinheim, 2010, p. 153.
31. M. Malhotra, M. Gautam, J. K. Radhakrishnan, V. Kapoor, S. Verma, U. Kumar, A. Kumar, G. Gupta, A. Goyal and S. Sitharaman, *Bull. Mater. Sci.*, 2005, **28**, 97.
32. M. V. Kovalenko, E. Kaufmann, D. Pachinger, J. Roither, M. Huber, J. Stangl, G. Hesser, F. Schäffler and W. Heiss, *J. Am. Chem. Soc.*, 2006, **128**, 3516.
33. G. G. Devyatyh, A. N. Moiseev, A. P. Kotkov, *et al.*, in Nanotechnology and Photonic Crystals, Proceedings of the 2nd Inter-Regional Seminar, Kaluga, 15–17 March 2004, ed. N. E. Bauman, Publishing House of Moscow State Technical University, Kaluga, 2004, pp. 249–252.
34. I. Kostov and I. Mincheva-Stefanova, *Sulfide Minerals*, Mir, Moscow, 1984, p. 280.
35. I. V. Vinarov, *Tsvet. Met.*, 1992, **3**, 46.
36. K. K. Lepesov, V. S. Zaretskas, L. F. Kozin, A. K. Lobster T. Kurdyumova, N. A. Baimbetov, L. G. Bastin and A. Zubkova, Method for amalgamation of metals and alloys that exhibit weak interaction with mercury, *USSR Pat.*, 1 133 311.
37. F. C. Lu, P. E. Berteau and D. J. Clegg. The toxicity of mercury in man and animals. *Mercury contamination in man and his environment. International Atomic Energy Agency*, Vienna, 1972, 67–85.

38. H. G. Seiler and H. Sigel (eds), *Handbook on Toxicity of Inorganic Compounds*, Marcel Dekker, New York, 1988, pp. 419–436.
39. J. K. Pietrowski and D. O. Coleman, *Environmental Hazards of Heavy Metals, Summary Evaluation of Lead, Cadmium and Mercury, London, Monitoring and Assessment*, Research Centre, 1980.
40. L. D. Hylander and M. Meili, *Sci. Total Environ.*, 2003, **304**, 13.
41. K. Kutzsche, *Neue Hütte*, 1989, **34**, 189.
42. (a) L. F. Kozin, Private communication; (b) M. Simon, P. Jönk, G. Wühl-Couturier and S. Halbach, *Mercury, Mercury Alloys and Mercury Compounds*, Ullmann's Encyclopedia of Industrial Chemistry, Wiley-VCH, Weinheim, 2006.
43. UNEP, Ad Hoc Open-Ended Working Group on Mercury, Bangkok, 12–16 November 2007.
44. Methods of mercury disposal: new methods, *Ecol. Bull. Russ.*, 2007, **3**, 8.
45. L. F. Kozin, *Amalgam Metallurgy*, Tekhnika, Kiev, 1970.
46. A. D. Akhmetov and V. A. Bednenko, Mercury pollution of the environment: emissions into the atmosphere, restoration of territories, impact on health, in *Proceedings of the International Seminar, Astana, 28 May–1 June 2007, Abstracts*, 2007, pp. 18–19.
47. M. H. Mendelsohn and H. S. Huang, *US Pat.*, 5 900 042, 1999.
48. G. B. Budkevich, V. Ya. Momot and I. I. Sirenko, *et al.*, *Zh. Prikl. Khim.*, 1974, **47**, 2217.
49. G. B. Budkevich, V. J. Momot, I. I. Sirenko, J. A. Tarasenko and I. A. Sheka, *Ukr. Khim. Zh.*, 1974, **40**, 364.
50. G. B. Budkevich, V. Y. Momot, I. I. Sirenko and Y. A. Tarasenko, Method for mercury removal from solutions, *USSR Pat.*, 1972, 382 720.
51. Mercury removal from sulfuric acid solutions, *Jpn. Pat. Appl.*, 56-48588.
52. Removing mercury from concentrated sulfuric acid solutions, *Jpn. Pat. Appl.*, 60-131809.
53. K. Reppenhagen and H. Winkler, Removing mercury from spent sulfuric acid, *Ger. Pat. Appl.*, 1987, 3 616 792.
54. T. D. Radzhabov, Method for mercury vapor removal from mercury-containing gases, *USSR Pat.*, 1991, 1 698 306.
55. V. E. Hrapunov, R. A. Isakova and L. S. Chelohsaev, *Tsvet. Met.*, 1988, **3**, 29.
56. I. A. Sheka, V. A. Trihlev, I. I. Sirenko and Y. A. Tarasenko, Method for mercury removal from gas, *USSR Pat.*, 1976, 536 838.
57. Quesksilberabscheidung aus Gasstromen, *East Ger. Pat.*, 257 594.
58. V. Marousek, P. Novak, M. Hejtmanek and J. Kalal, Extraction of mercury from solutions and air, *Czech. Pat.*, 1986, 235 768.
59. V. N. Kobahidze, V. V. Kobahidze and M. G. Zaitsev, Method for mercury removal from mercury-containing gases, *USSR Pat.*, 1973, 382 719.
60. I. I. Sirenko, I. A. Sheka and Yu. A. Tarasenko, *Ukr. Khim. Zhurn.*, 1974, **40**, 686.
61. Method for mercury recovery, *USSR Pat.*, 405 308.

62. J. Markovs, R. T. Maurer, A. S. Zarchy and E. S. Holmes, *US Pat.*, 5 354 357, 1994.
63. Russian Federation Patent 2,327,536 Method for disposal of mercury-containing wastes.
64. C. R. Palmer and R. G. Sandberg, *Precious Metals: Mining, Extraction and Processing, Proceedings of International Symposium, AIME Annual Meeting, Los Angeles, California, 27–29 February 1984*, TMS, Warrendale, PA, 1984, pp. 227–239.
65. Mercury sorbent recovery, *Fr. Pat. Appl.*, 2 615 756.
66. R. G. Sandberg, *US Bureau of Mines Circular*, No. 9059, US Department of the Interior, Washington, DC, 1986, pp. 19–25.
67. V. F. Kamenev and L. V. Fadeeva, *Tsvet. Met.*, 1985, **10**, 49.
68. K. Inoue, Y. Bada M. Goto, *et al.*, in *Proceedings of the First International Conference on Hydrometallurgy, Beijing, 1988, ICHM 88*, Oxford, 1989, pp. 571–575.
69. I. Dodreusky, A. Dimin, V. Nenov, *et al.*, in *Proceedings of the First International Conference on Hydrometallurgy, Beijing, 1988, ICHM 88*, Beijing, Oxford, 1989, p. 666.
70. V. F. Fedonina, M. O. Lishevskaya and N. S. Torocheshnikov, *et al.*, *Khim. Prom.*, 1972, **1**, 72.
71. L. E. Postolov, T. E. Mitchenko and V. G. Ovchinnikov, *Khim. Tekhnol.*, 1992, **182**, 15.
72. L. E. Postolov, T. E. Mitchenko and V. A. Skrypnik, *et al.*, *Khim. Prom.*, 1984, **136**, 35.
73. S. M. Rustamov, I. I. Zeinalova, F. T. Makhmudov and Z. Z. Bashirova, *J. Water Chem. Technol.*, 1993, **15**, 41.
74. M. W. Grossman and W. A. George, *US Pat.*, 4 879 010, 1989.
75. S. P. Kounaves and J. Buffle, *J. Electrochem. Soc.*, 1986, **133**, 245.
76. Method for mercury recovery from solutions, *USSR Pat.*, 1 668 483.
77. B. P. Giri, *US Pat.*, 6 221 128, 2001.
78. D. L. Ball and D. A. D. Baoteng, *US Pat.*, 5 013 358, 1991.
79. Method for mercury recovery, *USSR Pat.*, 195 112.
80. A. V. Malkov, N. P. Tarasov and J. Belloni, *All-Union Conference on Theoretical and Applied Radiation Chemistry, Obninsk, 23–25 October 1990, Abstracts of Reports*, Moscow, 1990, pp. 184–185.
81. M. E. Goldstone, C. Atkinson, P. W. W. Kirk and J. N. Lester, *Sci. Total Environ.*, 1990, **95**, 271.
82. M. Ya. Grushko, *Harmful Inorganic Compounds in Industrial Effluents*, Khimiya, Leningrad, 1979, p. 160.
83. G. E. Kislinskaya and L. I. Fedoryako, *Khim. Tekhnol.*, 1989, **2**, 23.
84. M. Iwamoto, S. N. Jonson and H. Miyamoto, *Mercury in Liquids Compressed Gases, Molten Salts and Other Elements*, Solubility Data Series, ed. A. S. Kertes and H. L. Clever, Pergamon Press, New York, 1986, vol. 29.
85. S. Okouchi and S. Sasaki, *Bull. Chem. Soc. Jpn.*, 1981, **54**, 2513.
86. V. K. Chviruk and N. V. Koneva, *Ukr. Khim. Zh.*, 1975, **41**, 1162.

87. A. A. Shokol and L. F. Kozin, *Ukr. Khim. Zh.*, 1959, **25**, 374.
88. V. K. Chviruk and N. V. Koneva, *Zh. Prikl. Khim.*, 1978, **51**, 743–746, 747–751.
89. V. K. Chviruk, N. V. Koneva and A. Yu. Noel', *Zh. Prikl. Khim.*, 1974, **47**, 2546.
90. F. G. Gavryuchenkov, M. I. Kurochkin, A. A. Potekhin and V. A. Rabinovich, *Chemistry Reference Book, translated from German*, Khimiya, Leningrad, 1975. p. 576.
91. R. A. Lidin, L. L. Andreeva and V. A. Molochko, *Constants of Inorganic Substances*, Drofa, Moscow, 2006.
92. S. Jensen and A. Jernelov, *Mercury Contamination in Man and His Environment*, Technical Report Series, No. 137, International Atomic Energy Agency, Vienna, 1972, pp. 43–66.
93. A. V. Holden, *Mercury Contamination in Man and His Environment*, Technical Report Series, No. 137, International Atomic Energy Agency, Vienna, 1972, pp. 143–168.
94. N. V. Lazarev and I. D. Gadaskina (eds), *Harmful Substances in Industry, Inorganic and Organoelement Compounds*, Khimiya, Leningrad, 1977.
95. Ya. M. Grushko, *Harmful Inorganic Compounds in Industrial Emissions into the Atmosphere*, Khimiya, Leningrad, 1987.
96. A. Stock and F. Cucuel, *Naturwissenschaften*, 1934, **6**, 390.
97. V. N. Kumok, O. M. Kuleshova and L. A. Karabin, *Solubility Products*, Nauka, Siberian Branch, Novosibirsk, 1983, p. 266.
98. F. P. W. Winteringham, *Mercury Contamination in Man and His Environment*, Technical Report Series, No. 137, International Atomic Energy Agency, Vienna, 1972, pp. 1–3.
99. S. A. Ostroumov, S. D. Hushkhvatova, V. N. Danilov and V. V. Ermakov, *Environ. Chem.*, 2008, **17**, 84.
100. V. G. Huhryansky, A. Ya. Tsyganenko and N. V. Pavlenko, *Chemistry of Biogenic Elements*, Vysha Shkola, Kiev, 1990.
101. R. Fuge, N. J. G. Pearce and W. T. Perkins, *Nature*, 1992, **357**, 369.
102. J. U. Skaare, N. H. Markussen, G. Norheim, S. Haugen and G. Holt, *Environ. Pollut.*, 1990, **66**, 309.
103. *The Maximum Permissible Concentration (MPC) of Pollutants in the Atmosphere of Inhabited Areas*, approved by Chief State Medical Officer of the USSR, 27 August 1984, No. 3086-84.
104. A. Reiner, *Neue Hütte*, 1985, **30**, 337.
105. Method for mercury removal from gases, *USSR Pat.*, 1 161 157.
106. C. O. Gale and E. L. Hill, Apparatus and method for continuous retorting of mercury from ores and others mercury contaminated materials, *US Pat.*, 6 268 590, 2001.
107. A. M. Zykov, S. N. Anichkov, S. P. Kolesnikov and E. V. Shuvalov, Mercury emissions into the atmosphere resulting from combustion of coals at Russian thermal power stations, in *Proceedings of the International Seminar, Astana, 28 May–1 June 2007, Abstracts*, 2007, pp. 30–31.

108. A. G. Kocharyan, I. P. Lebedeva and E. M. Gusev, *Environ. Chem.*, 2008, **17**, 207.
109. M. Horvat, in *Dynamics of Mercury Pollution on Regional and Global Scales*, ed. N. Pirrone and K. R. Mahaffey, Springer, New York, 2006.
110. W. F. Fitzgerald and G. A. Gill, *Anal. Chem.*, 1979, **51**, 1714.
111. W. L. Clevenger, B. W. Smith and J. D. Winefordner, *Crit. Rev. Anal. Chem.*, 1997, **27**, 1.
112. H. Morita, H. Tanaka and S. Shimomura, *Spectrochim. Acta*, 1995, **50B**, 6944.
113. US Environmental Protection Agency, *Method 1631, Revision E: Mercury in Water by Oxidation, Purge and Trap, and Cold Vapor Atomic Fluorescence Spectrometry*, 2002, http://water.epa.gov/scitech/methods/cwa/metals/mercury/upload/2007_07_10_methods_method_mercury_1631.pdf (last accessed 10 March 2013).
114. US Environmental Protection Agency, *Mercury Study Report to Progress, Volume III: Fate and Transport of Mercury in the Environment*, December 1997, http://www.epa.gov/ttn/oarpg/t3/reports/volume3.pdf.

CHAPTER 13

Demercurization Processes in Different Sectors of Industry

LEONID F. KOZIN,[a] STEVE C. HANSEN[b] AND NIKOLAI F. ZAKHARCHENKO[a]

[a] V. I. Vernadsky Institute of General and Inorganic Chemistry, National Academy of Sciences of Ukraine, Kiev, Ukraine; [b] Consultant

13.1 Introduction

In this chapter, special attention is given to methods of demercurization of chlor-alkali plants, recycling of spent fluorescent lamps, decontamination of industrial wastewater and demercurization of contaminated spaces to reduce airborne mercury levels dramatically to acceptable levels. Mercury forms mono- and divalent compounds. The monovalent compounds are poorly soluble in water, whereas divalent mercury compounds are characterized by high solubility (with the exception of mercury sulfide). Mercury compounds are mostly unstable and decompose under the influence of heat and some even under the action of light. Mercury forms numerous complexes with organic molecules, and also with inorganic ions. The properties of mercury compounds – the ability to dissolve in water and other environments, resistance to thermal stresses – are important when determining the method of chemical demercurization.[1]

13.2 Demercurization of a Chlor-Alkali Plant

An example of demercurization of a chlor-alkali plant is the work done at the Pavlograd Khimprom (Chemical Industry) in the city of Pavlograd in

Mercury Handbook: Chemistry, Applications and Environmental Impact
By Leonid F Kozin and Steve Hansen

Published by the Royal Society of Chemistry, www.rsc.org

the Republic of Kazakhstan.[2] During the period of operation from 1975 to 1993, the mercury-cell chlor-alkali plants lost over 1000 t of mercury. The first phase of demercurization included dismantling and disposal of the processing equipment, manual collection of metallic mercury and disassembly of the mercury-contaminated production complex. Mercury infiltrated the land on which the complex stood. Therefore, the surface layer of heavily polluted soil down to a thickness of 1 m was removed to repositories and thereby isolated from the atmosphere and groundwater. Mercury extraction installations were constructed for lightly contaminated materials.

Akhmetov and Bednenko[2] noted that in the spring of 1999, when the electrolysis shop was opened, intensive evaporation of metallic mercury spills began. The plant territory was declared an emergency zone and remained as such for 2 months until complete tearing down of the electrolysis room and the completion of manual collection of 17 t of bulk mercury that had been spilled. This is less than 1.7% of the overall spilled mercury, which amounted to more than 1000 t of metallic mercury, with a present-day cost of millions of dollars.

The same fate befell the Kiev chlor-alkali production facility in the Darnitskiy district of Kiev, Ukraine. In the opinion of the authors, these figures testify to the poor level of planning at the mercury-cell chlor-alkali plant. In the electrolysis shops of these chlor-alkali plants, it was essential to provide sloping floors and ceilings (all-welded metal) impervious to mercury and with airtight traps, to develop non-wettable concrete impervious to mercury vapor and automated removal of mercury from the traps, *etc.* Enormous losses of metallic mercury in the chlor-alkali industry have brought about catastrophic effects on the ecology of dozens of square kilometers, from the surface of the land down to subsurface waters. This led to the dismantling of factories and the creation of 'eternal' landfills in which the mercury content is higher than that of mercury-bearing ores and which will forever 'breathe' poisonous mercury vapors, bringing death to warm-blooded animals and humans. It should be noted that the maximum allowable concentration (MAC) of mercury in air is 0.0003 mg m^{-3} and the maximum allowable concentration in the soil is 2.1 mg kg^{-1}. Hence it is easy to calculate what an enormous area of the Earth's surface and what amount of land (soil) and groundwater have been turned into dead zones by 1000 t of 'lost' mercury at the Pavlograd Khimprom. Moreover, the mercury-cell chlor-alkali industry is large scale and ranks third in world consumption of mercury, which, even with a considerable reduction of production in 2005, for example, consumed 500 t of mercury.[3] The composition of wastewater from chlorine and caustic soda production is as follows (mg dm^{-3}):

Hg^{20}	45–60
Hg^0	15
Na_2CO_3	0.4–0.8
NaOH	4.5–5.8
NaCl	145–10 000

Fe[20]	4.9–5.6
NaClO	20.0
Available chlorine	10.0
Suspended particles	~1000

with a p of 11–12.[4–6]

An example of 'careful' handling of mercury and amalgam is the amalgam metallurgy used to obtain high-purity metals.[7–11] In this case, sealed electrolyzers are placed on special catch trays provided for the collection of amalgams in case of theoretically possible accidents at the electrolyzers. The equipment is fabricated from Plexiglas sheet (thickness 20–40 mm) for the walls and block Plexiglas (thickness 100 mm) for the floors of electrolyzers. The operating lifetime of electrolyzers prior to the appearance of the spiderweb cracking arrangement of Plexiglas was 10 years at an operating temperature of 40–45°C. However, during operations at the high-purity metal shop at the Chimkent Lead Plant (CLP), not a single accident has occurred since 1962 in the production of high-purity metals (cadmium Kd-0000, indium In-0000, thallium Tl-0000) in electrolyzers, each containing 500–600 kg of mercury, and also high-purity lead Pb-000 and Pb-0000 and bismuth Bi-000 and Bi-0000.[8–11] Plexiglas electrolyzers were used to obtain high-purity cadmium, indium and thallium and contained 500–600 kg of mercury each, while electrolyzers used for high-purity lead and bismuth each contained 1300–1400 kg of metallic mercury in the form of amalgams. Furthermore, the content of mercury vapor in the working premises of the shops was two orders of magnitude below the MAC. Therefore, the pure metals department at the CLP, which was distinguished by an unusually high purity, was considered by factory management to be a sanatorium and therefore highly skilled craftsmen and skilled workers who had received overexposure in other lead production shops were sent there to work in the pure metals department. This example shows that large amounts of mercury can be processed without harming the environment.[8–11]

13.3 Recycling of Fluorescent Lamps

The second highest mercury-consuming industry is the manufacture of fluorescent lamps. In the production of fluorescent lamps, each of which contains 1–10 mg of mercury, up to 120–150 tons of mercury were consumed in 2005. If we consider that the operating lifetime of fluorescent lamps will not exceed 1 year owing to cracking and low mechanical durability, then it becomes clear that the production of fluorescent lamps is a source of environmental hazards.

13.3.1 Thermal Demercurization of Fluorescent Lamps

For this reason, the world is developing technologies for demercurizing discarded fluorescent lamps.[12–20] For example, let us examine the technology of the thermal demercurization of fluorescent lamps.[19] A demercurization flowsheet is shown in ref. 17.

Fluorescent lamps and the waste material from their production are comminuted and are transported for demercurization in a reactor containing a 0.96–3.43 g L^{-1} solution of elemental sulfur in propylene carbonate. This solution circulates in a closed loop at 50–100 °C. Propylene carbonate, a sulfur solvent, has the following physiochemical properties: molecular weight, 102.09 g mol^{-1}; melting temperature, 70 °C; boiling point, 240 °C; density, 1.204 g cm^{-3}; and viscosity at 25 °C, 3 $mm^2\,s^{-1}$.

The reactor used for implementing this method contains a load and crush unit, coupled with a demercurization unit and centrifuge. The centrifuge separates the mercuric sulfide formed at high speed by the reaction

$$S + Hg^{\circ} \rightarrow HgS\downarrow \tag{13.1}$$

with precipitation of HgS into a crystal sediment.

The solution proceeds from the centrifuge to sump. The working solution preparation unit is connected to a pump. The unit is equipped with a filter to dispose of released gases that contain mercury vapor.

A fluorescent lamp is placed in the load and crush unit, which is linked by an emission-preventing airlock with the demercurization unit where it is crushed, and glass fragments flow into the demercurization unit. Here, the process of binding mercury into sulfide occurs *via* mixing in the working solution. Mercuric sulfide, phosphor and fine glass dust are washed off with excess solution from the glass fragments and, together with spent solution, proceed to the sump. Glass fragments proceed to the centrifuge for final disposal of spent solution C, which also enters the sump and dried glass fragments, III, are unloaded for further processing. Clarified solution proceeds to the working solution preparation unit, where it is thermostated and saturated with sulfur. Then working solution A is fed by a circulation pump into the demercurizer and is discharged into a gas filter which connects the load and crush unit with the atmosphere and prevents the ingress of mercury vapor into the crush unit. For the complete removal of mercury vapor from the load, discharged gases are circulated through them. Excess gas is dispelled into the atmosphere. Solution B and the mercury sulfide present therein, having passed through the filter, proceed to unit 2. Precipitate from the sump is unloaded as needed.

The loss of working solution during HgS precipitation carried away by glass fragments, phosphor and glass dust after centrifuging and rinsing made up 0.026–0.037% of the weight of the demercurized scrap. Overall, our results imply that at 50–100 °C the proposed technology[17] allows complete binding of free-form and amalgamated metallic mercury[7] into mercuric sulfide.

From the above, it follows that in order to improve environmental safety, pollution from mercury and its compounds should not be allowed in the environment and fluorescent lamps should be handled with great care. Spent fluorescent lamps are regarded as toxic mercury-containing wastes of the first hazard class and are buried at hazardous waste sites. According to the

All-Union Institute of Secondary Resources (VIVR), the disposal cost of lamps, for example, at the Krasny Bor site, in the Leningrad region, cost 513 rubles (~US$19) per 1000 fluorescent lamps in 1980 and since 1995 the price has risen to US$500 per 1000 lamps. Moreover, disposal of fluorescent lamps is accompanied by continuous expropriation of land and excludes the reclamation of secondary resources such as mercury, metal (aluminum, tin, tungsten, nickel), phosphors and glass fragments. Complete recycling of spent fluorescent lamps leads to the prevention of environmental pollution by mercury and its compounds and also results in macroeconomic turnover of the above-mentioned secondary resources. The average annual number of spent fluorescent lamps (in millions) is ~160 in Russia, ~40 in Ukraine, ~12 in Kazakhstan and ~9 in Belarus. The entire world now produces about 1.5 billion fluorescent lamps per year and uses about 4000 t of mercury in their production.

Attempts are being made to develop improved technologies for processing fluorescent lamps. The VIVR developed a pyrometallurgical method for processing the lamps. At the core of the technology are the processes of crushing the fluorescent tubes and thermal distillation of mercury. However, the interaction of mercury with metals and organics at high temperatures (400–600 °C) forms a mercury-containing resinous mass. Moreover, mercury is 'spread' by the given unit. There have been a multitude of other attempts to develop technologies for the recycling of spent and defective fluorescent tubes[21–23] which have failed in their practical implementation. These processes have proven inferior to the technologies developed in more recent years.[16–20]

13.3.2 Vibration–Pneumatic Demercurization Method

The Ecotrom-2 unit[16] is distinguished by high performance and low specific energy output ratio (per recycled standard low-pressure discharge lamp). Therefore, with overall electricity consumption practically equal to known thermal mercury lamp recycling units, Ecotrom-2 is much more productive (the productivity of a thermal unit is typically 180–200 tubes per hour, whereas the productivity of Ecotrom-2 is 1200 tubes per hour). The unit cost of water and compressed air needed to operate the Ecotrom-2 unit is also almost two times less.

The vibration–pneumatic demercurization unit Ecotrom-2 consists of two main units:

1. A lamp disintegration device, including a loading node, a pneumatic–vibrational separator with a crusher and a centrifugal separator (with an emission purification efficiency of 95–97%).
2. A multistage system for purifying waste gases: a hose filter (with a purifying efficiency of 99.96%), adsorbers (activated carbon) and a gas blower with a compressor.

The compressor creates a vacuum within the demercurization unit (from 5–8 kPa in the lamp loading zone to 19–23 kPa prior to gas blowing), which eliminates the possibility of dust and gas emissions in the production room, permits complete capture of fluorescent lamp dust and reduces the mercury content in waste gases to less than 0.0001 $mg\,m^{-3}$.

Fluorescent lamps are processed by the Ecotrom-2 demercurization unit as follows (see ref. 16). Lamps that arrive at the plant in special containers are directed to the loading node. They are then fed through an accelerating tube *via* the high-vacuum force existing within the unit that sends them to a separator in which they are crushed into glass with a fineness of less than 8 mm. The lamp bases are separated from the glass on a vibrating screen and proceed to a special collector (process container) which, when filled, is directed to the demercurization/roasting electric furnace where the bases are demercurized. Effluent gases from the furnace are diverted into the existing purification system. Phosphors are separated from the glass *via* blowing off in a vibrational countercurrent air/glass system. Phosphors that have been cleared of glass fragments then proceed to the storage hopper. The bulk of the phosphor (95–97%) is collected in a centrifugal separator and is accumulated in conveniently transportable metallic barrels with a polyethylene liner bag and a sealed lid. The phosphor that is not collected in the centrifugal separator precipitates into the receiver of the hose filter and is then packed into the above-mentioned containers. Mercury-containing phosphors and sorbent are sent for further processing (to obtain metallic mercury).

13.3.3 Hydrometallurgical Treatment for Fluorescent Lamp Recycling

We have also developed a highly effective hydrometallurgical technology for processing spent fluorescent lamps, allowing the recovery of metallic mercury, tin–lead–aluminum scrap and concentrated components of phosphor. The process diagram for fluorescent lamp conversion is presented in ref. 24.

As is shown, the process for converting spent fluorescent lamps includes mechanical crushing of the fluorescent lamps under a layer of solvent dissolution of mercury *via* redox reaction with an Fe(III) complex, electroreduction of mercury in an electrolyzer with a fluidized cathode, acidification of mercury, electrorefining of mercury to high purity and sorting of fragments including separation and flushing of glass fragments and oxides, sulfides and basic salts and metallic scrap. Process water is sent for ion-exchange treatment to remove trace mercury. Secondary removal of mercury from the water is then directed to the service tank. As mercury is accumulated in ion-exchange columns, ions are regenerated *via* standard techniques. The proposed technology is highly productive and permits the conversion of products that comply with the appropriate MACs for mercury.[4,7,25,26] Moreover, mercury recycling is close to 100%.

It follows from the above that in many industries and in nature, mercury is dispersed and, as was noted in Chapter 12, 'spreads' throughout the planet. In macroquantities, mercury has a toxic effect on the human body.[4,7,25–34]

Consequently, when working with mercury it is necessary to perform thoroughly precautionary measures that prevent even small traces of mercury from entering not only the human body, but also the environment. Techniques for working with mercury that ensure the safety of the unit's service personnel have been described in many studies and manuals.[7,25,35–39]

13.4 Removal of Mercury from Industrial Wastewater

Various groups[30,35,40–42] have examined methods for the complete removal of mercury from industrial wastewater and utility emissions with a purification rate of 97–99.5% and thereby ensured water with mercury contents below the MAC of 0.001 mg L^{-1} for drinking water,[4] no more than 0.005 mg L^{-1} for utility facilities[4] and 0.01 mg m^{-3} for purified air masses of industrial plants.[26,28–31] It is common for industrial wastewater to contain mercury concentrations that exceed the MAC by tens or hundreds of times. This includes waste produced by a wide range of industries, including pharmaceuticals, textiles, paper, paints and varnishes and instrument making. It also includes wastewater from the production of chlorine, caustic soda, tanning products and explosive materials, and further from the metallurgical and electrochemical industries.

Combined methods are used to remove mercury from industrial water. The process diagram for treating wastewater from chlorine and alkali production is displayed in ref. 24. The process diagram combines the stages for adjusting the pH to 2.5–4 (1) using HCl (2), oxidation of metallic mercury by elemental chlorine and adsorption in a solution, *via* activated carbon (3), or modified by cellulose complexing agents, to a mercury concentration of 0.1 mg dm^{-3}, filtration of solid suspended particles (4), dechlorination on activated carbon (5) and ion-exchange treatment (6 and 7).

Ion-exchange treatment of industrial waters employs mercury sorption *via* the ion exchange resin Mtilon T,[5,44,45] the synthetic cation KU-2-8 in H-form or klinoptilolite, aminated muriatic methylamine,[6] anions of type AB-17, VP-1AP, RMT, RNH and sulfur-containing cations of type KC, F-0 = 243 and RG = KSHL with sulfhydryl functional groups.[24,43] Ion-exchange methods allow the polishing and treatment of industrial wastewater to mercury concentrations of 0.01–0.001 mg dm^{-3},[5,6,24,44,45] though according to other data [43] this figure may be as high as 0.005 mg dm^{-3}.

Mercury in ion-exchange resins is in the form of complex compounds and, for example, reaches 120 g L^{-1} in the ion-exchange resin VP-1AP. In this material, mercury is bound in stable $R{-}NCH_3HgCl_3^-$ complexes on its surface and in pores and is extracted only by alkaline solutions of sodium sulfide containing 5–12% Na_2S + 4% NaOH.[43] Mercury is desorbed from the anion in the form of mercury disulfide, Na_2HgS_2. It is forced out of the mercury disulfide solution *via* electrolytic precipitation by powdered aluminum or granulated zinc or it is recovered by $SnCl_2$, hydrazine hydrate, formaldehyde and $FeSO_4$ at a concentration of 8–25 g -dm^{-3} of mercury disulfide solution. When $FeSO_4$ is applied as a regenerant solution, mercury is converted to mercury sulfide according to the reaction

$$Na_2HgS_2 + FeSO_4 \rightarrow Na_2SO_4 + HgS + FeS \qquad (13.2)$$

and is removed *via* filtration in a mixture with iron sulfide. In all remaining cases cited above, mercury is separated out in the form of fine metallic mercury, which is difficult to remove from solution. The process diagram for regenerating VP-1AP is presented in ref. 43. The diagram includes the regeneration stage of the VP-1AP anions *via* a sulfide–alkaline solution, flushing the sulfide ions from the anions with a 5% solution of NaOH and water (up to a pH of 7). Subsequently, the anions are sent to a device for presorption treatment, precipitation of mercury from the regenerant solution in the form of mercury sulfide *via* iron(II) sulfate treatment and separation of mercury-containing sludge. The sludge is then sent for conversion in furnaces for thermal regeneration of mercury sludge.[43] The mercury recycling rate of the specified process is 98%. The wastewater treatment yield *via* ion exchange with VP-1AP ion is $0.005\,mg\,Hg\,L^{-1}$ of treated wastewater.[43] This concentration complies with the mercury MAC established for utility water facilities but does not comply with the international standard MAC of $0.0001\,mg\,L^{-1}$.[4,25,26] Consequently, techniques are being developed for advanced removal of mercury from wastewater.

13.5 Demercurization of Workplaces and Plants

As noted above, mercury is widely used in science, engineering and industry. Mercury contamination of production facilities occurs during emergencies, accidental spills and improper handling of mercury. Therefore, in this section we have consolidated the results obtained from the development of methods for demercurizing facilities, the most important of which is the demercurization of workspaces without contaminating them with common impurities. Normally, demercurization is carried out using a number of solutions possessing oxidizing properties in relation to mercury.[25,35–39] These include a 20% solution of iron(III) chloride, a 1–5% solution of potassium permanganate, a 4–5% solution of mono- or dichloramine in carbon tetrachloride, then a 4–5% solution of sodium polysulfide and a 2.5% solution prepared from a mixture containing 15–20% of ethylenediaminetetraacetic acid (EDTA) and 85–80% thiosulfate.

The demercurization methodology has been described.[35–38] Prior to demercurization, metal equipment is thoroughly cleaned and removed from the room owing to the high corrosiveness of iron(III) chloride. Then the visible mercury is removed from contaminated surfaces by mechanical means (by use of vacuum or in traps). Next, the contaminated surface is treated with the appropriate solution, for example, with an iron(III) chloride solution based on a ratio of 1 bucket of solution to an area of 25 m^2. According to Pugachevich,[36] the surface covered with solution is wiped with a brush several times and is left for 1–2 days until completely dry. Then the demercurized surface is thoroughly cleansed with clean water. To remove drops of mercury, Jaeger[39] recommended sprinkling the floor with zinc dust, copper powder or activated carbon treated with iodine. Activated carbon that has not been treated with iodine has low activity in relation to mercury. The products of the reaction between mercury and the listed reagents are removed.

These methods are not 'clean' in relation to mercury and they create impurity sources. Moreover, these proposed methods do not allow demercurization of the walls and ceilings of workspaces. When working with mercury, vapors infiltrate and are absorbed by the surface layers of paint, walls and ceilings with concentrations lower than the MAC. To demercurize workspaces with an airborne mercury vapor content exceeding 80 MAC norms due to accumulated and 'spread' mercury, we used the following technique. We removed the furniture and equipment from the room, thoroughly sealed the windows and ventilation channels and prepared materials to seal the doors. We humidified the room using heating devices and boiling water and thereby humidified the walls. Then we placed 50 g of potassium permanganate in each of 2–3 flasks with volumes of ~2 L and filled them with concentrated hydrochloric acid. A chlorine-releasing reaction occurred. The doors were then thoroughly sealed. The released chlorine reached the walls and infiltrated the porous surface layers of the walls and ceiling, where it interacted with the absorbed mercury, and formed calomel and mercury(II) chloride according to the following reactions:

$$Cl_2 + 2Hg^0 \rightarrow Hg_2Cl_2 \tag{13.4}$$

$$Cl_2 + Hg^0 \rightarrow HgCl_2 \tag{13.5}$$

These reactions proceed at an enormously high rate. The standard electric potential of the half-reaction $Cl_2 + 2e \rightleftharpoons 2Cl^-$ is $E^\circ_{Cl_2/Cl^-} = +1.3595$ V.[46] Moreover, in the presence of moisture (moist walls, ceiling, floor), chlorine undergoes a disproportionation reaction (Kuznetsov reaction):

$$Cl_2 + H_2O \rightleftharpoons HClO + HCl \tag{13.6}$$

with the formation of a strong oxidant, hypochloric acid. The standard potentials of the half-reactions

$$HClO + H^+ + e \rightleftharpoons {}^1\!/_2Cl_2 + H_2O \tag{13.7}$$

$$HClO + H^+ + 2e \rightleftharpoons Cl^- + H_2O \tag{13.8}$$

equal to $E^\circ_{HClO/{}^1\!/_2Cl_2} = +1.63$ V and $E^\circ_{HClO/Cl^-} = +1.49$ V, respectively,[46] demonstrate the higher oxidizing power of hypochloric acid in comparison with chlorine.

The equilibrium constants of the reactions between HClO and mercury:

$$Hg^0 + HClO + HCl \rightarrow HgCl_2 + H_2O \tag{13.9}$$

and

$$Hg^0 + HClO + HCl \rightarrow Hg_2Cl_2 + H_2O \qquad (13.10)$$

are 4.82×10^{-37} and 1.06×10^{-43}, respectively, demonstrating the equilibrium shift of these reactions towards the formation of mercury(II) chloride and calomel, which have practically zero volatility.

After applying such a method to treat a workspace with a mercury MAC equal to 80 norms and after the space had been hermetically sealed for 3 days, the mercury content in the air after ventilation was reduced to 0.05 MAC. After the room had been whitewashed with white titanium paint and cleaned, the mercury content in the air was 0.01 MAC ($1.0\,ng\,m^{-3}$). Consequently, the mercury vapor content in the air of the room was reduced 8000-fold after demercurization. The thermal method of demercurization, which combines heating of the surface to 200–250 °C with sorbed mercury and vacuum venting of mercury vapor, gives only a 40–50-fold reduction in mercury content in air.[36,47]

The positive side of this proposed method of workspace demercurization is the conversion of metallic mercury into an inactive state and the creation, under a layer of fresh white titanium paint, of an oxidizing environment consisting of adsorbed active oxygen (HClO) and traces of chlorine that continue to spread into the plaster and pores in the walls, thereby neutralizing the metallic mercury that had accumulated over the years.

Methods of personal protection and personal preventive measures when working with mercury have been described in detail.[25,35,38] Work must be performed in starched protective clothing and leather or rubber footwear protected by poly(vinyl chloride) covers. After work, protective work clothing is left in the workclothes room.

After working with mercury, one must thoroughly wash the face and hands with warm water and soap and take a shower.[36] Also when working with mercury, special attention must be given to the condition of the mouth cavity. Teeth should be brushed twice per day (morning and night) and affected teeth should be treated in a timely manner.[36] After work, one must rinse the mouth cavity with a 1% solution of Berthollet's salt or a 5% solution of potassium permanganate. Food rich in vitamins and pectin must be consumed. In case of worsening health, irritability, insomnia, bleeding gums, *etc.*, a doctor must be consulted. Special attention should be paid to a healthy lifestyle, such as a daily 30–50 min morning exercise routine ending with hydrotherapeutic exercises, regular calisthenics, sports or mountain walking. Such activities increase the body's resistance and strengthen the nervous system.

References

1. A. G. Morachevskii, I. A. Shesterkin, V. B. Busse-Machukas, *et al.*, *Natrii. Svoistva, proizvodstvo, primenenie (Sodium. Properties, Production, Application)*, ed. A. G. Morachevskii, St. Petersburg, Khimiya, 1992.

2. A. D. Akhmetov and V. A. Bednenko, Mercury pollution of the environment, emissions into the atmosphere, restoration of territories, impact on health, in *Proceedings of the International Seminar, 28 May–1 June 2007, Astana*, 2007, 18–19.
3. United Nations Environment Programme, *Report on Current Supply and Demand for Mercury, Including Projections Considering the Phase-out of Primary Mercury Mining*, UNEP (DTIE)/Hg/OEWG 2/6, UNEP, Nairobi, 2008.
4. Ya. M. Grushko, *Harmful Inorganic Compounds in Industrial Effluents*, Khimiya, Leningrad, 1979.
5. V. F. Fedonina, M. O. Lishevskaya and N. S. Torocheshnikov, *et al.*, *Khim. Prom.*, 1972, **1**, 72–73.
6. S. M. Rustamov, I. I. Zeynalova, F. T. Makhmudov and Z. Z. Bashirova, Khim. Tekhnol. Vody (translated in *J. Water Chem. Technol.*), 1993, **15**, 378–382.
7. L. F. Kozin, *Physicochemistry and Metallurgy of High-Purity Mercury and Its Alloys*, Naukova Dumka, Kiev, 1992.
8. L. F. Kozin, A. G. Morachevskiy and I. A. Sheka, *Ukr. Khim. Zh.*, 1989, **55**, 495–509.
9. L. F. Kozin and A. G. Morachevskiy, *Physicochemistry and Metallurgy of High-Purity Lead*, Metallurgiya, Moscow, 1991.
10. L. F. Kozin and S. V. Volkov, *Chemistry and Technology of High-Purity Metals and Metalloids, Volume 1. Chemical and Electrochemical Methods for Ultra-purification*, Naukova Dumka, Kiev, 2002.
11. L. F. Kozin, *Amalgam Metallurgy*, Tekhnika, Kiev, 1970.
12. UNEP, *Assessment of Mercury Releases from the Russian Federation, Arctic Council's Action Plan to Prevent Pollution of the Arctic)*, Danish Environmental Protection Agency.
13. W. E. Walles and L. C. Mulford, *US Pat.*, 5 478 540, 1995.
14. A. Muratov, *Nedelya*, 1986, (18), April 28.
15. Russian Ministry of Natural Resources, *Certificate of Environmental Compliance with Requirements of GOST R ISO 14001-98 Russian Ministry of Natural Resources No. 00000309 as of 02/28/2003*, 2003.
16. N. A. Stolyarova, N. A. Chekhlan and E. A. Egorova, Utilization of fluorescent lamps on the demercurization equipment «Ecotrom-2» http://ea.donntu.edu.ua:8080/jspui/bitstream/123456789/15694/1/Stolyarova%20N.%20A.%20.pdf. Accessed 13 July 2013.
17. A. P. Alekseev, N. G. Dmitrieva, A. M. Zorkal'tsev, V. V. Leonenko VV, and G. A. Safonov, Method for processing of electroluminescent light sources and a device for its implementation, *USSR Pat.*, 2 044 087.
18. O. V. Grigor'vich and S. V. Sergeevich, Demercurator, *USSR Pat.*, 2 295 583.
19. V. N. Kovalev, A. N. Gorbenko, V. N. Timofeev and V. V. Ten, Method for thermal demercurization of mercury-containing materials and device for its implementation, *USSR Pat.*, 2 171 304.

20. Y. G. Dyachenko, A. A. Zakharov, A. G. Ljashenko, E. M. Ain, A. V. Agueyev and A. G. Gorobec, A device for fractionation in a unit for mercury-containing fluorescent lamps recycling, *USSR Pat.*, 2 191 638.
21. V. Grigorivich, V. S. Sergeevich and O. Vladimirovich, Method for disposal of mercury-containing wastes, *USSR Pat.*, 2 327 536.
22. C. R. Palmer and R. G. Sandberg, in *Precious Metals, Mining, Extraction and Process, Proceedings of International Symposium, AIME Annual Meeting, Los Angeles, CA., 27–29 February 1984*, AIME, Warrendale, PA, 1984, pp. 227–239.
23. *Fr. Pat. Appl.*, 2 615 756.
24. L. E. Postolov, T. E., Mitchenko and V. G. Ovchinnikov, *Khim. Tekhnol.*, 1992, (2) (182), 15.
25. H. G. Seiler and H. Sigel (eds), *Handbook on Toxicity of Inorganic Compounds*, Marcel Dekker, New York, 1988, pp. 419–436.
26. N. V. Lazarev and I. D. Gadaskina (eds), *Harmful Substances in Industry, Inorganic and Organoelement Compounds*, Khimiya, Leningrad, 1977.
27. J. O'M. Bockris (ed.), *Environmental Chemistry*, Khimiya, Moscow, 1982.
28. International Atomic Energy Agency, *Mercury Contamination in Man and His Environment*, Tech. Rep. Ser., No. 137, IAEA, Vienna, 1972, p. 177.
29. L. Friberg and J. Vostal, *Mercury in the Environment*, CRC Press, Cleveland, OH, 1982.
30. I. M. Trachtenberg and M. N. Korshun, *Mercury and Its Compounds in the Environment*, Vysha Shkola, Kiev, 1990.
31. Ya. M. Grushko, *Harmful Inorganic Compounds in Industrial Emissions into the Atmosphere*, Khimiya, Leningrad, 1987.
32. V. G. Huhryansky, A. Ya. Tsyganenko and N. V. Pavlenko, *Chemistry of Biogenic Elements*, Vysha Shkola, Kiev, 1990.
33. *The Maximum Permissible Concentration (MPC) of Pollutants in the Atmosphere of Inhabited Areas*, Approved by Chief State Medical Officer of the USSR on 27 August 1984, No. 3086-84.
34. V. N. Kurnosov, *Maximum Allowable Concentrations of Atmospheric Pollutants*, Medgiz, Moscow, 1961, pp. 54–71.
35. *Interindustrial Safety Rules in Production and Use of Mercury*, POT RM-009-99, DEAN Publishing House, St. Petersburg, 2001, p. 64.
36. P. P. Pugachevich, *Working with Mercury Under Laboratory and Industrial Conditions*, Khimiya, Moscow, 1972.
37. S. M. Mel'nikov, *Metallurgy of Mercury*, Metallurgiya, Moscow, 1971.
38. S. M. Mel'nikov, *Safety in Metallurgy of Mercury*, Metallurgiya, Moscow, 1974.
39. H. Jaeger (ed.), *Electrometallurgy of Aqueous Solutions, Industrial Electrochemistry Reference Book*, translation from German, Metallurgiya, Moscow, 1966.
40. T. O. Mamutovich, B. I. Ibrahimovic, A. A. Abildaevich and S. K. Sydykovich, Method for recycling of mercury-containing polymetallic raw materials, *USSR Pat.*, 1 620 499.

41. C. Nunez and F. Espiell, *Met. Trans.*, 1986, **B17**, 443–448.
42. C. Nunez and F. Espiell, *Met. Trans.*, 1985, **B15**, 229–233.
43. L. E. Postolov, T. E. Mitchenko, V. A. Skripnik and A. M. Roitenberg, *Khim. Tekhnol.*, 1984, **4**, 35.
44. Z. A. Rogovin, M. O. Lishevskaya, V. V. Voronenko and B. F. Fedyunina, *Khim.Prom.*, 1966, No. 7, 512.
45. Method for wastewater treatment from metals, *USSR Pat.*, 230 738.
46. V. Latimer, *The Oxidation States of the Elements and Their Potentials in Aqueous Solutions*, Izdatel'stvo Inostrannaya Literatura, Moscow, 1954, p. 400.
47. H. B. Masters, *Min. Annu. Rev.*, 1988, June, 91–95.

CHAPTER 14

Safety and Health Practices for Working with Metallic Mercury[†]

WOODHALL STOPFORD

Duke University Medical Center, Durham, NC 27710, USA

14.1 Toxicity of Metallic Mercury

Acute mercury toxicity can occur when mercury is inhaled in milligram quantities. Such exposures are invariably associated with work in a confined, mercury-contaminated space with little ventilation or by heating of mercury such as in home amalgamation efforts, welding of contaminated surfaces or spills of hot mercury. After a delay of a few hours, a metal fume fever can result, with symptoms of nausea, abdominal cramps, diarrhea, muscle aches, fever and an elevated white blood cell count.

With higher exposures, one can also see symptoms of pulmonary irritation with chest tightness, a cough and shortness of breath. X-rays of the chest in this disorder disclose an interstitial pneumonitis and pulmonary function tests show restrictive changes and also a diffusion defect for oxygen. Survivors can develop chronic shortness of breath and interstitial fibrosis of the lungs. Possibly more common than pulmonary toxicity is inflammation of the mouth. Shortly after an acute exposure, the mouth and gums can become red and sore. Within a few days, patients can experience a metallic taste together with further inflammation of the gums, loosening of the teeth, ulcers of the mouth and a blue line at the gum margins. Occasionally, a tremor is noted and, less commonly, bloody diarrhea and transient liver and kidney abnormalities.

[†]Provided by Bethlehem Apparatus, Inc. Hellertown, PA.

Mercury Handbook: Chemistry, Applications and Environmental Impact
By Leonid F Kozin and Steve Hansen

Published by the Royal Society of Chemistry, www.rsc.org

14.2 Toxicity from Chronic Exposure

In industry, the earliest finding that might be noted from chronic exposure to moderate levels of metallic mercury is that of a tremor. It is initially seen as a fine, postural tremor noted only when the arms are outstretched. With greater exposures or exposures of longer duration, the tremor increases in amplitude and becomes coarse. In addition, it appears to be aggravated by intentional activities such as writing or picking up a cup or coffee. As the severity of the tremor increases, it can be interrupted by clonic-like jerks of one or more extremities. In the severest cases, these jerks progress to involve the entire body.

Sometimes, associated with a moderate tremor there is difficulty in performing fine movements, incoordination, difficulty with gait and even hoarseness associated with ataxia of the vocal cords.

If mercury exposure continues to the point of the development of a significant tremor, a state of erethism can sometimes be found. Affected workers become easily upset and embarrassed, irritable and sometimes quarrelsome. They lose self-confidence and often have a feeling as if they are being watched. If they think they are being watched, their tremor activity might increase in severity. They often have difficulties with sleeping or have nightmares. Some tend to be drowsy and fall asleep on the job. Often there is depression and memory loss associated with mercury exposure. Rarely there are hallucinations, delusions or mania. All findings and complaints of an individual who manifests eresthismic symptoms can be resolved over a period of months without any further intervention if the individual is removed from exposure to mercury.

With extreme exposures to metallic mercury, one can find concentric construction of the visual fields, poor night vision and red–green color blindness, problems that can be resolved by treatment with chelating agents. A more common eye problem is the finding of mercurialentis, *e.g.*, a brownish discoloration of the anterior capsule of the lens. Such a defect occurs only after chronic exposure to mercury and can be present without any evidence of mercurialism.

Individuals who have severe mercurialism can also have symptomatic defects of other sensory organs. These include a high frequency hearing loss, symptoms of vertigo, ringing in the ears, loss of balance and a partial loss of smell sense. In workers with severe mercurialism secondary to inorganic mercury exposure, one can sometimes find disorders of the spinal cord and peripheral nerves, including a syndrome resembling amyotrophic lateral sclerosis (ALS), disorders of the sensory nerves and a Parkinsonism-like syndrome of stiffness and rigidity of the extremities.

A common manifestation of chronic exposure to excessive levels of mercury vapor is gum disease. The gums initially become swollen and boggy and later retract. In individuals who have pre-existing pyorrhea, evidence of infection can be aggravated. In severe cases, there can be loosening of the teeth with bony reabsorption of the jaw. The gum disease can be brought under control with good oral hygiene even if exposure to mercury continues. Individuals who have jaw involvement can improve if they are taken out of exposure and treated with braces.

Once inorganic mercury enters the body, it is primarily stored in the proximal tubules of the kidney. It is not unexpected, therefore, that the earliest signs of kidney damage would be manifested by a disorder of the proximal tubules. These tubules are involved with reabsorption of nutrients and substances that are normally filtered into the urine, but are usually conserved by the body. With proximal tubular disease secondary to mercury, one can find abnormal amounts of glucose, phosphate, amino acids and low molecular weight proteins in the urine.

Unfortunately, studies of renal tubular disease are difficult and often proteinuria from a glomerular injury is the first manifestation of kidney toxicity secondary to excessive absorption of mercury. On biopsy of an individual with such a problem, one finds evidence of a membranous glomerulonephritis in association with proximal tubular injury.

Inorganic mercury is unusual in that it is lost relatively rapidly from the body. In a worker with early manifestations of mercury toxicity of any organ, improvement and complete resolution of the problem are almost invariably noted when the worker is removed from exposure for a period of time. If excessive exposure continues in the face of obvious clinical problems or if there are recurrent exposures and toxicity, manifestations of mercurialism can become chronic. Manifestations can include tremors, paralysis, loss of memory and chronic renal disease. Over the past 20 years, cases of chronic inorganic mercurialism have been reported only sporadically and have only been noted after exposures that were severe, uncontrolled and prolonged.

14.3 Surveillance Programs for Industry

A pre-employment examination program is needed to identify and restrict certain individuals from potential exposure to mercury. Exposure is contraindicated in those individuals who have problems that might be either aggravated with mercury exposure or confused with findings of mercurialism and thus hinder the effectiveness of a medical monitoring program. Problems which would restrict employment include

- alcoholism
- chronic kidney disease
- known allergy to mercury.

If protein is found in a urine specimen, but there is evidence of chronic kidney disease, the individual can be employed once the underlying problem has been corrected. Individuals with evidence of gingivitis (gum inflammation) should be under the care of a dentist prior to employment. Tremor and psychological problems should be documented.

There has always been difficulty in correlating measured exposure levels to mercury vapor with biological measurements of mercury absorption except when data are analyzed on a group basis. Most studies that correlate biological levels with air levels are based on area vapor measurements. However, when there is a potential exposure to metallic mercury, personal contamination can result with the formation of a microenvironment of mercury vapor around the worker's breathing zone that is several times higher than that of the general work environment. If work clothes are brought home, contamination of the home can result in exposures that continue for a longer period than the work day, and also exposure to family members.

Environmental monitoring should include daily area mercury vapor levels at all work sites by direct measurement of mercury vapor concentration and periodic time-weighted average breathing zone measurements. Measurements can be made with a gold foil monitor or a portable personal sampling pump connected to Hopcalite tube.

Urine mercury levels are useful for estimating both exposures and body burden since most inorganic mercury that is absorbed is deposited in the kidneys prior to excretion. Urine mercury levels tend to vary from person to person with similar exposures and also from time to time in the same individual. In order to make urine mercury determinations more reflective of an individual's exposure, efforts have been made to decrease the variability in measurements. By making urine mercury determinations on the same day each week and the same time of day, the variability seen in any one individual's urine mercury determinations can be decreased. Corrections for urine concentration by using either the excretion of creatinine or urine specific gravity can also decrease the variability between urine mercury determinations. Corrections for variability in urine concentration can be avoided by using for analysis the first voided specimen on rising.

Determinations of mercury in whole blood are better indicators of current exposure than urine mercury determinations. In humans, the initial half-life for loss of mercury from blood is also fairly short, $\sim$5 days. There is an excellent correlation between blood mercury levels and average area mercury vapor levels or breathing zone mercury vapor levels.

Some individuals can excrete milligrams of mercury per liter of urine without any evidence of ill-effect, whereas others might excrete less than 300 $\mu g\,L^{-1}$ and have evidence of adverse effects from mercury exposure. Because of this variation in susceptibility, only a detailed medical monitoring program can identify those individuals who are having adverse effects from mercury exposure. Such an examination should include a complete history, emphasizing neurological and psychological complaints, and also complete physical examination with emphasis on the oropharyngeal and neurological components of the examination. Certain physiological studies are indicated. Periodic tests of strength can be made with a simple grip gauge. The severity of a tremor can be documented. An important part of the medical examination is a close assessment of kidney function. One simple way is to assess total protein excretion quantitatively. Urine glucose and albumin levels can easily be checked with an indicator strip. Periodic blood tests to assess kidney function should be performed.

14.4 Preventive Measures

Metallic mercury that is spilled on a floor surface is available not only for vaporization, but also for tracking into all parts of the facility with resultant exposure to workers who might normally not be exposed to mercury vapor. To prevent this type of contamination, floors should be sealed with an epoxy sealer and washed periodically with trisodium phosphate solution to remove all small particles of mercury. More obvious spills of mercury should be vacuumed into a water trap at the time of the occurrence of the spill. The vacuum should be appropriately exhausted or filtered to prevent further mercury exposure. An alternative method of picking up mercury spills is to use a copper pad filled with zinc filings to amalgamate the mercury.

In areas where the working surface cannot be sealed, either a solution of calcium polysulfide in a wetting agent or a 20% solution of ferric chloride can be sprayed on to the surface. This treatment will adequately suppress mercury vapor production until the surface is disturbed.

Smoking should not be allowed in mercury exposure areas. Furthermore, tobacco products should not be kept in shop areas. Cigarettes can readily absorb mercury vapor and therefore should not be brought into an area where mercury is used. Furthermore, cigarettes can be readily contaminated by being laid down on a work surface or by being smoked prior to washing one's hands. The cigarette can then readily volatilize any mercury present, giving excessive exposure not only to the smoker, but also to anyone nearby.

In addition to creating an excessive microenvironmental level of mercury vapor around a worker's face, personal contamination can also lead to excessive absorption through the skin. To help prevent skin contamination, not only should adequate laboratory garments be worn, but also either gloves should be worn or thiosulfate-containing barrier cream should be applied to protect the hands. Hands should be washed before smoking or drinking. With exposure to levels of mercury vapor less than $0.5\,mg\,m^{-3}$, a silver-impregnated dust mask can be worn. With higher exposures, a chemical cartridge respirator (*e.g.*, Mersorb-MGA) should be used. Accumulation of mercury-containing dust on the outside of the mercury vapor absorptive masks can also lead to the rapid breakthrough of these masks and excessive mercury exposure to workers above and beyond what they would receive without wearing the mask. A frequently changed dust filter in front of the cartridge can prevent this problem. When working in an environment with a very high concentration of mercury vapor, a full face mask should be worn to prevent excessive absorption of mercury vapor through the cornea with resultant incapacities of the cornea and lens.

Underwear and socks can absorb mercury excreted in sweat and, in turn, make this mercury available for reabsorption through either the skin or the respiratory tract. Although relatively minor, reabsorption of mercury from sweat can be prevented by changing underwear and showering at the end of each working day.

One of the major ways of decreasing workers' exposure to mercury is by adequate ventilation. In a stagnant environment, even at 0 °C, a dangerous exposure situation to mercury vapor can result. As the temperature increases, the vapor pressure of mercury increases rapidly such that at 30 °C there is an approximately sixfold greater vapor level than at 0 °C under the same conditions. Adequate general ventilation is mandatory. During the summer months, increased ventilation by opening all doors and windows is helpful. Under some circumstances, it is easier to use air conditioning to keep the temperature down and vapor levels under control without any drastic changes in ventilation. High-volume local exhaust ventilation of point sources of mercury vapor, such as contaminated ovens, can greatly decrease the overall ventilation requirements of a facility and also prevent excessive levels of mercury vapor in local work areas. Makeup air requirements can be decreased by filtering recirculated air through mercury-absorbing activated charcoal filters.

In situations where engineering controls are inadequate to control a mercury vapor exposure, administrative controls can be used to decrease worker exposure and, thus, worker ill health. Excessive biological levels of mercury are not an indication of mercury poisoning. Although biological results can be used as an indication of excessive absorption and, thus, the need for improved work practices, personal hygiene, engineering controls, personal protective equipment and safety measures, they should only be used with caution as a basis for administrative actions. If a medical examination discloses evidence both of absorption of mercury and of abnormalities that might be related to this absorption, administrative action should be taken. The employee should be put into a low-exposure area until the abnormality has cleared and evidence of excessive absorption has disappeared. If abnormalities persist, further medical evaluation is needed to look for other etiologic reasons for the abnormalities.

Mercury can present an environmental hazard through inappropriate discharges of process or wash water, venting of contaminated air or disposal of mercury-contaminated solid wastes. Waste water contaminated with mercury can be adequately decontaminated prior to discharge with activated charcoal. Other systems are available based on sulfite deposition of mercury or by reverse osmosis. Mercury-absorbing activated charcoal can also be used to treat vented air contaminated with high levels of mercury vapor. Work room air should be vented away from intake vents. In order to avoid disposing of mercury-contaminated solid wastes, such wastes should be sent to a facility equipped to recover mercury.

Metallic mercury exposures, when excessive, can produce various neurologic, oropharyngeal and renal problems that may be resolved spontaneously once exposure has stopped. Chronic problems can develop with unusual exposures that are prolonged beyond the first clinical manifestation of mercurialism. In this situation, a number of findings, including constriction of visual fields, a Parkinsonism-like syndrome and evidence of combined motor neuron disease can develop. However, these findings can be reversed by the use of an appropriate therapy.

Often there can be personal contamination of a worker with exposure through the skin and airways that cannot be detected by routine air measurements. For this reason, a biological and medical monitoring program is required to detect early manifestations of mercurialism and intervene at the time when only administrative controls are required to resolve the problems. There are large variations in a worker's susceptibility such that biological levels of mercury by themselves should not be used as a basis for moving a worker to a low-exposure situation. With adequate work practices, personal protection, housekeeping and engineering controls, chronic exposures can be kept to a minimum and most of the hazards to workers eliminated.

Further Reading

H. B. Elkins, L. D. Pagnotto and H. L. Smith, *Am. Ind. Hyg. Assoc. J.*, 1974, **35**, 559.

L. J. Goldwater, The Harben Lectures, *J. R. Inst. Public Health Hyg.*, 1964, **27**, 279.

J. K. Piotrowski, B. Trojanowska and E. M. Mogilnicka, *Int. Arch. Occup. Environ. Health*, 1975, **35**, 245.

M. Randall and H. B. Humphrey, *New Process for Controlling Mercury Vapor*, US Bureau of Mines Information Circular No. 7206, US Bureau of Mines, Washington, DC, 1942.

A. O. Rathje, *Am. Ind. Hyg. Assoc. J.*, 1969, **29**, 126.

J. Shurgan and L. Harris, *Am. Ind. Hyg. Assoc. J.*, 1977, **38**, 146.

W. Stopford, Industrial exposure to mercury, in *The Biogeochemistry of Mercury in the Environment*, ed. J. O. Nriagu, Elsevier/North Holland, Amsterdam, 1979, pp. 368–397.

W. Stopford, S. D. Bundy, L. J. Goldwater and J. A. Bittikofer, *Am. Ind. Hyg. Assoc. J.*, 1978, **39**, 378.

APPENDIX I

Phase Diagrams and Intermetallic Compounds in Binary Amalgam Systems

AI.1 Binary Mercury Phase Diagrams

Most metals in contact with mercury a peritectic reaction or a series of peritectic reactions called a peritectic cascade. Few systems do not show any peritectic reactions. With several exceptions, notably lithium, zinc, cadmium, indium and thallium, mercury forms intermetallic compounds with little or no solid solubility.

Alkali, alkaline earth, lanthanide and actinide metals react strongly with mercury and form a large number of intermetallic compounds. The intermetallic phases formed in these systems generally have very narrow ranges of homogeneity. Transition metals (Ti, Zr, Hf, Mn, Ni, Cu, Rh, Pd, Pt, Cd, Zn, Ag and Au) typically form peritectic cascades and intermetallic compounds. Metalloids such as aluminum, antimony and bismuth form degenerate eutectic systems. Gallium forms a simple monotectic reaction with mercury. Lead and tin also show intermetallic solid solutions and peritectic reaction sequences. The halogen–mercury systems form a syntectic reaction ($L_1 + L_2 \rightarrow HgX$) with $X = F$, Cl, Br and I.

A complete set of binary mercury phase diagrams is available as follows:

C. Guminski, Contributions of electrochemistry to the knowledge of amalgams, *Pol. J. Chem.*, 2004, **78**, 1733–1751.

AI.2 Intermetallic Phases in Binary Amalgam Systems

Intermetallic phases with crystal structure and lattice parameters where available are given in the following tables, with relevant literature citations.

Mercury Handbook: Chemistry, Applications and Environmental Impact
By Leonid F Kozin and Steve Hansen

Published by the Royal Society of Chemistry, www.rsc.org

Ag–Hg

Phase	*Crystal structure*	*Lattice parameters (nm)*	*Density ($g\ cm^{-3}$)*	*Prototype*
ζ 43–46% Hg	Hexagonal close packed	$a = 0.2964$, $c = 0.4831$		Mg
γ 56–57% Hg	Cubic	$a = 1.0046$	13.65^{calc}, 13.48^{meas}	

P. Anderson and S. J. Jensen, Chemical composition and crystal structure of the γ-phase in the silver–mercury system, *Scand. J. Dent. Res.*, 1971, **79**, 466–471.

M. R. Baren, The Ag–Hg (silver–mercury) system, *J. Phase Equilib.*, 1996, **17**, 122–128.

C. Cipriani, G. P. Bernardini, M. Corazza, G. Mazzetti and V. Moggi, *Eur. J. Mineral.*, 1993, **5**, 903–914.

C. W. Fairhurst and J. B. Cohen, The crystal structures of two compounds found in dental amalgam: Ag_2Hg_3 and Ag_3Sn, *Acta Crystallogr. B*, 1972, **28**, 371–377.

S. J. Jensen, Neutron diffraction study of the γ-phase in the silver–mercury system, *Scand. J. Dent. Res.*, 1972, **80**, 162–165.

E. Seeliger and A. Mücke, Para-schachnerite, $Ag_{1.2}Hg_{0.8}$, and schachnerite, $Ag_{1.1}Hg_{0.9}$, from Landsbergite near Obermoschel, Pfalz, *N. Jahrb. Mineral. Abhandl.*, 1972, **117**, 1–18.

S. Stenbeck, X-ray analysis of alloys of mercury with silver, gold and tin, *Z. Anorg. Chem.*, 1933, **214**, 16–18.

Am–Hg

J. Maly, The amalgamation behavior of heavy elements, *J. Inorg. Nucl. Chem.*, 1969, **31**, 1007–1017.

M. F. Tikhonov, V. Z. Neponmyashchii, S. V. Kalinina, A. D. Khokhlov, V. I. Bulkin and B. M. Filin, Preparation of the Am amalgam, *Radiokhimiya*, 1986, **28**, 804–809.

Au–Hg

Phase	*Crystal structure*	*Lattice parameters (nm)*	*Density ($g\ cm^{-3}$)*	*Prototype*
α_1 16–23% Hg	Hexagonal	$a = 0.8736$, $c = 0.9577$		
ζ (Au_3Hg) 21–26% Hg	Hexagonal	$a = 0.2914$, $c = 0.4803$		Mg
Au_2Hg	Hexagonal	$a = 0.7019$, $c = 1.0184$	16.67^{calc}	Au_2Hg
Au_6Hg_5	Hexagonal	$a = 1.308$, $c = 1.720$		N_6Nb_5
Au_5Hg_8	Cubic	$a = 0.992$		Cu_5Zn_8

A. F. Berndt and J. D. Cummins, The crystal structure of the Au_2Hg phase, *Acta Crystallogr. B*, 1970, **26**, 864–867.

T. Lindahl, The crystal structure of Au_6Hg_5, *Acta Chem. Scand.*, 1970, **24**, 946–952.

T. B. Massalski, The lattice spacings of close-packed hexagonal Au–In, Au–Cd and Au–Hg alloys, *Acta Metall.*, 1957, **5**, 541–547.
H. Okamoto and T. B. Massalski, *Bull. Alloy Phase Diagrams*, 1989, **10**, 50–58.
C. Rolfe and W. Hume-Rothery, The constitution of alloys of gold and mercury, *J. Less-Common Met.*, 1967, **13**, 1–10.
S. Stenbeck, X-ray analysis of alloys of mercury with silver, gold and tin, *Z. Anorg. Chem.*, 1933, **214**, 16–18.

Ba–Hg

Phase	*Structure*	*Lattice parameters (nm)*	*Density ($g\,cm^{-3}$)*	*Protoype*
Ba_7Hg_{31}	Tetragonal	$a=1.0778$, $c=1.0189$		Ba_7Cd_{31}
BaHg	Cubic	$a=0.4133$		CsCl
$BaHg_2$	Orthorhombic	$a=0.5144$, $b=0.8072$, $c=0.8717$		$CeCu_2$
$BaHg_6$				
$BaHg_{11}$	Cubic	$a=0.95871$		$BaHg_{11}$
$BaHg_{13}$	Cubic			
Ba_2Hg	Tetragonal	$a=0.4200$, $c=1.519$		Ba_2Cd
$Ba_{14}Hg_{51}$	Hexagonal			

E. Biehl and H.-J. Deiseroth, Preparation, structural relations and magnetism of amalgams MHg_{11} (M: K, Rb, Ba, Sr), *Z. Anorg. Allg. Chem.*, 1999, **625**, 1073–1080.
G. Bruzzone and F. Merlo, The barium–mercury system, *J. Less-Common Met.*, 1975, **39**, 271–276.
C. Guminski, The Ba–Hg (barium–mercury) system, *J. Phase Equilib.*, 2000, **21**, 173–178.
F. Merlo, The crystal structure of Ca_3Cd_2, Ba_2Cd and Ba_2Hg, *J. Less-Common Met.*, 1976, **50**, 275–278.

Br–Hg

Phase	*Crystal structure*	*Lattice parameters (nm)*	*Density ($g\,cm^{-3}$)*	*Prototype*
α-Hg_2Br_2	Orthorhombic			
β-Hg_2Br_2	*tI*16	$a=0.4666$, $c=1.1133$		Hg_2Cl_2
$HgBr_2$	*oC*12			$HgBr_2$

H. Braekken, The crystal structure of $HgBr_2$, *Z. Kristallogr.*, 1932, **81**, 152–154.
E. Dorm, Intermetallic distances in Hg(I) halides Hg_2F_2, Hg_2Cl_2, Hg_2Br_2, *J. Chem. Soc., Chem. Commun.*, 1971, 466–467.
R. Dworsky and K. Komarek, The Hg–Br system, *Monatsh. Chem.*, 1970, **101**, 976–983.
C. Guminski, The Br–Hg (bromine–mercury) system, *J. Phase Equilib.*, 2000, **21**, 539–543.

R. J. Havighurst, Parameters in crystal structure: mercurous halides, *J. Am. Chem. Soc.*, 1926, **48**, 2113–2125.

H. E. Swanson, N. T. Gilfrich and M. I. Cooke, *Standard XRD Powder Patterns*, Circular 539, Pt. 7, National Bureau of Standarrds, Gaithersburg, MD, 1957, pp. 33–35.

Ca–Hg

Phase	*Crystal structure*	*Lattice parameters (nm)*	*Density* $(g\ cm^{-3})$	*Protoype*
CaHg	Cubic	$a = 0.3758$		CsCl
$CaHg_2$	Trigonal	$a = 0.4894$, $c = 0.3571$		$CeCd_2$
$CaHg_3$	Hexagonal	$a = 0.6635$, $c = 0.502$		Ni_3Sn
$CaHg_{8-11}$	Cubic			$BaHg_{11}$
Ca_2Hg	Orthorhombic	$a = 0.786$, $b = 0.489$, $c = 0.987$		Co_2Si (Ni_2Si)
Ca_3Hg	Orthorhombic	$a = 0.816$, $b = 1.015$, $c = 0.6823$		Fe_3C
Ca_3Hg_2	Tetragonal	$a = 0.8476$, $c = 0.4197$		U_3Si_2
Ca_5Hg_3	Tetragonal	$a = 0.8183$, $c = 1.470$		Cr_5B_3
$Ca_{11-x}Hg_{54-x}$	Hexagonal	$a = 1.339$, $c = 0.9615$		

G. Bruzzone and F. Merlo, The calcium–mercury system, *J. Less-Common Met.*, 1973, **32**, 237–241.

A. V. Tkachuk and A. Mar, Alkaline-earth metal mercury intermetallics $A_{11-x}Hg_{54+x}$ (A = Ca, Sr), *Inorg. Chem.*, 2008, **47**, 1313–1318.

Cd–Hg

Phase	*Crystal structure*	*Lattice parameters (nm)*	*Density* $(g\ cm^{-3})$	*Prototype*
ω 29–98% Hg	Body-centered tetragonal	$a = 0.3965$, $c = 0.2869$		In
ω′ 33–42% Hg (Cd_2Hg)	Body-centered tetragonal	$a = 0.3965$, $c = 0.8607$		$MoSi_2$
ω′ 58–71% Hg ($CdHg_2$)	Body-centered tetragonal	$a = 0.3966$, $c = 0.8607$		$MoSi_2$

T. Claeson, H. L. Luo, T. R. Anatharaman and M. E. Merriam, Order–disorder transformations at 2:1 composition in the Cd–Hg system, *Acta Metall.*, 1996, **14**, 285–290.

R. Kubiak, A. Pietraszko and K. Lukaszewicz, X-ray investigation of $Cd_{65}Hg_{35}$ and $Cd_{55}Hg_{45}$, *Acta Crystallogr. A*, 1978, **34**, S179.

B. Predel and W. Schwermann, Order–disorder transition in the Cd–Hg system, *J. Inst. Met.*, 1971, **99**, 209–212.

V. V. Prytkin, L. M. Lityagina and E. V. Biblik, The effect of temperature and pressure on the lattice parameters of the ω-phase in the Cd–Hg system, *Metallofizika*, 1986, **8**, 76–80.

Ce–Hg

Phase	*Crystal structure*	*Lattice parameters (nm)*	*Density (g cm^{-3})*	*Prototype*
CeHg	Cubic	$a = 0.3816$		CsCl
$CeHg_2$	Trigonal	$a = 0.4942$, $c = 0.3540$		$CeCd_2$
$CeHg_3$	Hexagonal	$a = 0.6755$, $c = 0.4957$		Ni_3Sn
$Ce_{11}Hg_{45}$	Cubic	$a = 2.1857$		$Sm_{11}Cd_{45}$

A. Iandelli and A. Palenzona, Crystal chemistry of intermetallic compounds, in *Handbook on the Physics and Chemistry of Rare Earths*, ed. K. A. Gschneidner Jr and L Eyring, North-Holland, Amsterdam, 1979, pp. 1–54.

F. Merlo and M. L. Fornasini, Crystal structure of the $R_{11}Hg_{45}$ compounds (R = La, Ce, Pr, Nd, Sm, Gd, Yb, U), *J. Less-Common Met.*, 1979, **64**, 221–231.

Cl–Hg

Phase	*Crystal structure*	*Lattice parameters (nm)*	*Density (g cm^{-3})*	*Prototype*
α-Hg_2Cl_2				
β-Hg_2Cl_2		$a = 0.4482$, $c = 1.091$		βHg_2Cl_2
γ-Hg_2Cl_2	Orthorhombic	$a = 0.423$, $b = 0.454$, $c = 1.044$		
$HgCl_2$		$a = 1.2768$, $b = 0.59756$, $c = 0.43347$		$PbCl_2$

N. J. Calos, C. H. L. Kermard and D. R. Lindsay, The structure of Hg_2Cl_2 derived from neutron powder data, Z. *Kristallogr.*, 1989, **187**, 305–307.

E. Dorm, Intermetallic distances in Hg(I) halides Hg_2F_2, Hg_2Cl_2, Hg_2Br_2, *J. Chem. Soc., Chem. Commun.*, 1971, 466–467.

C. Guminski, The Cl–Hg (chlorine–mercury) system, *J. Phase Equilib.*, 1994, **15**, 101–107.

V. Subramamian and K. Serf, $HgCl_2$, a redetermination, *Acta Crystallogr. B*, 1980, **36**, 2132–2135.

S. J. Yosim and S. W. Mayer, The Hg–$HgCl_2$ system, J. *Phys.Chem. Solids*, 1960, **64**, 909–911.

Cs–Hg

Phase	*Crystal structure*	*Lattice parameters (nm)*	*Density (g cm^{-3})*	*Prototype*
Cs_2Hg_{27}	Cubic	$a = 1.6557$	12.47^{calc}	Cs_2Hg_{27}
Cs_3Hg_2	Cubic	$a = 1.0913$	11.27^{calc}	Cs_3Hg_{20}
Cs_5Hg_1	Tetragonal	$a = 1.1803$, $c = 1.0814$	9.87^{calc}	Rb_5Hg_{19}
CsHg	Orthorhombic	$a = 0.8727$, $b = 0.5488$, $c = 0.9082$	8.19	KHg_2
CsHg	Triclinic	$a = 0.7154$, $b = 0.7470$, $c = 0.7635$ $\alpha = 107.82°$, $\beta = 103.34°$, $\gamma = 90.95°$	5.91	KHg

H.-J. Deiseroth and A. Strunck, Square Hg_4 clusters in the compound CsHg, *Angew. Chem. Int. Ed. Engl.*, 1987, **26**, 687–688.

H.-J. Deiseroth, A. Strunck and W. Bauhofer, $RbHg_2$ and $CsHg_2$ – preparation, crystal structures and electrical conductivities, *Z. Anorg. Allg. Chem.*, 1988, **558**, 128–136.

H.-J. Deiseroth, A. Strunck and W. Bauhofer, CsHg, an unusual variant of the CsCl structure. Preparation, crystal structure and physical properties, *Z. Anorg. Allg. Chem.*, 1989, **575**, 31.

C. Hoch and A. Simon, Cs_2Hg_{27}, the mercury-richest amalgam with close relationship to the Bergman phases, *Z. Anorg. Allg. Chem.*, 2008, **634**, 853–856.

E. Todorov and S. C. Sevov, Synthesis and structure of the alkali-metal amalgams A_3Hg_{20} (A = Rb, Cs), K_3Hg_{11}, Cs_5Hg_{19} and A_7Hg_{31} (A = K, Rb), *J. Solid State Chem.*, 2000, **149**, 419–427.

Cu–Hg

Phase	*Crystal structure*	*Lattice parameters (nm)*	*Density (g cm^{-3})*	*Prototype*
Cu_7Hg_6	Rhombohedral	$a = 0.94024$ $\alpha = 90.425°$		

H. J. Bernhardt and K. Schmetzer, Belendorffite, a new copper amalgam dimorphous with kolymite, *Neues Jahrb. Mineral. Monatsh.*, 1992, (1), 21 28.

M. M. Carnasciali and G. A. Costa, Cu_xHg_y: a puzzling compound, *J. Alloys Compd.*, 2001, **317–318**, 491–496.

T. Lindahl and S. Westman, Structure of rhombohedral γ brass-like phase in Cu–Hg system, *Acta Chem. Scand.*, 1969, **23**, 1181–1190.

E. A. Markova, N. M. Chernitsova, Yu. S. Borodaev, L. S. Dubakina and O. E. Yushko-Zakharova, The new mineral kolymite, Cu_7Hg_6, *Int. Geol. Rev.*, 1982, **24**, 233–237.

Dy–Hg

Phase	*Crystal structure*	*Lattice parameters (nm)*	*Density (g cm^{-3})*	*Prototype*
DyHg	Cubic	$a = 0.3676$		CsCl
$DyHg_2$	Trigonal	$a = 0.4816$, $c = 0.3466$		$CeCd_2$
$DyHg_3$	Hexagonal	$a = 0.6543$, $c = 0.4880$		Ni_3Sn

A. Iandelli and A. Palenzona, Atomic size of rare earths in intermetallic compounds, MX compounds of CsCl type, *J. Less-Common Met.*, 1965, **9**, 1–6.

A. Iandelli and A. Palenzona, Crystal chemistry of intermetallic compounds, in *Handbook on the Physics and Chemistry of Rare Earths*, ed. K. A. Gschneidner Jr and L Eyring, North-Holland, Amsterdam, 1979, pp. 1–54.

A. Palenzona, MX_3 intermetallic phase of the rare earths with Hg, In, Tl, Pb, *J. Less-Common Met.*, 1966, **10**, 290–292.

Er–Hg

Phase	Crystal structure	Lattice parameters (nm)	Density (g cm^{-3})	Prototype
ErHg	Cubic	$a=0.3645$		CsCl
$ErHg_2$	Trigonal	$a=0.4790$, $c=0.3442$		$CeCd_2$
$ErHg_3$	Hexagonal	$a=0.6505$, $c=0.4866$		Ni_3Sn

A. Iandelli and A. Palenzona, Atomic size of rare earths in intermetallic compounds, MX compounds of CsCl type, *J. Less-Common Met.*, 1965, **9**, 1–6.

A. Palenzona, MX_3 intermetallic phase of the rare earths with Hg, In, Tl, Pb, *J. Less-Common Met.*, 1966, **10**, 290–292.

Eu–Hg

Phase	Crystal structure	Lattice parameters (nm)	Density (g cm^{-3})	Prototype
EuHg	Cubic	$a=0.3880$		CsCl
$EuHg_2$	Trigonal	$a=0.4978$, $c=0.3710$		$CeCd_2$
$EuHg_3$	Hexagonal	$a=0.6794$, $c=0.5074$		Ni_3Sn
$Eu_{14}Hg_{51}$	Hexagonal	$a=1.357$, $c=0.974$		$Gd_{14}Ag_{51}$

D. M. Bailey and G. R. Kline, *Acta Crystallogr. B*, 1971, **27,** 650.

S. J. Lyle and W. A. Westal, A Mössbauer spectroscopic study of the Eu–Hg system, *J. Less-Common Met.*, 1984, **99**, 265–272.

A. Iandelli and A. Palenzona, Atomic size of rare earths in intermetallic compounds, MX compounds of CsCl type, *J. Less-Common Met.*, 1965, **9**, 1–6.

A. Iandelli and A. Palenzona, Crystal chemistry of intermetallic compounds, in *Handbook on the Physics and Chemistry of Rare Earths*, ed. K. A. Gschneidner Jr and L Eyring, North-Holland, Amsterdam, 1979, pp. 1–54.

F–Hg

Phase	Crystal structure	Lattice parameters (nm)	Density (g cm^{-3})	Prototype
Hg_2F_2		$a=0.3673$, $c=1.0884$		CaF_2
HgF_2		$a=0.55373$		

E. Dorm, *J. Chem. Soc., Chem. Commun.*, 1971, 466–467.

H. E. Swanson, M. C. Morris, R. P. Stinchfield and E. H. Evans, *Standard X-Ray Diffraction Powder Patterns*, NBS Monograph 25, National Bureau of Standards, Gaithersburg, MD, 1963, Sect. 2, p. 25.

Ga–Hg

No intermetallic compounds.

Gd–Hg

Phase	*Crystal structure*	*Lattice parameters (nm)*	*Density (g cm^{-3})*	*Prototype*
GdHg	Cubic	$a = 0.3719$		CsCl
$GdHg_2$	Trigonal	$a = 0.4854$, $c = 0.3496$		$CeCd_2$
$GdHg_3$	Hexagonal	$a = 0.6591$, $c = 0.4889$		Ni_3Sn
$Gd_{14}Hg_{51}$	Hexagonal			$Gd_{14}Ag_{51}$
$Gd_{11}Hg_{45}$	Cubic	$a = 2.1551$		$Sm_{11}Cd_{45}$

C. Guminski, The Gd–Hg (gadolinium–mercury) system, *J. Phase Equilib.*, 1995, **16**, 181–185.

A. Iandelli, Intermetallic and metalloid Gd compounds, *An. Accad. Naz. Lincei Rend. Cl. Sci. Fis. Mat. Nat.*, 1960, **29**, 62–69.

A. Iandelli and A. Palenzona, Crystal chemistry of intermetallic compounds, in *Handbook on the Physics and Chemistry of Rare Earths*, ed. K. A. Gschneidner Jr and L Eyring, North-Holland, Amsterdam, 1979, pp. 1–54.

H. R. Kirchmayr, Compounds of Y, Sm and Gd with Hg, *Acta Phys. Austriaca*, 1964, **18**, 193–204.

F. Merlo and M. L. Fornasini, Crystal structure of the $R_{11}Hg_{45}$ compounds (R = La, Ce, Pr, Nd, Sm, Gd, Yb, U), *J. Less-Common Met.*, 1979, **64**, 221–231.

A. Palenzona, MX_3 intermetallic phase of the rare earths with Hg, In, Tl, Pb, *J. Less-Common Met.*, 1966, **10**, 290–292.

Hf–Hg

Phase	*Crystal structure*	*Lattice parameters (nm)*	*Density (g cm^{-3})*	*Prototype*
Hf_2Hg	Tetragonal	$a = 0.3345$, $c = 1.1496$		$MoSi_2$

F. Kurka and P. Ettmayer, Die Kristallstruktur von Hf_2Hg, *Monatsh. Chem.*, 1967, **98**, 2414–2418.

Ho–Hg

Phase	*Crystal structure*	*Lattice parameters (nm)*	*Density (g cm^{-3})*	*Prototype*
HoHg	Cubic	$a = 0.3660$		CsCl
$HoHg_2$	Trigonal	$a = 0.4798$, $c = 0.3470$		$CeCd_2$
$HoHg_3$	Hexagonal	$a = 0.6526$, $c = 0.4872$		Ni_3Sn

A. Iandelli and A. Palenzona, Atomic size of rare earths in intermetallic compounds, MX compounds of CsCl type, *J. Less-Common Met.*, 1965, **9**, 1–6.

A. Iandelli and A. Palenzona, Crystal chemistry of intermetallic compounds, in *Handbook on the Physics and Chemistry of Rare Earths*, ed. K. A. Gschneidner Jr and L Eyring, North-Holland, Amsterdam, 1979, pp. 1–54.

H. R. Kirchmayr, Lattice constants and structures of compounds HoHg, $HoHg_2$ and $HoHg_3$, *Monatsh. Chem.*, 1964, **95**, 1667–1670.
A. Palenzona, MX_3 intermetallic phase of the rare earths with Hg, In, Tl, Pb, *J. Less-Common Met.*, 1966, **10**, 290–292.

I–Hg

Phase	*Crystal structure*	*Lattice parameters (nm)*	*Density (g cm^{-3})*	*Prototype*
Hg_2I_2		$a=0.4924$, $c=1.1633$		Hg_2Cl_2
α-HgI_2		$a=0.4374$, $c=1.2435$		
β-HgI_2		$a=0.4702$, $b=0.7432$, $c=1.3872$		$HgBr_2$
γ-HgI_2		$a=0.422$, $c=2.370$		

H. Oppermann, On the constitution barogram of the HgI_2–I_2 system, *Z. Anorg. Allg. Chem.*, 1989, **576**, 229–234.
C. Guminski, The Hg–I (mercury–iodine) system, *J. Phase Equilib.*, 1997, **18**, 206–215.

In–Hg

Phase	*Crystal structure*	*Lattice parameters (nm)*	*Density (g cm^{-3})*	*Prototype*
ε 78–94 at.% In_2Hg	Face-centered cubic	$a=0.4694$		Cu
InHg	Rhombohedral	$a=0.4846$ $\alpha=43.24°$		CuPt–$L1_1$
$InHg_4$	Face-centered orthorhombic	$a=1.0872$, $b=0.4847$, $c=0.3522$		γ-Pu

T. Claeson and M. F. Merriam, New phase in the mercury–indium system, *J. Less-Common Met.*, 1966, **11**, 186–190.
B. R. Coles, M. F. Merriam and Z. Fisk, The phase diagram of the mercury–indium alloy system, *J. Less-Common Met.*, 1963, **5**, 41–48.
T. X. Mahy and B. C. Giessen, A new representative of the γ-Pu structure type: the crystal structure of β($Hg_{0.80}In_{0.20}$), *J. Less-Common Met.*, 1979, **63**, 257–264.
H. Okamoto, *Phase Diagrams of Binary Indium Alloys*, ASM, Metals Park, OH, 1991, p. 129.
M. Segnini and B. C. Giessen, The crystal structure of HgIn, *Acta Crystallogr. B*, 1972, **28**, 320–321.
C. Tyzack and G. V. Raynor, *Trans. Faraday Soc.*, 1954, **50**, 675–684.

K–Hg

Phase	*Structure*	*Lattice parameters (nm)*	*Density (g cm^{-3})*	*Prototype*
KHg_{11}	Cubic	$a=0.9630$	12.46^{calc}	$BaHg_{11}$
K_3Hg_{11}		$a=0.5122$, $b=1.0063$, $c=1.4782$	10.13^{calc}	α-La_3Al_{11}

Phase	*Structure*	*Lattice parameters (nm)*	*Density (g cm^{-3})*	*Prototype*
KHg_6	Orthorhombic			
KHg_7				
K_7Hg_{31}	Hexagonal	$a = 1.0850$ m, $c = 1.0210$	10.36^{calc}	Ba_7Hg_{31}
K_2Hg_7	Hexagonal	$a = 0.67175$, $c = 0.64133$		K_2Hg_7
α-KHg_2	Orthorhombic	$a = 0.810$, $b = 0.516$, $c = 0.877$	7.88^{calc}, 7.95^{meas}	KHg_2
β-KHg_2				Mod. AlB_2
$K_{29}Hg_{48}$	Cubic			
K_5Hg_7	Orthorhombic	$a = 1.006$, $b = 1.945$, $c = 0.834$		K_5Hg_7
KHg	Triclinic	$a = 0.659$, $b = 0.676$, $c = 0.706$ $\alpha = 106°5'$, $\beta = 101°52'$, $\gamma = 92°47'$	5.41^{calc}, 5.47^{meas}	KHg

E. Biehl, Thesis, University of Siegen, 1998.

E. Biehl and H.-J. Deiseroth, Preparation, structural relations and magnetism of amalgams MHg_{11} (M: K, Rb, Ba, Sr), *Z. Anorg. Allg. Chem.*, 1999, **625**, 1073–1080.

E. Biehl, H.-J. Deiseroth, K_2Hg_7 und Rb_2Hg_7, zwei Vertreter eines neuen Strukturtyps binärer intermetallischer Verbindungen, *Z. Anorg. Allg. Chem.*, 1999, **625**, 1337–1342.

H.-J. Deiseroth, Alkali metal amalgams: a group of unusual alloys, *Prog. Solid State Chem.*, 1997, **25**, 73–123.

E. J. Duwell and N. C. Baenziger, The crystal structures of KHg and KHg_2, *Acta Crystallogr.*, 1955, **8**, 705.

E. J. Duwell and N. C. Baenziger, The crystal structure of K_5Hg_7, *Acta Crystallogr.*, 1960, **13**, 476–479.

E. Maey, *Z. Phys. Chem.*, 1899, **29**, 119–138.

E. Todorov and S. C. Sevov, Synthesis and structure of the alkali-metal amalgams A_3Hg_{20} (A = Rb, Cs), K_3Hg_{11}, Cs_5Hg_{19} and A_7Hg_{31} (A = K, Rb), *J. Solid State Chem.*, 2000, **149**, 419–427.

La–Hg

Phase	*Crystal structure*	*Lattice parameters (nm)*	*Density (g cm^{-3})*	*Prototype*
LaHg	Cubic	$a = 0.3864$		CsCl
$LaHg_2$	Trigonal	$a = 0.4960$, $c = 0.3650$		$CeCd_2$
$LaHg_3$	Hexagonal	$a = 0.6816$, $c = 0.4971$	12.30^{calc}	Ni_3Sn
$La_{11}Hg_{45}$	Cubic	$a = 2.1997$		γ-Brass
$La_{13}Hg_{58}$	Hexagonal	$a = 1.567$, $c = 1.548$		$Pu_{13}Zn_{58}$
$LaHg_6$	Orthorhombic	$a = 0.9763$, $b = 2.886$, $c = 0.5004$		

G. Bruzzone and F. Merlo, The lanthanum–mercury system, *J. Less-Common Met.*, 1976, **44**, 259–265.

C. Guminski, The La–Hg (lanthanum–mercury) system, *J. Phase Equilibria*, 1995, **16**, 86–192.

A. Iandelli and A. Palenzona, Crystal chemistry of intermetallic compounds, in *Handbook on the Physics and Chemistry of Rare Earths*, ed. K. A. Gschneidner Jr and L Eyring, North-Holland, Amsterdam, 1979, pp. 1–54.

F. Merlo and M. L. Fornasini, Crystal structure of the $R_{11}Hg_{45}$ compounds (R = La, Ce, Pr, Nd, Sm, Gd, Yb, U), *J. Less-Common Met.*, 1979, **64**, 221–231.

Li–Hg

Phase	*Crystal structure*	*Lattice parameters (nm)*	*Density (g cm^{-3})*	*Prototype*
$LiHg_3$	Hexagonal	$a = 0.6253$, $c = 0.4804$		Ni_3Sn
$LiHg_2$				
LiHg (38–62 at.% Li at 375 °C)	Cubic	$a = 0.3315$	9.28^{meas} (52.1 at.% Li)	CsCl
Li_3Hg	Cubic	$a = 0.6548$		Li_3Bi

G. Grube and W. Wolf, *Z. Elektrochem.*, 1935, **41**, 675–679.

E. Zintl and A. Schneider, Röntgenanalyse der Lithium-Amalgame, *Z. Elektrochem.*, 1935, **41**, 771–774.

E Zintl and G. Brauer, *Z. Phys. Chem.*, 1933, **B20**, 245–271.

G. J. Zukowsky, *Z. Anorg. Allg. Chem.*, 1911, **71**, 403–418.

Lu–Hg

Phase	*Crystal structure*	*Lattice parameters (nm)*	*Density (g cm^{-3})*	*Prototype*
LuHg	Cubic	$a = 0.3607$		CsCl
$LuHg_3$	Hexagonal	$a = 0.6467$, $c = 0.4851$		Ni_3Sn

C. Guminski, The Hg–Lu (mercury–lutetium) system, *J. Phase Equilib.*, 1995, **16**, 276.

A. Iandelli and A. Palenzona, Atomic size of rare earths in intermetallic compounds, MX compounds of CsCl type, *J. Less-Common Met.*, 1965, **9**, 1–6.

A. Iandelli and A. Palenzona, Crystal chemistry of intermetallic compounds, in *Handbook on the Physics and Chemistry of Rare Earths*, ed. K. A. Gschneidner Jr and L Eyring, North-Holland, Amsterdam, 1979, pp. 1–54.

A. Palenzona, MX_3 intermetallic phase of the rare earths with Hg, In, Tl, Pb, *J. Less-Common Met.*, 1966, **10**, 290–292.

Mg–Hg

Phase	*Crystal structure*	*Lattice parameters (nm)*	*Density (g cm^{-3})*	*Prototype*
Mg_3Hg	Hexagonal	$a = 0.483$, $c = 0.862$		Na_3As
Mg_5Hg_2				
Mg_2Hg	Orthorhombic	$a = 0.6219$, $b = 0.4617$, $c = 0.8799$		$CoSi_2$

Phase	*Crystal structure*	*Lattice parameters (nm)*	*Density (g cm^{-3})*	*Prototype*
Mg_5Hg_3	Hexagonal	$a=0.8260$, $c=0.5931$		Mn_5Si_3
MgHg	Cubic	$a=0.3449$		CsCl
$MgHg_2$		$a=0.3838$, $c=0.8799$		$MoSi_2$

G. Brauer and W. Hauke, Crystal structure of intermetallic compounds MgAu and MgHg, *Z. Phys. Chem. B*, 1936, **33**, 304–310.

G. Brauer and R. Rudolph, X-ray analysis of magnesium amalgams, *Z. Anorg. Chem.*, 1941, **248**, 405–524.

G. Brauer, H. Nowotny and R. Rudolph, X-ray investigation of the Mg–Hg system, *Z. Metallkd.*, 1947, **38**, 81–84.

J. L. C. Daams and J. H. N. van Vucht, The Mg–Au–Hg system, *Philips J. Res.*, 1984, **39**, 275–292.

A. A. Nayeb-Hashemi and J. B. Clark, The Hg–Mg (mercury–magnesium system), *J. Phase Equilib.*, 1987, **8**, 65–70.

Mn–Hg

Phase	*Crystal structure*	*Lattice parameters (nm)*	*Density (g cm^{-3})*	*Prototype*
α-MnHg	Cubic	$a=0.3318$		CsCl
β-MnHg	Tetragonal	$a=0.3298$, $c=0.3912$		MnHg
γ-MnHg				
Mn_2Hg_5	Tetragonal	$a=0.9758$, $c=0.2998$	12.85^{meas}, 13.00^{calc}	Mn_2Hg_5

J. E. deWet, Intermetallic phases in the Mn Hg system, *Angew. Chem.*, 1955, **67**, 208.

J. E deWet, A new metallic layer structure, *Nature*, 1957, **180**, 1412–1413.

J. E deWet, Crystal structure of Mn_2Hg_5, *Acta Crystallogr.*, 1961, **14**, 733–738.

E. Lihl, Investigations on the amalgams of Mn, Fe, Co, Ni and Cu, *Z. Metallkd.*, 1953, **44**, 160 166.

E. Lihl, On the structure of the Hg–Mn system, *Monatsh. Chem.*, 1955, **86**, 186–190.

Z. Moser and C. Gumiński, The Hg–Mn system, *J. Phase Equilib.*, 1993, **14**, 726–733.

M. Ohashi, The magnetic and thermal properties of the intermetallic compound MnHg, *J. Phys. Soc. Jpn.*, 1965, **20**, 911–914.

Na–Hg

Phase	*Crystal structure*	*Lattice parameters (nm)*	*Density (g cm^{-3})*	*Protoype*
$Na_{11}Hg_{52}$	Hexagonal	$a=3.9703$, $c=0.96810$	12.04^{calc}	$Na_{11}Hg_{52}$
$NaHg_2$	Hexagonal	$a=0.5029$, $c=0.3230$	9.96^{calc}, 9.94^{meas}	Modified AlB_2
α-NaHg	End-centered orthorhombic	$a=0.7184$, $b=1.0784$, $c=0.5198$		Distorted CsCl

Phase	*Crystal structure*	*Lattice parameters (nm)*	*Density ($g\,cm^{-3}$)*	*Protoype*
β-NaHg	Rhombohedral	$a = 0.5071$, $c = 1.2668$		Distorted NaTl
γ-NaHg	Cubic	$a = 0.5129$		NaTl
Na_3Hg_2	Tetragonal	$a = 0.84587$, $c = 0.77078$	5.66^{calc}	Na_3Hg_2
α-Na_8Hg_3	Hexagonal	$a = 0.5433$, $c = 0.7795$		Au_8Al_3
β-Na_8Hg_3	Monoclinic	$a = b = c = 0.7676$ $\beta = 90.7°$		Distorted Li_3Bi
γ-Na_8Hg_3	Cubic	$a = 0.7674$		Li_3Bi
α-Na_3Hg	Hexagonal	$a = 0.5438$, $c = 0.9808$	3.56^{calc} (34 °C)	Na_3As
β-Na_3Hg	Rhombohedral	$a = 0.5404$, $c = 1.3420$	3.95^{calc} (42 °C)	Modified Li_3Bi

H.-J. Deiseroth and M. Rochnia, Temperature dependent phase transitions of sodium-rich amalgams, *Z. Anorg. Allg. Chem.*, 1992, **616**, 35–38.

H.-J. Deiseroth and M. Rochnia, β-Na_3Hg: a solid with a quasi-fluid sodium substructure between 36 and 60 °C, *Angew. Chem. Int. Ed. Engl.*, 1993, 32, 1494–1495.

H.-J. Deiseroth and M. Rochnia, Temperature-dependent phase conversions of sodium-rich amalgams, *Z. Anorg. Allg. Chem.*, 1994, **620**, 1736–1740.

H.-J. Deiseroth and D. Toelstede, Na_8Hg_3: an alkali metal rich amalgam with isolated mercury anions?, *Z. Anorg. Allg. Chem.*, 1990, **587**, 103–109.

H.-J. Deiseroth and D. Toelstede, Na_3Hg – the most sodium-rich amalgam in the system sodium–mercury, *Z. Anorg. Allg. Chem.*, 1992, **615**, 43–48.

H.-J. Deiseroth, A. Stupperich, R. Pankaluoto and N. E. Christensen, NaHg: a variant of the cesium chloride structure – structural relations and electronic structure, *Z. Anorg. Allg. Chem.*, 1991, **597**, 41–50.

H.-J. Deiseroth, E. Biehl and M. Rochnia, Sodium amalgams: phase diagram, structural chemistry and thermodynamic data, a summary of recent developments, *J. Alloys Compd.*, 1997, **246**, 80–90.

C. Hoch and A. Simon, $Na_{11}Hg_{52}$: complexity in a polar metal, *Angew. Chem. Int. Ed.*, 2012, **51**, 3262–3265.

E. Maey, *Z. Phys. Chem.*, 1899, **29**, 119.

J. W. Nielsen and N. C. Baenziger, The crystal structures of $NaHg_2$, NaHg and Na_3Hg_2, *Acta Crystallogr.*, 1954, **7**, 277–282.

M. Rochnia and H.-J. Deiseroth, Polymorphism of NaHg: the first experimental observation of a reversible, temperature driven B2 (CsCl)–B32 (NaTl) phase transition, *Croat. Chem. Acta*, 1995, **68**, 701–708.

A. V. Tkachuk and A. Mar, Redetermination of Na_3Hg_2, *Acta Crystallogr. E*, 2006, **62**, i129–i130.

Nd–Hg

Phase	*Crystal structure*	*Lattice parameters (nm)*	*Density (g cm^{-3})*	*Prototype*
NdHg	Cubic	$a=0.3780$		CsCl
$NdHg_2$	Trigonal	$a=0.4904$, $c=0.3520$		$CeCd_2$
$NdHg_3$	Hexagonal	$a=0.6695$, $c=0.4929$		Ni_3Sn
$Nd_{11}Hg_{45}$	Cubic	$a=2.1716$		$Sm_{11}Cd_{45}$

C. Guminski, The Hg–Nd (mercury–neodymium) system, *J. Phase Equilibria*, 1995, **16**, 448–453.

A. Iandelli and A. Palenzona, Crystal chemistry of intermetallic compounds, in *Handbook on the Physics and Chemistry of Rare Earths*, ed. K. A. Gschneidner Jr and L Eyring, North-Holland, Amsterdam, 1979, pp. 1–54.

F. Merlo and M. L. Fornasini, Crystal structure of the $R_{11}Hg_{45}$ compounds (R = La, Ce, Pr, Nd, Sm, Gd, Yb, U), *J. Less-Common Met.*, 1979, **64**, 221–231.

Ni–Hg

Phase	*Crystal structure*	*Lattice parameters (nm)*	*Density (g cm^{-3})*	*Protoype*
NiHg	Tetragonal	$a=0.422$, $c=0.314$		AuCu
$NiHg_2$	Tetragonal	$a=0.456$, $c=0.283$		$PtHg_2$
$NiHg_3$	Cubic	$a=0.3005$		
$NiHg_4$	Cubic	$a=0.3004$		$PtHg_4$

A. Baranski and Z. Galus, An electrochemical study of the equilibria in the nickel–mercury system, *J. Electroanal. Chem. Interfacial Electrochem.*, 1973, **46**, 289–305.

L. F. Bates and J. H. Prentice, *Proc. Phys. Soc.*, 1939, **51** 419.

F. Lihl, Investigation of the amalgams of the metals manganese, iron, cobalt, nickel and copper, *Z. Metallkd.*, 1953, **44**, 160.

F. Lihl and H. Nowotny, The structure of $NiHg_4$, *Z. Metallkd.*, 1953, **44**, 358–360.

M. Puselj and Z. Ban, Preparation and crystal structure of NiHg, *Z. Naturforsch.*, 1977, **32B**, 497.

O–Hg

Phase	*Crystal structure*	*Lattice parameters (nm)*	*Density (g cm^{-3})*	*Prototype*
HgO	Triclinic	$a=0.665$, $b=0.554$, $c=0.701$ $\alpha=90.15°$, $\beta=90.45°$, $\gamma=90.95°$		HgO
HgO	Orthorhombic	$a=0.66074$, $b=0.55254$, $c=0.35215$		HgO
HgO	Tetragonal	$a=0.82941$, $b=0.71121$		
HgO	Cubic	$a=0.534$		ZnS
HgO	Hexagonal	$a=0.3578$, $c=0.8685$		HgS
α-HgO_2	Rhombohedral	$a=0.44702$, $b=0.54592$, $c=0.35192$ $\beta=108.45°$		α-HgO_2

Phase	Crystal structure	Lattice parameters (nm)	Density (g cm^{-3})	Prototype
β-HgO_2		$a = 0.608$, $b = 0.601$, $c = 0.480$		β-HgO_2
Hg_2O	Decomposes at 100 °C			

C. Guminski, The Hg–O (mercury-oxygen) system, *J. Phase Equilib.*, 1999, **20**, 85–88.

Pb–Hg

Phase	Crystal structure	Lattice parameters (nm)	Density (g cm^{-3})	Prototype
Pb_2Hg	Face-centered tetragonal	$a = 0.3520$, $c = 0.4512$		CuAu

M. Ellner and B. Predel, The structure of TlSn(h) and its crystal chemical relationship to similar phases, *Z. Metalkd.*, 1975, **66**, 503–506.
C. Tyzack and G. V. Raynor, The lattice spacings of Pb-rich substitutional solid solutions, *Acta Crystallogr.*, 1954, **7**, 505–510.

Pd–Hg

Phase	Crystal structure	Lattice parameters (nm)	Density (g cm^{-3})	Prototype
$PdHg_4$ or Pd_5Hg_{21}				Complex γ-brass
Pd_2Hg_5	Tetragonal	$a = 0.9463$, $c = 0.3031$		Mn_2Hg_5
PdHg	Tetragonal	$a = 0.3026$, $c = 0.3702$		AuCu

A. Barański, A. Kryska and Z. Galus, On the electrochemical properties of the Pd + Hg system, *J. Electroanal. Chem.*, 1993, **349**, 341–354.
H. Bittner and H. Nowotny, Investigation of the system Pd–Hg, *Monatsh. Chem.*, 1952, **83**, 287–288.
H. Bittner and H. Nowotny, Further magnetic measurements of Hume–Rothery phases, *Monatsh. Chem.*, 1952, **83**, 1308–1313.
J. D. Cummins and A. F. Berndt, A single crystal study of Pd–Hg and γ-Sn–Hg, *J. Less-Common Met.*, 1969, **19**, 431–432.
P. Ettmayer, Die Kristallstruktur von Pd_2Hg_5, *Monatsh. Chem.*, 1965, **96**, 884–888.
K. Terada and F. W. Cagle Jr, The solid solution of mercury in palladium, *Acta Crystallogr.*, 1961, **14**, 1299.

Po–Hg

Phase	Crystal structure	Lattice parameters (nm)	Density (g cm^{-3})	Prototype
PoHg	Ordered fcc	$a = 0.6250$		NaCl

W. G. Witteman, A. L. Giorgi and D. T. Vier, The preparation and identification of some intermetallic compounds of polonium, *J. Phys. Chem.*, 1960, **64**, 434–440.

Pr–Hg

Phase	Crystal structure	Lattice parameters (nm)	Density (g cm^{-3})	Prototype
PrHg	Cubic	$a=0.3799$		CsCl
$PrHg_2$	Trigonal	$a=0.4918$, $c=0.3539$		$CeCd_2$
$PrHg_3$	Hexagonal	$a=0.6724$, $c=0.4937$		Ni_3Sn
$Pr_{11}Hg_{45}$	Cubic	$a=2.1786$		$Sm_{11}Cd_{45}$

A. Iandelli and A. Palenzona, Crystal chemistry of intermetallic compounds, in *Handbook on the Physics and Chemistry of Rare Earths*, ed. K. A. Gschneidner Jr and L Eyring, North-Holland, Amsterdam, 1979, pp. 1–54.

F. Merlo and M. L. Fornasini, Crystal structure of the $R_{11}Hg_{45}$ compounds (R = La, Ce, Pr, Nd, Sm, Gd, Yb, U), *J. Less-Common Met.*, 1979, **64**, 221–231.

Pt–Hg

Phase	Crystal structure	Lattice parameters (nm)	Density (g cm^{-3})	Prototype
$PtHg_4$	Body-centered cubic	$a=0.61808$		$PtHg_4$
$PtHg_2$	Tetragonal	$a=0.4675$, $c=0.2918$	15.6^{calc}	$PtHg_2$
PtHg	Tetragonal	$a=0.4193$, $c=0.3817$		AuCu ($L1_0$)

E. Bauer, H. Nowotny and A. Stempfl, X-ray investigation in the Pt–Hg system, *Monatsh. Chem.*, 1953, **84**, 211–212.

E. Bauer, H. Nowotny and A. Stempfl, X-ray investigation in the Pt Hg system, *Monatsh. Chem.*, 1953, **84**, 692–700.

S. K. Lahiri, J. Angilello and M. Natan, Precise lattice parameter determination of $PtHg_4$, *J. Appl. Crystallogr.*, 1982, **15**, 100–101.

G. D. Robbins and C. G. Enke, Investigation of the compound formed at a Pt–Hg interface, *J. Electroanal. Chem.*, 1969, **23**, 343–349.

G. R. Souza, I. A. Pastre, A. V. Benedetti, C. A. Ribeiro and F. L. Fertonani, Solid state reactions in the platinum–mercury system, *J. Thermal Anal. Calorim.*, 2007, **88**, 127–132.

Pu–Hg

Phase	Crystal structure	Lattice parameters (nm)	Density (g cm^{-3})	Prototype
Pu_5Hg_{21} or $Pu_{11}Hg_{45}$	Cubic	$a=2.178$	13.90^{calc}	γ-Brass or $Sm_{11}Cd_{45}$
$PuHg_3$	Hexagonal			Probably Ni_3Sn

A. F. Berndt, A gamma phase in the Pu–Hg system, *J. Less-Common Met.*, 1966, **11**, 216–219.

A. S. Coffinberry and F. H. Ellinger, Intermetallic compounds of Pu, in *Peaceful Uses of Atomic Energy*, United Nations, New York, 1956, vol. 9, pp. 138–146.

C. Gumiński, The Hg–Pu system, *J. Equil. Diagr. Diffusion*, 2005, **26**, 77–79.

Rb–Hg

Phase	Structure	Lattice parameters (nm)	Density (g cm^{-3})	Prototype
RbHg	Triclinic			KHg
$RbHg_{11}$	Cubic	$a=0.9707$	12.48^{calc}	$BaHg_{11}$
Rb_3Hg_{20}	Cubic	$a=1.0737$	11.45^{calc}	
$RbHg_2$	Orthorhombic	$a=0.8449$, $b=0.5300$, $c=0.8981$	8.06	KHg_2
Rb_7Hg_{31}				K_7Hg_{31}
$Rb_{15}Hg_{16}$	Tetragonal	$a=1.6653$, $c=1.8134$		$Rb_{15}Hg_{16}$
Rb_2Hg_7	Hexagonal	$a=0.68436$, $c=0.65774$		
Rb_5Hg_{19}	Tetragonal	$a=1.1561$, $c=1.0510$		Defect $BaAl_4$

E. Biehl and H.-J. Deiseroth, Rb_5Hg_{19}: Eine neue, geordnete Defektvariante des $BaAl_4$-Strukturtyps, *Z. Anorg. Allg. Chem.*, 1999, **625**, 389–394.

E. Biehl and H.-J. Deiseroth, Preparation, structural relations and magnetism of amalgams MHg_{11} (M: K, Rb, Ba, Sr), *Z. Anorg. Allg. Chem.*, 1999, **625**, 1073–1080.

E. Biehl and H.-J. Deiseroth, K_2Hg_7 und Rb_2Hg_7, zwei Vertreter eines neuen Strukturtyps binärer intermetallischer Verbindungen, *Z. Anorg. Allg. Chem.*, 1999, **625**, 1337–1342.

H.-J. Deiseroth and A. Strunck, Hg_8 ('mercubane') clusters in $Rb_{15}Hg_{16}$, *Angew. Chem. Int. Ed. Engl.*, 1989, **28**, 1251–1252.

H.-J. Deiseroth, A. Strunck and W. Bauhofer, $RbHg_2$ and $CsHg_2$ – preparation, crystal structures and electrical conductivities, *Z. Anorg. Allg. Chem.*, 1988, **558**, 128–136.

E. Todorov and S. C. Sevov, Synthesis and structure of the alkali-metal amalgams A_3Hg_{20} (A = Rb, Cs), K_3Hg_{11}, Cs_5Hg_{19} and A_7Hg_{31} (A = K, Rb), *J. Solid State Chem.*, 2000, **149**, 419–427.

Rh–Hg

Phase	Crystal structure	Lattice parameters (nm)	Density (g cm^{-3})	Prototype
$RhHg_{4.63}$				
$RhHg_5$				
$RhHg_2$	Tetragonal	$a=0.4551$, $c=0.2998$		$PtHg_2$

P. Ettmayer and B. Mathis, Crystal structure of $RhHg_2$, *Monatsh. Chem.*, 1967, **58**, 505–506.

G. Jangg, R. H. Kirchmayr and H. B. Mathis, Investigations in the Hg–Rh system, *Z. Metallkd.*, 1967, **58**, 724–726.

H. Nowotny, University of Vienna, unpublished research.
E. Y. Yonashiro and F. L. Fertonani, Thermogravimetry applied to the study of reaction of Hg with Pt–Rh, *Thermochim. Acta*, 2002, **383**, 153–160.

S–Hg

Phase	Crystal structure	Lattice parameters (nm)	Density ($g\ cm^{-3}$)	Prototype
α-HgS	Trigonal	$a=0.41488$, $c=0.95039$		α-HgS
β-HgS	Cubic	$a=0.58514$		ZnS
γ-HgS	Hexagonal	$a=0.6861$, $c=1.4077$		
HgS	Cubic	$a=0.5070$		NaCl

P. Auvray and E. Genet, Affinement of the crystal structure of cinnabar α-HgS, *Bull. Soc. Fr. Mineral. Cristall.*, 1973, **96**, 218–219.
T. Huang and A. L. Ruoff, Pressure-induced phase transition of HgS, *J. Appl. Phys.*, 1983, **54**, 5459–5461.
A. N. Mariano and E. P. Warekois, High pressure phases of some compounds of groups II–VI, *Science*, 1963, **142**, 672–673.
T. Ohmiya, Thermal expansion and the phase transformation in HgS, *J. Appl. Crystallogr.*, 1974, **7**, 396–397.
R. W. Potter II and H. L. Barnes, Phase relations in the binary Hg–S, *Am. Mineral.*, 1978, **63**, 1143–1152.
H. E. Swanson, R. K. Fuyat and G. M. Ugrinic, *NBS Circ. 539*, 1953, **IV**, 17–23.

Sc–Hg

Phase	Crystal structure	Lattice parameters (nm)	Density ($g\ cm^{-3}$)	Protoype
ScHg	Cubic	$a=0.3480$		CsCl
$ScHg_3$	Hexagonal	$a=0.6356$, $c=0.4759$		$MgCd_3$

E. Laube and H. Nowotny, Die Kristallstrukturen von ScHg, $ScHg_3$, YCd, YHg und YHg_3, *Monatsh. Chem.*, 1963, **94**, 851–858.

Se–Hg

Phase	Crystal structure	Lattice parameters (nm)	Density ($g\ cm^{-3}$)	Prototype
α-HgSe	Cubic	$a=0.60864$		ZnS
β-HgSe	Hexagonal	$a=0.432$, $c=0.968$		HgS
γ-HgSe	Cubic	$a=0.5360$		NaCl
δ-HgSe	Body-centered tetragonal	$a=0.5112$, $c=0.2721$		β-Sn

T.-L. Huang and A. L. Ruoff, High-pressure-induced phase transitions of mercury chalcogenides, *Phys. Rev. B*, 1985, **31** 5976–5983.
A. N. Mariano and E. P. Warekois, High pressure phases of some compounds of groups II–VI, *Science*, 1963, **142**, 672–673.
H. P. Singh and B. Dayal, Lattice parameters and thermal expansion of ZnTe and HgSe, *Acta Crystallogr. A*, 1970, **26**, 363–364.

Sm–Hg

Phase	*Crystal structure*	*Lattice parameters (nm)*	*Density (g cm^{-3})*	*Prototype*
SmHg	Cubic	a = 0.3744		CsCl
$SmHg_2$	Trigonal	a = 0.4877, c = 3.515		$CeCd_2$
$SmHg_3$	Hexagonal	a = 0.6632, c = 0.4900		Ni_3Sn
$Sm_{11}Hg_{45}$	Cubic	a = 2.1651		$Sm_{11}Cd_{45}$

A. Iandelli and A. Palenzona, Crystal chemistry of intermetallic compounds, in *Handbook on the Physics and Chemistry of Rare Earths*, ed. K. A. Gschneidner Jr and L Eyring, North-Holland, Amsterdam, 1979, pp. 1–54.
H. R. Kirchmayr, Compounds of Y, Sm and Gd with Hg, *Acta Phys. Austriaca*, 1964, **18,** 193–204.
F. Merlo and M. L. Fornasini, Crystal structure of the $R_{11}Hg_{45}$ compounds (R = La, Ce, Pr, Nd, Sm, Gd, Yb, U), *J. Less-Common Met.*, 1979, **64,** 221–231.

Sn–Hg

Phase	*Crystal structure*	*Lattice parameters (nm)*	*Density (g cm^{-3})*	*Prototype*
β 95–98% Sn	Hexagonal (probably)	a = 0.32415, c = 0.30065		
γ 89–94% Sn	Hexagonal	a = 0.32109, c = 0.29888		BiIn
δ 84–87% Sn	Orthorhombically distorted hexagonal	a = 0.5551, b = 0.3179, c = 0.2983		
$HgSn_4$				

Y.-W. Yen, J. Gröbner, R. Schmid-Fetzer and S. C. Hansen, Thermodynamic assessment of the Hg–Sn system, *J. Phase Equilib.*, 2003, **24**, 151–167.
J. D. Cummins and A. F. Berndt, A single crystal study of palladium–mercury and γ mercury–tin, *J. Less-Common Met.*, 1969, **19,** 431–432.
G. V. Raynor and J. A. Lee, The tin-rich intermediate phases in the alloys of tin with cadmium, indium and mercury, *Acta Metall.*, 1954, **2**, 616–620.

Sr–Hg

Phase	Crystal structure	Lattice parameters (nm)	Density (g cm^{-3})	Protoype
$SrHg_8$		$a=1.3328$, $b=0.49128$, $c=2.6446$	12.98^{calc}	$SrHg_8$
SrHg	Cubic	$a=0.3930$		CsCl
Sr_2Hg				
Sr_3Hg	Orthorhombic	$a=0.8523$, $b=1.108$, $c=0.7405$		Fe_3C
$SrHg_3$	Heagonal	$a=0.6906$, $c=0.5106$		Ni_3Sn
$SrHg_2$	Body-centered orthorhombic	$a=0.4985$, $b=0.7754$, $c=0.8550$		$CeCu_2$
Sr_3Hg_2	Tetragonal	$a=0.8883$, $c=0.4553$		U_3Si_2
$Sr_{13}Hg_{58}$	Hexagonal	$a=1.594$, $c=1.579$		$Gd_{13}Zn_{58}$
$SrHg_{11}$	Cubic	$a=0.95099$		$BaHg_{11}$
$Sr_{11-x}Hg_{54+x}$	Hexagonal	$a=1.3602$, $c=0.9818$		

E. Biehl and H.-J. Deiseroth, Preparation, structural relations and magnetism of amalgams MHg_{11} (M: K, Rb, Ba, Sr), *Z. Anorg. Allg. Chem.*, 1999, **625**, 1073–1080.

G. Bruzzone and F. Merlo, The strontium–mercury system, *J. Less-Common Met.*, 1974, **35**, 153–157.

C. Druska, T. Doert and P. Böttcher, Refinement of the crystal structure of Sr_3Hg_2, *Z. Anorg. Allg. Chem.*, 1996, **622**, 401–404.

C. Gumiński, The Hg–Sr system, *J. Phase Equilib. Diffus.*, 2005, **26**, 81–86.

A. V. Tkachuk and A. Mar, Alkaline-earth metal mercury intermetallics $A_{11-x}Hg_{54+x}$ (A = Ca, Sr), *Inorg. Chem.*, 2008, **47**, 1313–1318.

A. V. Tkachuk and A. Mar, In search of the elusive amalgam $SrHg_8$: a mercury-rich intermetallic compound with augmented pentagonal prisms, *Dalton Trans.*, 2010, **39**, 7132–7135.

Tb–Hg

Phase	Crystal structure	Lattice parameters (nm)	Density (g cm^{-3})	Prototype
TbHg	Cubic	$a=0.3690$		CsCl
$TbHg_2$	Trigonal	$a=0.4833$, $c=0.3487$		$CeCd_2$
$TbHg_3$	Hexagonal	$a=0.6565$, $c=0.4887$		Ni_3Sn
$Tb_{11}Hg_{45}$	Cubic			

C. Guminski, The Hg–Tb (mercury–terbium) system, *J. Phase Equilib.*, 1995, **16**, 193–196.

A. Iandelli and A. Palenzona, Atomic size of rare earths in intermetallic compounds, MX compounds of CsCl type, *J. Less-Common Met.*, 1965, **9**, 1–6.

A. Iandelli and A. Palenzona, Crystal chemistry of intermetallic compounds, in *Handbook on the Physics and Chemistry of Rare Earths*, ed. K. A. Gschneidner Jr and L Eyring, North-Holland, Amsterdam, 1979, pp. 1–54.

H. R. Kirchmayr and W. Lugscheider, Structure of Tb–Hg and Yb–Hg systems, *Z. Metallkd.*, 1968, **59**, 296–297.

A. Palenzona, MX_3 intermetallic phase of the rare earths with Hg, In, Tl, Pb, *J. Less-Common Met.*, 1966, **10**, 290–292.

Te–Hg

Phase	*Crystal structure*	*Lattice parameters (nm)*	*Density (g cm^{-3})*	*Prototype*
α-HgTe	Cubic	$a = 0.6453$		ZnS
β-HgTe	Hexagonal	$a = 0.445$, $c = 0.989$		HgS
γ-HgTe	Cubic	$a = 0.583$		NaCl
δ-HgTe	Body-centered tetragonal	$a = 0.5524$, $c = 0.2973$		β-Sn
ε-HgTe		$a = 0.3339$, $b = 0.3611$, $c = 0.3284$		Distorted CsCl

T.-L. Huang and A. L. Ruoff, High-pressure induced phase transitions of mercury chalcogenides, *Phys. Rev. B*, 1983, **31**, 5976–5983.

A. N. Mariano and E.P. Warekois, High pressure phases of some compounds of groups II–VI, *Science*, 1963, **142**, 672–673.

R. C. Sharma, Y. A. Chang and C. Gumiński, The Hg–Te system, *J. Phase Equilib.*, 1995, **16**, 338–348.

A. Werner, H. D. Hochheimer, K. Strössner and A. Jayaraman, High pressure X-ray diffraction studies on HgTe and HgS to 20 GPa, *Phys. Rev. B*, 1983, **28**, 3330–3334.

Th–Hg

Phase	*Crystal structure*	*Lattice parameters (nm)*	*Density (g cm^{-3})*	*Prototype*
$ThHg_3$	Hexagonal	$a = 0.6716$, $c = 0.4902$	14.39 ^{calc}	Ni_3Sn
$ThHg_2$	Hexagonal	$a = 0.4822$, $c = 0.7438$		$CaIn_2$
ThHg				
Th_2Hg	Tetragonal	$a = 0.7696$, $c = 0.5902$		$CuAl_2$

P. Chiotti, V. V. Akhachinskii, I. Ansara and M. H. Rand, *The Chemical Thermodynamics of Actinide Elements and Compounds, Part 5, The Actinide Binary Alloys*, International Atomic Energy Agency, Vienna, 1981, p. 34.

R. E. Domagala, R. E. Elliott and W. Rostocker, The system Hg–Th, *Trans. AIME*, 1958, **212**, 393–395.

P. Ettmayer, Beitrag zum System Quecksilber–Thorium, *Monatsh. Chem.*, 1965, **96**, 443–449.

R. Ferro, The crystal structures of $ThHg_3$, $ThIn_3$, $ThTl_3$, $ThSn_3$ and $ThPb_3$, *Acta Crystallogr.*, 1958, **11**, 737–738.

A. Palenzona, Th_2Hg: another representative of the $CuAl_2$-type structure, *J. Less-Common Met.*, 1986, **125**, L5–L6.

Ti–Hg

Phase	Crystal structure	Lattice parameters (nm)	Density (g cm^{-3})	Protoype
TiHg	Tetragonal	$a = 0.3009$, $c = 0.4041$		AuCu ($L1_0$)
γ-Ti_3Hg	Cubic	$a = 0.51888$		Cr_3Si
δ-Ti_3Hg	Cubic	$a = 0.41654$		$AuCu_3$ ($L1_2$)
$TiHg_3$				

J. L. Murray, *The Hg–Ti System, Phase Diagrams of Binary Ti Alloys*, ASM, Metals Park, OH, 1987, p. 140.

P. Pietrokowsky, A cursory investigation of intermediate phases in the systems Ti–Zn, Ti–Hg, Zr–Zn, Zr–Cd and Zr–Hg by X-ray powder diffraction method, *J. Met.*, 1954, **6**, 219–226.

E. Vielhaber and H. L. Luo, *Solid State Commun.*, 1967, **5**, 221–223.

Tl–Hg

Phase	Crystal structure	Lattice parameters (nm)	Density (g cm^{-3})	Prototype
γ-(Hg_5Tl_2) 20–30% Tl	Face-centered cubic	$a = 0.4628–0.468$		Cu

W. Gierlotka, J. Sopousek and K. Fitzner, Thermodynamic assessment of the Hg–Tl system, *CALPHAD*, 2006, **30**, 425–430.

R. St. Amand and B. C. Giessen, On the metastable system Hg–Tl, *J. Less-Common Met.*, 1978, **58**, 161–172.

Tm–Hg

Phase	Crystal structure	Lattice parameters (nm)	Density (g cm^{-3})	Prototype
TmHg	Cubic	$a = 0.3632$		CsCl
$TmHg_3$	Hexagonal	$a = 0.6491$, $c = 0.4856$		Ni_3Sn

C. Guminski, The Hg–Tm (mercury–thulium) system, *J. Phase Equilib.*, 1995, **16**, 459.

A. Iandelli and A. Palenzona, Atomic size of rare earths in intermetallic compounds, MX compounds of CsCl type, *J. Less-Common Met.*, 1965, **9**, 1–6.

A. Iandelli and A. Palenzona, Crystal chemistry of intermetallic compounds, in *Handbook on the Physics and Chemistry of Rare Earths*, ed. K. A. Gschneidner Jr and L Eyring, North-Holland, Amsterdam, 1979, pp. 1–54.

A. Palenzona, MX_3 intermetallic phase of the rare earths with Hg, In, Tl, Pb, *J. Less-Common Met.*, 1966, **10**, 290–292.

U–Hg

Phase	*Crystal structure*	*Lattice parameters (nm)*	*Density (g cm^{-3})*	*Prototype*
UHg				
UHg_2	Hexagonal	$a=0.4976$, $c=0.3218$		AlB_2
UHg_3	Hexagonal	$a=0.3320$, $c=0.4875$	14.88^{calc}	Ni_3Sn
$U_{11}Hg_{45}$	Cubic	$a=2.1720$		$Sm_{11}Cd_{45}$

B. R. T. Frost, The U–Hg system, *J. Inst. Met.*, 1953–54, **82**, 456–462.

C. Gumiński, The Hg–U system, *J. Phase Equilib.*, 2003, **24**, 461–468.

T. S. Lee, P. Chiotti and J. T. Mason, Phase relations in the U–Hg system, *J. Less-Common Met.*, 1979, **66**, 33–40.

F. Merlo and M. L. Fornasini, Crystal structure of the $Hg_{45}R_{11}$ compounds (R = La, Ce, Pr, Nd, Sm, Gd, U), *J. Less-Common Met.*, 1979, **64**, 221–231.

R. E. Rundle and A. S. Wilson, The structures of some metal compounds of U, *Acta Crystallogr.*, 1949, **2**, 148–150.

Y–Hg

Phase	*Crystal structure*	*Lattice parameters (nm)*	*Density (g cm^{-3})*	*Prototype*
YHg	Cubic	$a=0.3677$		CsCl
YHg_2	Trigonal	$a=0.4779$, $c=0.3471$		$CeCd_2$
YHg_3	Hexagonal	$a=0.6528$, $c=0.486$		Ni_3Sn
$Y_{11}Hg_{45}$	Cubic			$Sm_{11}Cd_{45}$

H. R. Kirchmayr, Compounds of Y, Sm and Gd with Hg, *Acta Phys. Austriaca*, 1964, **18**, 193–204.

E. Laube and H. Nowotny, Die Kristallstrukturen von ScHg, $ScHg_3$, YCd, YHg und YHg_3, *Monatsh. Chem.*, 1963, **94**, 851–858.

Yb–Hg

Phase	*Crystal structure*	*Lattice parameters (nm)*	*Density (g cm^{-3})*	*Prototype*
Yb_2Hg				
YbHg	Cubic	$a=0.3731$		CsCl
$YbHg_2$	Trigonal	$a=0.4896$, $c=0.3534$		$CeCd_2$
$YbHg_3$	Hexagonal	$a=0.6596$, $c=0.5021$		Ni_3Sn
$Yb_{14}Hg_{51}$	Cubic	$a=1.341$, $b=0.961$		$Gd_{14}Ag_{51}$

A. Iandelli and A. Palenzona, Atomic size of rare earths in intermetallic compounds, MX compounds of CsCl type, *J. Less-Common Met.*, 1965, **9**, 1–6.

A. Iandelli and A. Palenzona, Crystal chemistry of intermetallic compounds, in *Handbook on the Physics and Chemistry of Rare Earths*, ed. K. A. Gschneidner Jr and L Eyring, North-Holland, Amsterdam, 1979, pp. 1–54.

H. R. Kirchmayr and W. Lugscheider, Structure of Tb–Hg and Yb–Hg systems, *Z. Metallkd.*, 1968, **59**, 296–297.

F. Lihl, *Phase Diagrams of Yb–Hg and Tb–Hg*, US Government Report AD-658216, 1967.
F. Merlo and M. L. Fornasini, Crystal structure of the $R_{11}Hg_{45}$ compounds (R = La, Ce, Pr, Nd, Sm, Gd, Yb, U), *J. Less-Common Met.*, 1979, **64**, 221–231.
A. Palenzona, MX_3 intermetallic phase of the rare earths with Hg, In, Tl, Pb, *J. Less-Common Met.*, 1966, **10**, 290–292.

Zn–Hg

Phase	Crystal structure	Lattice parameters (nm)	Density ($g\ cm^{-3}$)	Prototype
γ-(Zn_3Hg) 70–77% Zn	Orthorhombic	$a = 0.2708$, $b = 0.4696$, $c = 0.5471$		β′-Cu_3Ti
β-(Zn_2Hg or Zn_3Hg_2) 56–67% Zn				
$ZnHg_3$	Hexagonal			

E. Cohen and P. J. H. van Ginneken, Die Zinkamalgame und der Clarkische Normalelement, *Z. Phys. Chem.*, 1911, **75,** 437–493.
R. Kubiak, M. Wolcyrz and W. Zacharko, New phase in the Hg–Zn system: Hg_3Zn, *Cryst. Res. Technol.*, 1988, **23**, K57–K59.
M. Pušclj, Z. Ban and A. Drašner, On the crystal structure of $HgZn_3$, *Z. Naturforsch.*, 1982, **37B**, 557–559.

Zr–Hg

Phase	Crystal structure	Lattice parameters (nm)	Density ($g\ cm^{-3}$)	Prototype
ZrHg	Tetragonal	$a = 0.315$, $c = 0.417$		AuCu
Zr_3Hg	Cubic	$a = 0.55583$		β-W
$ZrHg_3$	Cubic	$a = 0.43652$		AuCu

E. E. Havinga, H. Damsma and M. H. Van Maaren, Oscillatory dependence of superconductive critical temperature on number of valency electrons in Cu_3Au-type alloys, *J. Phys. Chem. Solids*, 1970, **31**, 2653–2662.
P. Pietrokowsky, A cursory investigation of intermediate phases in the systems Ti–Zn, Ti–Hg, Zr–Zn, Zr–Cd and Zr–Hg by X-ray powder diffraction method, *J. Met.*, 1954, **6**, 219–226.
E. Vielhaber and H.-L. Luo, New A-15 phases, *Solid State Commun.*, 1967, **5,** 321–323.

APPENDIX II

Density and Surface Tension of Binary Amalgams

Bi–Hg

x_{Bi} *(mole fraction)*	T *(K)*	*Density* $(g\,cm^{-3})$	σ $(mN\,m^{-1})$	*Ref.*
0.10	373	12.98	409	1
	393	12.94	429	1
	424	12.88	444	1
	497	12.73	437	1
	567	12.59	421	1
	638	12.45	402	1
0.19	370	12.63	419	1
	478	12.44	428	1
	557	12.30	422	1
	628	12.17	410	1
0.34	420	12.01	412	1
	487	11.91	415	1
	511	11.87	416	1
	557	11.80	411	1
	628	11.69	404	1
0.47	438	11.54	380	1
	448	11.53	390	1
	477	11.48	405	1
	518	11.42	404	1
	557	11.36	400	1
	626	11.27	396	1
	671	11.20	392	1
0.70	490	10.81	386	1
	515	10.77	390	1
	575	10.69	386	1
	636	10.61	382	1
	668	10.57	379	1
0.83	533	10.40	382	1
	571	10.36	380	1
	618	10.30	377	1
	673	10.23	373	1

Mercury Handbook: Chemistry, Applications and Environmental Impact
By Leonid F Kozin and Steve Hansen

Published by the Royal Society of Chemistry, www.rsc.org

1. Kh. I. Ibragimov and V. S. Savvin, Surface tension of the Hg–M (M = Cd, In, Sn, Tl, Pb, Bi) amalgams, *Inorg. Mater.*, 1996, **32**, 963–970.

Cd–Hg

x_{Cd} (mole fraction)	T (K)	Density (g cm^{-3})	σ (mN m^{-1})	Ref.
0.26	373	12.37	507	1
	423	12.26	497	1
	473	12.15	486	1
	523	12.04	474	1
	573	11.93	462	1
0.52	430	11.03	527	1
	480	10.94	560	1
	520	10.86	556	1
	578	10.76	545	1
	678	10.57	527	1
0.74	519	9.623	574	1
0.74	553	9.570	587	1
	580	9.527	587	1
	615	9.473	599	1
	659	9.404	586	1
0.85	506	9.055	538	1
	578	8.951	595	1
	622	8.888	601	1
	691	8.789	597	1

1. Kh. I. Ibragimov and V. S. Savvin, Surface tension of the Hg–M (M = Cd, In, Sn, Tl, Pb, Bi) amalgams, *Inorg. Mater.*, 1996, **32**, 963.

Cs-Hg

x_{Cs} (mole fraction)	T (K)	σ (mN m^{-1})	Ref.
0	295	462.0	2
0.000080	295	385.8	2
0.000160	295	378.4	2
0.000241	295	375.3	2
0.000421	295	372.1	2
0.000562	295	369.0	2
0.000722	295	366.4	2
0.000903	295	354.2	2
0.00112	295	363.4	2
0.00136	295	362.0	2
0.00162	295	360.4	2
0.00197	295	359.1	2
0.00233	295	357.2	2
0.00317	295	356.0	2
0.00357	295	353.8	2
0.00403	295	353.0	2
0.00457	295	351.7	2
0.00520	295	350.7	2
0.00598	295	349.7	2

x_{Cs} *(mole fraction)*	T *(K)*	σ *(mN m^{-1})*	*Ref.*
0.00678	295	348.6	2
0.00776	295	347.6	2
0.00885	295	346.5	2
0.01000	295	345.0	2

1. V. B. Lazarev, Y. I. Malov and G. A. Sharpataya, Work function and surface tension of concentrated cesium amalgams, *Dokl. Akad. Nauk SSSR*, 1968, **178**, 355.
2. P. P. Pugachevich and O. A. Timofeevicheva, Experimental study of the surface tension of metallic solutions. II. Surface tension of very dilute alkali-metal amalgams at 22°, *Russ. J. Phys. Chem.*, 1959, **33,** 350.

In–Hg

x_{In} *(mole fraction)*	T *(K)*	*Density (g cm^{-3})*	σ *(mN m^{-1})*	*Ref.*
0	293		483	1
0.076	293		489	1
0.126	293		500	1
0.140	293		488	1
0.146	293		500	1
0.207	293		500	1
0.294	293		520	1
0.429	293		534	1
0.642	293		555	1
0.756	294		556	1
0	298	13.534	485.1	4
0.163	298	12.533	495.9	4
0.368	298	11.224	529.9	4
0.636	298	9.470	563.3	4
0	373		471	1
0.074	373		479	1
0.123	373		487	1
0.140	373		477	1
0.143	373		486	1
0.207	373		485	1
0.236	374		487	1
0.294	373		509	1
0.428	373		507	1
0.429	373		524	1
0.569	373		525	1
0.641	373		546	1
0.756	373		547	1
0	433		460	1
0.075	433		469	1
0.122	433		475	1
0.140	433		466	1
0.147	433		478	1
0.236	433		477	1
0.297	433		500	1
0.428	433		499	1

x_{In} (mole fraction)	T (K)	Density ($g\,cm^{-3}$)	σ ($mN\,m^{-1}$)	Ref.
0.428	433		515	1
0.569	432		514	1
0.644	433		534	1
0.756	433		533	1
0.878	433		556	1
0	503		412	2
0.038	503		416	2
0.073	503		424	2
0.206	503		459	2
0.285	503		484	2
0.424	503		511	2
0.481	503		512	2
0.528	503		516	2
0.617	503		519	2
0.715	503		524	2
0.835	503		531	2
0.900	503		535	2
0.954	503		545	2
0.977	503		550	2
1.000	503		551	2

1. Y. Oguchi, T. Itami and M. Shimoji, Surface tension of liquid In–Hg amalgams, *Phys. Chem. Liq.*, 1981, **10**, 315.
2. Kh. I. Ibragimov, Interpretation of the surface tension of mercury in amalgam systems within the framework of the Mott theory, *Russ. J. Phys. Chem.*, 1980, **54**, 90.
3. Kh. I. Ibragimov and S. L. Aziev, The surface properties of indium–mercury melts, *Russ. J. Phys. Chem.*, 1976, **50**, 168 (VINITI Document No. 2874-75, deposited 9 October 1975).
4. D. A. Olsen and D. C. Johnson, The surface tension of mercury–thallium and mercury indium amalgams, *J. Phys. Chem.*, 1963, **67**, 2529.

K–Hg

x_K (mole fraction)	T (K)	σ ($mN\,m^{-1}$)	Ref.
0	295	462.0	1
0.00000249	295	426.0	1
0.00000722	295	420.3	1
0.0000194	295	414.0	1
0.0000550	295	406.9	1
0.000123	295	401.4	1
0.000236	295	396.3	1
0.000399	295	393.0	1
0.000594	295	389.8	1
0.000847	295	386.8	1
0.001242	295	383.9	1
0.001766	295	381.1	1
0.002368	295	378.6	1
0.002708	295	378.5	1

x_K *(mole fraction)*	T *(K)*	σ *(mN m⁻¹)*	*Ref.*
0	293	467	2
0.00025	293	404	2
0.0012	293	384	2
0.0021	293	390	2
0.0027	293	388	2
0.0035	293	385	2
0.0040	293	382	2
0.0050	293	380	2
0.0065	293	377	2

1. P. P. Pugachevich and O. A. Timofeevicheva, Experimental study of the surface tension of metallic solutions. II. Surface tension of very dilute alkali-metal amalgams at 22°, *Russ. J. Phys. Chem.*, 1959, **33**, 350.
2. P. P. Pugachevich, Experimental study of surface tension of potassium amalgam, *Dokl. Akad. Nauk SSSR*, 1951, **74**, 831.

Na–Hg

x_{Na} *(mole fraction)*	T *(K)*	σ *(mN m⁻¹)*	*Ref.*
0	295	463.1	2
0.00000080	295	460.4	2
0.00000421	295	459.2	2
0.00000843	295	458.5	2
0.0000106	295	457.8	2
0.0000148	295	456.9	2
0.0000184	295	456.3	2
0.0000231	295	455.7	2
0.0000273	295	454.4	2
0.0000313	295	454.0	2
0.0000353	295	452.6	2
0.0000405	295	452.5	2
0.0000485	295	450.2	2
0.0000568	295	448.2	2
0.0000682	295	445.3	2
0.0000802	295	444.3	2
0.0000931	295	442.6	2
0.000105	295	441.4	2
0.000117	295	440.5	2
0.000129	295	439.7	2
0.000148	295	439.0	2
0.000164	295	438.0	2
0.000183	295	437.4	2
0.000201	295	436.5	2
0.0002	298–623 K (20–350 °C)	454–0.260(*T*, °C)	4

1. A. Y. Lee, Effect of sodium concentration on the surface tension of mercury, *Ind. Eng. Chem. Prod. Res. Dev.*, 1968, **7**, 66.

2. P. P. Pugachevich and O. A. Timofeevicheva, Experimental study of the surface tension of metallic solutions. II. Surface tension of very dilute alkali-metal amalgams at 22°, *Russ. J. Phys. Chem.*, 1959, **33**, 350.
3. P. P. Pugachevich and O. A. Timofeevicheva, *Dokl. Akad. Nauk SSSR*, 1954, **94**, 285.
4. A. I. Pridantsev, L. A. Gavrilov, L. S. Kokorev, A. A. Smirnov and V. I. Petrovichev, Adhesion and surface tension of mercury and amalgams of sodium and magnesium, *Voprosy Teplofiziki Yadernykh Reaktorov*, 1969, (2), 41–48.

Pb–Hg

x_{Pb} *(mole fraction)*	T *(K)*	*Density (g cm^{-3})*	σ *(mN m^{-1})*	*Ref.*
0.29	423	12.59	470	1
0.40	423	12.34	478	1
0.53	523	11.89	471	1
0.58	523	11.79	472	1
0.67	523	11.61	469	1
0.53	573	11.82	463	1
0.58	573	11.72	466	1
0.67	573	11.54	463	1
0.78	573	11.30	460	1
0	623		383	2
0.095	623		402	2
0.196	623		410	2
0.293	623		427	2
0.403	623		442	2
0.500	623		452	2
0.532	623		453	2
0.583	623		456	2
0.651	623		454	2
0.677	623		457	2
0.783	623		453	2
0.912	623		451	2
1.0	623		443	2
0.53	673	11.67	449	1
0.58	673	11.58	454	1
0.67	673	11.41	455	1
0.78	673	11.17	451	1
0.91	673	10.87	446	1

Pb–Hg (continued)

	$\sigma = \sigma_0 - \alpha T$ *(mN m^{-1})*[3]		$\rho = \rho_0 - \beta T$ *(g cm^{-3})*[4]	
x_{Pb} *(mole fraction)*	σ_0	α	ρ_0	β
0	487	0.281	13.53	0.0024
0.041	487	0.247	13.52	0.0024
0.096	490	0.226	13.44	0.0023
0.234	502	0.203	13.15	0.0021
0.352	513	0.183	12.94	0.0020
0.389	514	0.173	12.83	0.0019

	$\sigma = \sigma_0 - \alpha T$ (mN m^{-1})[3]		$\rho = \rho_0 - \beta T$ (g cm^{-3})[4]	
x_{Pb} (mole fraction)	σ_0	α	ρ_0	β
0.441	513	0.154	12.71	0.0019
0.528	513	0.147	12.47	0.0018
0.577	509	0.139	12.33	0.0017
0.649	501	0.121	12.13	0.0017
0.665	496	0.106	12.05	0.0016
0.780	484	0.089	11.69	0.0015
0.787	479	0.072	11.68	0.0015
0.911	481	0.098	11.28	0.0013
1.000	469	0.086	11.05	0.0013

1. Kh. I. Ibragimov and V. S. Savvin, Surface tension of the Hg–M (M = Cd, In, Sn, Tl, Pb, Bi) amalgams, *Inorg. Mater.*, 1996, **32**, 963.
2. Kh. I. Ibragimov, Interpretation of the surface tension of mercury in amalgam systems within the framework of the Mott theory, *Russ. J. Phys. Chem.*, 1980, **54**, 90.
3. Kh. I. Ibragimov and V. S. Savvin, Surface tension and density of Hg–Pb melts, *Izv. Vyssh. Uchebn. Zaved. Tsvet. Metall.*, 1976, (4), 148–149.
4. M. P. Vukalovich, A. I. Ivanov, *et al.*, *Thermophysical Properties of Mercury*, Standards Publishing House, Moscow, 1971.

Rb–Hg[1]

x_{Rb} (mole fraction)	T (K)	σ (mN m^{-1})
0	298	475
0.0000048	298	405.1
0.0000068	298	401.0
0.000049	298	383.2
0.00015	298	374.5
0.00045	298	370.2
0.00061	298	369.2
0.00180	298	364.0
0.00670	298	355.5

1. Yu. I. Malov, Kh. I. Badakhov and V. B. Lazarev, *Elektrokhimiya*, 1971, **7**, 432 (translation: Work function and surface tension in dilute rubidium amalgams, *Sov. Electrochem.*, 1971, **7**, 416).

Sn–Hg

x_{Sn} (mole fraction)	T (K)	σ (mN m^{-1})	Ref.
0.051	523	539	1
0.097	523	537	1
0.151	523	533	1

x_{Sn} (mole fraction)	T (K)	σ ($mN\,m^{-1}$)	Ref.
0.218	523	528	1
0.280	523	519	1
0.344	523	515	1
0.414	523	504	1
0.516	523	488	1
0.621	523	471	1
0.710	523	454	1
0.785	523	444	1
0.823	523	439	1
0.850	523	434	1
0.893	523	428	1
0.941	523	424	1
0.027	597	391.4	2
0.055	597	394.4	2
0.107	597	399.2	2
0.150	597	407.3	2
0.177	597	412.5	2
0.217	597	419.5	2
0.286	597	431.7	2
0.378	597	452.0	2
0.473	597	473.4	2
0.576	597	491.8	2
0.650	597	503.2	2
0.702	597	511.0	2
0.769	597	520.2	2
0.831	597	527.6	2
0.878	597	531.7	2
0.925	597	534.6	2
0.962	597	536.5	2
0.048	623	533	1
0.097	623	531	1
0.151	623	527	1
0.218	623	522	1
0.280	623	512	1
0.341	623	505	1
0.414	623	494	1
0.516	623	475	1
0.618	623	456	1
0.780	623	436	1
0.782	623	423	1
0.820	623	417	1
0.847	623	411	1
0.893	623	405	1
0.944	623	398	1

1. Kh. I. Ibragimov and V. S. Savvin, Surface tension of the Hg–M (M = Cd, In, Sn, Tl, Pb, Bi) amalgams, *Inorg. Mater.*, 1996, **32**, 963.
2. Kh. I. Ibragimov, Interpretation of the surface tension of mercury in amalgam systems within the framework of the Mott theory, *Russ. J. Phys. Chem.*, 1980, **54**, 90.

Tl-Hg

x_{Tl} *(mole fraction)*	T *(K)*	*Density (g cm^{-3})*	σ *(mN m^{-1})*	*Ref.*
0.10	299	13.27	468	1
	329	13.20	464	1
	373	13.10	458	1
	471	12.89	439	1
	587	12.63	411	1
	680	12.42	386	1
0.31	303	12.77	471	1
	360	12.66	467	1
	396	12.60	464	1
	493	12.42	452	1
	580	12.26	437	1
	673	12.09	420	1
	729	11.99	408	1
0.47	373	12.35	472	1
	418	12.27	467	1
	523	12.08	455	1
	613	11.92	443	1
	723	11.73	426	1
0.69	488	11.78	464	1
	548	11.67	458	1
	599	11.59	452	1
	683	11.45	441	1
	723	11.38	435	1
0	298	13.534	485.1	5
0.0196	298	13.490	477.4	5
0.0393	298	13.443	474.3	5
0.0590	298	13.405	472.5	5
0.0836	298	13.353	471.0	5
0.1082	298	13.305	473.0	5
0.1279	298	13.258	475.0	5
0.1476	298	13.222	478.1	5
0.2465	298	13.021	479.0	5
0.3955	298	12.727	480.9	5
0.020	573		401	2
0.048	573		403	2
0.066	573		406	2
0.082	573		407	2
0.101	573		410	2
0.132	573		415	2
0.308	573		434	2
0.468	573		445	2
0.685	573		454	2
0.997	573		462	2

1. Kh. I. Ibragimov and V. S. Savvin, Surface tension of the Hg–M (M = Cd, In, Sn, Tl, Pb, Bi) amalgams, *Inorg. Mater.*, 1996, **32**, 963.
2. Kh. I. Ibragimov, Interpretation of the surface tension of mercury in amalgam systems within the framework of thc Mott theory, *Russ. J. Phys. Chem.*, 1980, **54**, 90.
3. Kh. I. Ibragimov and V.S. Savvin, The surface tension of melts in the thallium–mercury system, *Fiz.-Khim. Issled. Metall. Protsessov*, 1980, (8), 61–66.

4. S. L. Aziev and Kh. I. Ibragimov, Surface tension and density of the thallium–mercury system, *Russ. J. Phys. Chem.*, 1975, **49**, 952 (VINITI Document No. 621-75, deposited 10 March 1975).
5. D. A. Olsen and D. C. Johnson, The surface tension of mercury–thallium and mercury–indium amalgams, *J. Phys. Chem.*, 1963, **67**, 2529.

Zn–Hg

	σ ($mN\,m^{-1}$)							
x_{Zn} (*mole fraction*)	*298 K*	*323 K*	*373 K*	*423 K*	*473 K*	*523 K*	*573 K*	*623 K*
0	465	462	452	439	429	416	402	387
0.009	470	467	457	446	434	421	407	392
0.017	471	468	459	447	436	423	409	393
0.030	475	470	461	448	438	424	411	395
0.058	477	474	464	454	442	432	419	402
0.113			467	458	448	438	426	412
0.167			474	463	455	446	436	422
0.226				468	463	452	446	435
0.304				479	474	470	462	454
0.384					488	487	483	477
0.467						519	515	510
0.530						544	540	538
0.574							556	552
0.643							584	582
0.679							598	596
0.765								635
0.825								670

	Density ($g\,cm^{-3}$)							
x_{Zn} (*mole fraction*)	*298 K*	*323 K*	*373 K*	*423 K*	*473 K*	*523 K*	*573 K*	*623 K*
0.009	13.464	13.409	13.294	13.180	13.065	12.950	12.835	12.720
0.017	13.435	13.380	13.265	13.150	13.035	12.920	12.805	12.690
0.030	13.415	13.360	13.245	13.130	13.015	12.900	12.784	12.670
0.058	13.254	13.205	13.105	12.974	12.890	12.775	12.675	12.569
0.113			12.840	12.737	12.640	12.538	12.450	12.340
0.167			12.590	12.512	12.420	12.390	12.295	12.190
0.226				12.360	12.250	12.150	12.050	11.960
0.304				12.125	12.020	11.915	11.810	11.705
0.384					11.585	11.492	11.404	11.322
0.467						11.007	10.931	10.850
0.530						10.657	10.583	10.509
0.574							10.329	10.250
0.643							9.887	9.804
0.679							9.674	9.594
0.765								8.970
0.825								8.420

1. Kh. I. Ibragimov, A. G.-M. Nal'giev and B. B. Sagov, The surface properties of liquid mercury–zinc alloys, *Russ. J. Phys. Chem.*, 1975, **49**, 1097 (VINITI Document No. 1014-75, deposited 10 April 1975).

APPENDIX III

Inorganic and Organic Mercury Compounds

Tables of inorganic and organic mercury compounds hare been assembled. Crystal structure determination was used as the criterion for selecting compounds because it prevents the possibility of incorrectly assigning chemical formulae.

Amido Compounds

Name	*Formula*	*MW*	*Color*	*Crystal structure*	*Ref.*
Mercury(II) amidochloride	$HgNH_2Cl$	252.066		Orthorhombic	1
Mercury(II) amidobromide	$HgNH_2Br$	296.517		Orthorhombic	2
Iminomercury(II) bromide	Hg_2NHBr_2	576.003		Hexagonal	3
Mercury(I) nitrite	$Hg_2(NO_2)_2$	493.190		Monoclinic	4

1. W. N. Lipscomb, *Acta Crystallogr.*, 1951, **4**, 266.
2. L. Nijssen and W. N. Lipscomb, *Acta Crystallogr.*, 1952, **5**, 604.
3. K. Brodersen, *Acta Crystallogr.*, 1955, **8**, 723.
4. S. Ohba, F. Matsumoto, M. Ishihara and Y. Saito, *Acta Crystallogr.*, 1986, **C42**, 1.

Arsenides and Arsenates

Name	*Formula*	*MW*	*Color*	*Crystal structure*	*Ref.*
Mercury(I) orthoarsenate	α-$(Hg_2)_3(AsO_4)_2$	1481.376	Red–brown	Monoclinic	1
Mercury(I) orthoarsenate	β-$(Hg_2)_3(AsO_4)_2$	1481.376			
Mercury(I) diarsenate(V)	$(Hg_2)_2(As_2O_7)$	1064.197		Orthorhombic	2
	$[Hg_3](PO_4)Cl$	732.193			3

Mercury Handbook: Chemistry, Applications and Environmental Impact
By Leonid F Kozin and Steve Hansen

Published by the Royal Society of Chemistry, www.rsc.org

Name	*Formula*	*MW*	*Color*	*Crystal structure*	*Ref.*
	$[Hg_3](AsO_4)Br$	820.592			3
	$[Hg_3](AsO_4)Cl$	776.141			3
Mercury(II) arsenate	$Hg_3(AsO_4)_2$	879.606		Monoclinic	4
	$(Hg_3)_3(AsO_4)_4$	2360.982			5, 6
	$(Hg)_3[HgO_2]Cl_2$	905.264			7

1. B. Kamenar and B. Kaitner, *Acta Crystallogr.*, 1973, **B29**, 1666.
2. M. Weil, *Z. Anorg. Allg. Chem.*, 2004, **630,** 213.
3. M. Weil, *Z. Naturforsch.*, 2001, **56B**, 753.
4. A.-K. Larsson, S. Lidin, C. Stålhandske and J. Albertsson, *Acta Crystallogr.*, 1993, **C49**, 784.
5. M. Weil and R. Glaum, *J. Solid State Chem.*, 2001, **157**, 68.
6. A. L. Wessels, W. Jeitschko and M. H. Moller, 1997, *Z. Naturforsch.*, **52B**, 469.
7. S. V. Borisov, S. A. Magarill and N. V. Pervukhina, *J. Struct. Chem.*, 2003, **44,** 441.

Antimonides

Name	*Formula*	*MW*	*Color*	*Crystal structure*	*Ref.*
Mercury antimony oxide	$Hg_2Sb_2O_7$	957.292		Cubic	1

1. V. I. Sidey, P. M. Milyan, O. O. Semrad and A. M. Solomon, *J. Alloys Compd.*, 2008, **457**, 480.

Azides

Name	*Formula*	*MW*	*Color*	*Crystal structure*	*Ref.*
Mercury(I) azide Synonym: diazidodimercury	$Hg_2(N_3)_2$	485.220	White Light-sensitive	Monoclinic	1
Mercury(II) azide	α-$Hg(N_3)_2$	284.630		Orthorhombic	2
	β-$Hg(N_3)_2$	284.630			2

1. P. Nockemann, U. Cremer, U. Ruschewitz and G. Meyer, *Z. Anorg. Allg. Chem.*, 2003, **629**, 2079.
2. U. Müller, *Z. Anorg. Allg. Chem.*, 1973, **399**, 183.

Borates

Name	*Formula*	*MW*	*Color*	*Crystal structure*	*Ref.*
	α-HgB_4O_7	355.827			1
	β-HgB_4O_7	355.827		Orthorhombic	2 (high-pressure phase)

1. M. Weil, *Acta Crystallogr.*, 2003, **E59**, i40.
2. H. Emmea, M. Weil and H. Huppertz, *Z. Naturforsch.*, 2005, **60B,** 815.

Bromides and Bromates

Name	*Formula*	*MW*	*Color*	*Crystal structure*	*Ref.*
Mercury(I) bromide	Hg_2Br_2	560.988		Ferroelectric phase forms at 143 K	See Chapter 5 1, 2
Mercury(II) bromide	$HgBr_2$	360.398			See Chapter 5
Mercury(II) iminobromide	$Hg_2(NH)Br_2$	576.003		Hexagonal	3
Millon's base	Hg_2NBr	495.091	Tridymite	Hexagonal	4
	$Hg_2NOH \cdot 2H_2O$	451.221	Cristobalite	Cubic	5
Mercury(I) bromate	$Hg_2(BrO_3)_2$	656.984	Colorless	Monoclinic	6
Mercury(II) hydroxybromate (mercury bromate hydroxide)	$Hg(OH)BrO_3$	345.499			7, 8
Mercury(I,II) bromide oxide	$Hg_8O_4Br_3$	1908.446		Monoclinic	9

1. E. Dorm, *J. Chem. Soc. D*, 1971, 466–467.
2. M. E. Boiko, B. S. Zadokhin and K. Lukaszewicz, *Fiz. Tverd. Tela (Leningrad)*, 1993, **35**, 1483.
3. K. Brodersen, *Acta Crystallogr.*, 1955, **8**, 723.
4. L. Nijssen and W. N. Lipscomb, *Acta Crystallogr.*, 1954, **7**, 103.
5. W. N. Lipscomb, *Acta Crystallogr.*, 1951, **4**, 156.
6. E. Dorm, *et al.*, *Acta Chem. Scand.*, 1967, **21**, 1661.
7. A. Weiss, *et al.*, *Z. Naturforsch.*, 1960, **15B**, 678.
8. G. Bjornlund, *Acta Chem. Scand.*, 1971, **25**, 1645.
9. C. Stålhandske, *Acta Chem. Scand.*, 1987, **A41**, 576.

Cyanimides and Carbonates

Name	*Formula*	*MW*	*Color*	*Crystal structure*	*Ref.*
Mercury cyanamide	HgNCN(II)	240.615			1
Mercury carbodiimide	HgNCN(I)	240.615			2, 3
Mercury(I) carbonate	Hg_2CO_3	461.189	White	Monoclinic	4

1. X. Liu, P. Müller, P. Kroll and R. Dronskowski, *Inorg. Chem.*, 2002, **41**, 4259.
2. S. K. Deb and A. D. Yoffe, *Trans. Faraday Soc.*, 1958, **55**, 106.

3. M. Becker and M. Jansen, *Z. Anorg. Allg. Chem.*, 2000, **626**, 1639.
4. A. N. Christensen, P. Norby and J. C. Hanson, *Z. Kristallogr.*, 1994, **209**, 874.

Chlorides and Chlorates

Name	*Formula*	*MW.*	*Color*	*Crystal structure*	*Ref.*
Mercury(I) chloride	α-Hg_2Cl_2	472.088		Tetragonal	1, 2
	β-Hg_2Cl_2	472.088			2
Mercury(II) chloride	$HgCl_2$	271.498	White, light-sensitive	Tetragonal	See Chapter 5
Cesium mercury tetrachloride	Cs_2HgCl_4	608.212		Orthorhombic	3
	$2HgCl_2 \cdot HgO$ $(Hg_3Cl_3)OCl$			Cubic	4
Mercury chloride oxide	α-$HgCl_2 \cdot 2HgO$ α-$Hg_3Cl_2O_2$	704.674	Black	Monoclinic	5–7
Mercury chloride oxide	β-$Hg_3Cl_2O_2$	704.674	Dark red	Monoclinic	7–9
	$Hg(OH)ClO_3$	301.047			10
	$HgCl_2 \cdot 4HgO$	1268.060	Brown Black	Orthorhombic	11
Mercury(I) chlorate	$Hg_2(ClO_3)_2$	568.08		Monoclinic	12
Mercury oxychloride	Hg_3OCl	653.222		Monoclinic	13
Mercury(II) chlorate	$Hg(ClO_3)_2$	367.492	White		14
Mercury chloride oxide	$Hg_4O_2Cl_2$	905.264	Light yellow	Monoclinic	15–17
Mercury(II) hydroxide chlorate(V)	$Hg(OH)ClO_3$	301.047		Orthorhombic	10
	$[Hg_2]_3O_2Cl_2$	1306.444		Monoclinic	7
	$Hg_5O_4Cl_2$	1137.852		Orthorhombic	7
	$[Hg_2]_3HgO_3Cl_2$	1523.033		Orthorhombic	7
	$Hg_8O_4Br_3$	1908.446		Monoclinic	7

1. N. J. Calos, C. H. L. Kennard and R. L. Davis, *Z. Kristallogr.*, 1989, **187**, 305.
2. C. Guminski, *J. Phase Equilib.*, 1994, **15**, 101.
3. B. Bagautdinov, J. Luedecke, M. Schneider and S. van Smaalen, *Acta Crystallogr.*, 1998, **B54**, 626
4. D. Grdenic and S. Scavnicar, *Nature*, 1953, **172**, 584.
5. S. Scavnicar, *Acta Crystallogr.*, 1955, **8**, 379.
6. K. Aurivillius, *et al.*, *Acta Crystallogr.*, 1974, **B30**, 1907.
7. S. V. Borisov, S. A. Magarill and N. V. Pervukhina, *J. Struct. Chem.*, 2003, **44**, 1018.
8. K. Aurivillius, *et al.*, *Acta Crystallogr.*, 1978, **B34**, 79.
9. M.A. Neuman, *et al.*, *J. Cryst. Mol. Struct.*, 1976, **6**, 177.

10. D. Göbbels and M.S. Wickleder, *Acta Crystallogr.*, 2004, **E60**, i40.
11. A. Weiss, G. Nagorsen and A. Weiss, *Z. Kristallogr.*, 1954, **9B**, 81.
12. D. Göbbels and G. Meyer, *Z. Anorg. Allg. Chem.*, 2003, **629**, 2446.
13. N. V. Pervukhina, G. V. Romanenko, S. A. Magarill, V. I. Vasiliev and S. V. Borisov, *J. Struct. Chem.*, 1999, **40**, 155.
14. J. Lamure, *et al.*, in *Nouveau Traité de Chimie Minérale*, ed. P. Pascal, Masson, Paris, 1962, vol. 5, p. 739.
15. K. Aurivilius and L. Folkmarsson, *Acta Chem. Scand.*, 1968, **22**, 2529.
16. S. Scavnicar, *Acta Crystallogr.*, 1956, **9**, 956.
17. K. Broderson, G. Gobel and G. Liehr, *Z. Anorg. Allg. Chem.*, 1989, **575**, 145.

Chromates

Name	*Formula*	*MW*	*Color*	*Crystal structure*	*Ref.*
Mercury(I) chromate	α-Hg_2CrO_4	517.172			1
Mercury(I) chromate	β-Hg_2CrO_4	517.172			1
Mercury(II) chromate	α-$HgCrO_4$	316.584		Monoclinic	2–4
Mercury(II) chromate	β-$HgCrO_4$	316.584		Orthorhombic	2
	$HgCrO_4 \cdot H_2O$	334.597		Triclinic	2
Mercury(II) chromate hemihydrate	$HgCrO_4 \cdot \frac{1}{2}H_2O$	325.590		Monoclinic	5, 6
Mercury chromate oxide	$Hg_3(CrO_4)O_2$ $HgCrO_4 \cdot 2HgO$	749.762			4, 7, 8
	$Hg_6Cr_2O_9$	1451.523	Orange	Orthorhombic	9
	$Hg_6Cr_2O_{10}$	1467.522	Red	Orthorhombic	9
Mercury(II) dichromate	$HgCr_2O_7$	416.575		Tetragonal	10
	$HgCr_2O_4$	368.578		Cubic	11
	$HgCr_2O_4$	368.578			12

1. M. Weil and B. Stöger, *Z. Anorg. Allg. Chem.*, 2006, **632**, 2131.
2. B. Stoger and M. Weil, *Z. Naturforsch.*, 2006, **61B** 708.
3. C. Stålhandske, *Acta Crystallogr.*, 1978, **B34**, 1968.
4. K. Aurivillius, *et al.*, *Acta Chem. Scand.*, 1961, **15**, 1932.
5. K. Aurivillius and C. Stålhandske, *Z. Kristallogr.*, 1975, **142**, 129.
6. K. Aurivillius, *Acta Chem. Scand.*, 1972, **26**, 2113.
7. G. Nogorsen, *et al.*, *Angew. Chem. Int. Ed. Engl.*, 1962, **1**, 115.
8. J. Lamure, *C. R. Hebd. Seances Acad. Sci.*, 1963, **256**, 3696.
9. M. Weil and B. Stöger, *J. Solid State Chem.*, 2006, **179**, 2479.
10. M. Weil, B. Stöger, E. Zobetz and E. J. Baran, *Monatsh. Chem.*, 2006, **137**, 987.
11. A.L. Wessels, R. Czekalla and W. Jeitschko, *Mater. Res. Bull.*, 1998, **33**, 95.
12. M. Weil and B. Stöger, *Acta Crystallogr.*, 2006, **E62**, i199.

Cyanides and Cyanates

Name	*Formula*	*MW*	*Color*	*Crystal structure*	*Ref.*
Mercury(II) cyanide	$Hg(CN)_2$	252.626		Monoclinic	1
Mercury(II) cyanide	$Hg(CN)_2$	252.626		Orthorhombic	1
Mercury(II) cyanide	$Hg(CN)_2$	252.625		Four high-pressure phases	2–4
Mercury cyanamide	$HgCN_2$	240.615		Orthorhombic	5
	HgNCN(I)	240.615			5, 6
	HgNCN(II)	240.615		Monoclinic	7
	$Hg_2(CN_2)Cl_2$	512.114			8
	$Hg_3(CN_2)_2Cl_2$	752.73			8
Potassium tetracyanomercurate	α-$K_2[Hg(CN)_4]$	382.858		Trigonal	9, 10
Potassium tetracyanomercurate	β-$K_2[Hg(CN)_4]$	382.858		Cubic	9, 10
Rubidium tetracyanomercurate	α-$Rb_2Hg(CN)_4$	475.598			9, 11
Rubidium tetracyanomercurate	β-$Rb_2Hg(CN)_4$	475.598			9, 11
Mercuric oxycyanide	$HgO \cdot Hg(CN)_2$	469.218		Orthorhombic	12
Mercury(II) oxycyanide	$HgO \cdot Hg(CN)_2$	469.218	White Explosive, sensitive to impact and heat	Orthorhombic	12
Mercury(II) cyanonitrate	$Hg(CN)NO_3$	288.613		Hexagonal	13
Mercury fulminate	$Hg(CNO)_2$	284.624		Orthorhombic	14

1. R. C. Seccombe and C. H. L. Kennard, *J. Organomet. Chem.*, 1969, **18**, 243.
2. J. Hvoslef, *Acta Chem. Scand.*, 1958, **12**, 1568.
3. D. M. Adams, *et al.*, *J. Phys. C: Solid State Phys.*, 1983, **16**, 3349.
4. P. T. T. Wong, *J. Chem. Phys.*, 1984, **80**, 5937.
5. M. Becker and M. Jansen, *Z. Anorg. Allg. Chem.*, 2000, **626** 1639.
6. X. Liu, Dissertation, Aachen University of Technology (RWTH), 2002.
7. X. Liu, P. Müller, P. Kroll and R. Dronskowski, *Inorg. Chem.*, 2002, **41**, 4259.
8. X. Liu and R. Dronskowski, *Z. Naturforsch.*, 2002, **57B**, 1108.
9. L. Wiehl and S. Haussühl, *Z. Kristallogr.*, 1981, **156**, 117.
10. T. Asaji and R. Ikeda, *J. Mol. Struct.*, 1995, **345**, 93.
11. P. Klüfers, H. Fuess and S. Haussühl, *Z. Kristallogr.*, 1981, **156**, 255.
12. S. Šćavničar, *Z. Kristallogr.*, 1963, **118**, 248.
13. C. Mahon and D. Britton, *Inorg. Chem.*, 1971, **10**, 2331.
14. W. Beck, J. Evers, M. Göbel, G. Oehlinger and T. M. Klapötke, *Z. Anorg. Allg. Chem.*, 2007, **633**, 1417.

Fluorides

Name	Formula	MW	Color	Crystal structure	Ref.
	Hg_2F_2	439.176			See Chapter 5
	$Hg(OH)F$	236.597		Orthorhombic	1, 2
	Hg_2PO_3F	499.149			3
	HgF_2	238.586			See Chapter 5
	HgF_4	276.582			4
Mercury hexafluoroniobate	Hg_3NbF_6	808.664			5
Mercury hexafluorotantalate	Hg_3TaF_6	896.706			5

1. D. Grdenic and M. Sikirica, Inorg. Chem., 1973, **12**, 544.
2. C. Stalhandske, *Acta Crystallogr.*, 1979, **B35**, 949.
3. M. Weil, M. Puchberger and E. J. Baran, *Inorg. Chem.*, 2004, **43,** 8330.
4. X. Wang, L. Andrews, S. Riedel and M. Kaupp, *Angew. Chem. Int. Ed.*, 2007, **46,** 8371.
5. I. D. Brown, R. J. Gillespie and K. R. Morgan, *Inorg. Chem.*, 1984, **23**, 4506.

Iodides and Iodates

Name	Formula	MW	Color	Crystal structure	Ref.
Mercury(I) iodide	Hg_2I_2	654.988			See 5.2.7
Mercury(II) iodide	α-HgI_2	454.398			
	β-HgI_2	454.398			See 5.2.8
Mercury arsenide iodide	$Hg_4As_2I_3$	1332.917	Black		1, 2
Mercury phosphide hexaiodide	$Hg_9P_5I_6$	2721.604		Monoclinic	3
Mercury(I,II) oxide iodide	Hg_2OI	544.083		Monoclinic	4, 5
Mercury(II) iodate	α-$Hg(IO_3)_2$	550.392			6
	β-$Hg(IO_3)_2$	550.392		Monoclinic	6
	$Hg(H_3IO_6)$	426.512			7
	$Hg_3(H_2IO_6)_2$	1051.598			7

1. H. Puff, *et al.*, *Z. Anorg. Allg. Chem.*, 1965, **341**, 217.
2. P. L. Labbé, *et al.*, *Z. Kristallogr.*, 1989, **187**, 117.
3. M. Ledésert, A. Rebbah and P. Labbé, *Z. Kristallogr.*, 1990, **192**, 223.
4. C. Stålhandske, K. Aurivillius and G.-I. Bertinsson**,** *Acta Crystallogr.*, 1985, **C41**, 167.
5. S. V. Borisov, S. A. Magarill and N. V. Pervukhina, *J. Struct. Chem.*, 2003, **44**, 1018.
6. M. Weil, *Z. Naturforsch.*, 2003, **58B,** 627.
7. T. J. Mormann and W. Jeitschko, *Z. Kristallogr.*, 2001, **216**, 1.

Iodomercurates

Name	*Formula*	*MW*	*Color*	*Crystal structure*	*Ref.*
Sodium iodomercurate	Na_2HgI_4	754.186			
Potassium iodomercurate	K_2HgI_4	786.402			
Cadmium iodomercurate	Cd_2HgI_4	933.028			1
Lead iodomercurate	Pb_2HgI_4	1122.606			1
Silver iodomercurate	α-Ag_2HgI_4	923.942			2
	β-Ag_2HgI_4	923.942			3
	δ-Ag_2HgI_4	923.942			4
	ε-Ag_2HgI_4	923.942			4
Copper iodomercurate	α-Cu_2HgI_4	835.298			2, 5
	β-Cu_2HgI_4	835.298			
	δ-Cu_2HgI_4	835.298			4
				Hexagonal – high pressure	6
Cesium iodomercurate	Cs_2HgI_4	974.016		Monoclinic	7
	Cs_3HgI_5	1233.830		Orthorhombic	7
	$Cs_2Hg_3I_8 \cdot H_2O$	1900.827		Monoclinic	7
Sulfur iodomercurate	$Hg_3S_2I_2$		Orange	Orthorhombic	8
Selenium iodomercurate	$Hg_3Se_2I_2$		Light red	Orthorhombic	8
Tellurium iodomercurate	Hg_3TeI_4			Cubic	9
Thallium iodomercurate	Tl_4HgI_6	1779.546	Yellow		10

1. J. A. A. Ketelaar, *Z. Phys., Chem.*, 1934, **B26**, 327.
2. S. Hull and D. A. Keen, *J. Phys.: Condens. Matter*, 2000, **12**, 3751.
3. K. W. Browall, J. S. Kasper and H. Wiedemeier, *J. Solid State Chem.*, 1974, **10**, 20.
4. S. Hull and D. A. Keen, *J. Phys.: Condens. Matter*, 2001, **13,** 5597.
5. L. Eriksson, P. Wang and P.-E. Werner, *Z. Kristallogr.*, 1991, **197**, 235.
6. H. Mikler, *Monatsh. Chem.*, 1989, **120**, 7.
7. R. Sjövall and C. Svensson, *Acta Crystallogr.*, 1988, **C44**, 207.
8. J. Beck and S. Hedderich, *J. Solid State Chem.*, 2000, **151**, 73.
9. H. Wiedemeier, M. A. Hutchins, Y. Grin, C. Feldmann and H. G. von Schnering, *Z. Anorg. Allg. Chem.*, 1997, **623**, 1843.
10. J. H. Kennedy, C. Schaupp, Y. Yang, Zhengming Zhang, T. Novinson and T. Hoffard, *J. Solid State Chem.*, 1990, **88**, 555.

Molybdates

Name	*Formula*	*MW*	*Color*	*Crystal structure*	*Ref.*
Mercury(I) molybdate	α-Hg_2MoO_4	561.116			1, 2
	β-Hg_2MoO_4	561.116			1
	$Hg_2Mo_2O_7$	705.053			3
	$Hg_2Mo_5O_{16}$	1136.864			3
	$HgMoO_4$	360.526		Monoclinic	4

1. T. J. Mormann and W. Jeitschko, *Inorg. Chem.*, 2000, **39**, 4219.
2. A. L. Wessels, R. Czekalla and W. Jeitschko, *Mater. Res. Bull.*, 1998, **33**, 95.
3. S. V. Borisov, S. A. Magarill, N. V. Pervukhina and N. A. Kryuchkova, *J. Struct. Chem.*, 2002, **43,** 293.
4. W. Jeitschko and A. W. Sleight, *Acta Crystallogr.*, 1973, **B29**, 869.

Niobates

Name	*Formula*	*MW*	*Color*	*Crystal structure*	*Ref.*
	$Hg_2Nb_2O_7$	698.992		Cubic	1

1. W. Sleight, *Inorg. Chem.*, 1968, **7**, 1704.

Nitrites, Nitrides and Nitrates

Name	*Formula*	*MW*	*Color*	*Crystal structure*	*Ref.*
Mercury(I) nitrite	$Hg_2(NO_2)_2$	493.19			1
	$[(Hg_2)_2O(NO_3)]NO_3 \cdot HNO_3$	1005.4		Orthorhombic	2
	$[(Hg_2)_5(OH)_4(NO_3)_2](NO_3)_4$	2446.0		Triclinic	2
	$[Hg_2(OHg)_2](NO_3)_2$	958.4		Monoclinic	2
Basic mercury nitrate	$Hg_8O_4(OH)(NO_3)_5$	1995.743			3
	$K_3[Hg(NO_2)_4]NO_3$	563.908			4

1. R. B. English, D. Röhm and C. J. H. Schutte, *Acta Crystallogr.*, 1985, **C41**, 997.
2. B. Kamenar, D. Matkovic-Calogovic and A. Nagl, *Acta Crystallogr.*, 1986, **C42**, 385.
3. M. Weil, *Z. Anorg. Allg. Chem.*, 2005, **631**, 1346.
4. D. Hall and R.V. Holland, *Inorg. Chim. Acta*, **3**, 1969, 235.

Oxides, Oxalates and Oxomercurates

Name	*Formula*	*MW.*	*Color*	*Crystal structure*	*Ref.*
Mercury(II) oxide	HgO	216.589		Triclinic	See Appendix I
	HgO	216.589		Orthorhombic	1
	HgO	216.589		Tetragonal	
	HgO	216.589		Cubic	
	HgO	216.589		Hexagonal	
	α-HgO_2	232.588		Rhombohedral	
	β-HgO_2	232.588			
	Hg_2O	417.179		Decomposes at 100 °C	
	Hg_2O_2NaI	583.072			1

Name	*Formula*	*MW.*	*Color*	*Crystal structure*	*Ref.*
Mercury(II) oxalate	C_2HgO_4	288.608			2
	Li_2HgO_2	246.47		Tetragonal	3
	Na_2HgO_2	278.568		Tetragonal	3
	K_2HgO_2	310.784		Tetragonal	3
	Rb_2HgO_2	403.524		Tetragonal	3
	Cs_2HgO_2	498.398		Tetragonal	3
	$BaHgO_2$	369.918		Hexagonal	4
	$BaHgO_2$	369.918		Rhombohedral	5
	$SrHgO_2$	320.211			6
	$CdHgO_2$	344.999		Monoclinic	7

1. K. Aurivillius, *Acta Chem. Scand.*, 1964, **18**, 1305.
2. A. N. Christensen, P. Norby and J. C. Hanson, *Z. Kristallogr.*, 1994, **209**, 874.
3. R. Hoppe and H. J. Röhrborn, *Z. Anorg. Allg. Chem.*, 1964, **329**, 110.
4. M. Soll and H. Müller-Buschbaum, *J. Less-Common Met.*, 1990, **162**, 169.
5. S. N. Putilin, S. M. Kazakov and M. Marezio, *J. Solid State Chem.*, 1994, **109,** 406.
6. S. N. Putilin, M. G. Rozova, D. A. Kashporov, E. V. Antipov and L. M. Kovba, *Russ. J. Inorg. Chem.*, 1991, **36,** 928.
7. T. Hansen and H. Müller-Buschbaum, *Z. Anorg. Allg. Chem.*, 1994, **620,** 1137.

Phosphorus compounds

Name	*Formula*	*MW*	*Color*	*Crystal structure*	*Ref.*
Mercury(II) pyrophosphate dihydrate	$Hg_2P_2O_7(H_2O)_2$	611.151	Colorless		1
Mercury(I) dihydrogen-phosphate	$Hg_2(H_2PO_4)_2$	595.152		Monoclinic	2
	$AgHg_2PO_4$	604.018		Orthorhombic	3
	$(Hg_2)_2(H_2PO_4)(PO_4)$	994.316	Yellowish	Monoclinic	4
Mercury(I) phosphate	α-$(Hg_2)_3(PO_4)_2$	1393.48	Orange	Monoclinic	5
	β-$(Hg_2)_3(PO_4)_2$	1393.48	Orange	Monoclinic	5
Mercury(II) polyphosphate	$Hg(PO_3)_2$	358.532			6
Mercury(II) phosphate	$Hg_3(PO_4)_2$	791.71		Monoclinic	7
	$(Hg_2)_2P_2O_7$	976.301	Light yellow	Monoclinic	5
Mercury(II) diphosphate	$Hg_2P_2O_7$	575.121			8
Mercury(II) hydrogenphosphate	$HgHPO_4$	2185.213	Colorless	Triclinic	9
	$(Hg_3)_3(PO_4)_4$	2057.198	Colorless	Trigonal	10
	$(Hg_3)_2(HgO_2)(PO_4)_2$	1626.068	Yellow	Monoclinic	10

1. M. Weil, *Monatsh. Chem.*, 2003, **134**, 1509.
2. B. A. Nilsson, *Z. Kristallogr.*, 1975, **141**, 321.
3. R. Masse, J.-C. Guitel and A. Durif, *J. Solid State Chem.*, 1978, **23**, 369.
4. M. Weil, *Z. Anorg. Allg. Chem.*, **626**, 2000, 1752.
5. M. Weil and R. Glaum, *Z. Anorg. Allg. Chem.*, 1999, **625**, 1752.
6. M. Weil and R. Glaum, *Acta Crystallogr.*, 2000, **C56**, 133.
7. K. Aurivillius and B. A. Nilsson, *Z. Kristallogr.*, 1975, **141**, 1.
8. M. Weil and R. Glaum, *Acta Crystallogr.*, 1997, **C53**, 1000.
9. E. Dubler, L. Beck, L. Linowsky and G. B. Jameson, *Acta Crystallogr.*, 1981, **B37**, 2214.
10. M. Weil and R. Glaum, *J. Solid State Chem.*, 2001, **157**, 68.

Rhenates

Name	*Formula*	*MW*	*Color*	*Crystal structure*	*Ref.*
	$HgReO_4$	450.793			1
	$Hg_5Re_2O_{10}$	1535.354			1
	$Hg_5Re_2O_{10}$	1535.354			1
	Hg_2ReO_5	667.382			1

1. S. V. Borisov, S. A. Magarill, N. V. Pervukhina and N. A. Kryuchkova, *J. Struct. Chem.*, 2002, **43,** 293.

Selenides and Selenates

Name	*Formula*	*MW*	*Color*	*Crystal structure*	*Ref.*
	HgSe	279.55			See Appendix I
Mercury(II) selenite	α-$HgSeO_3$	327.547			1
	β-$HgSeO_3$	327.547		Trigonal	2
	γ-$HgSeO_3$	327.547		Trigonal	2
Mercury(II) selenate monohydrate	$HgSeO_4 \cdot H_2O$	361.561		Monoclinic	3
	α-Hg_2SeO_3	528.137	Light yellow		4
	β-Hg_2SeO_3	528.137	Colorless		4
	γ-Hg_2SeO_3	528.137			4
Mercury(II) selenate(IV)	$HgSeO_4$	343.546		Orthorhombic	5
	$HgSeO_4 \cdot HgO$	560.135		Monoclinic	5
	$HgSeO_4 \cdot 2HgO$	1168.122		Trigonal	5
	$HgSeO_3 \cdot HgO \cdot 1/6H_2O$	563.136		Trigonal	6
	Hg_3SeO_6	776.724			5
	$Hg_3Se_3O_{10}$	998.64			7
	$Hg_4Se_4O_9$	1262.191			2

1. M. Koskenlinna and J. Valkonen, *Acta Crystallogr.*, 1995, **C51**, 1040.
2. M. Weil, *Solid State Sci.*, 2002, **4**, 1153.
3. C. Stålhandske, *Acta Crystallogr.*, 1978, **B34**, 1408.
4. M. Weil, *J. Solid State Chem.*, 2003, **172,** 35.
5. M. Weil, *Z. Naturforsch.*, 2002, **57B**, 1043.
6. M. Weil, *Acta Crystallogr.*, 2002, **C58**, i164.
7. M. Weil and Kolitsch, *Acta Crystallogr.*, 2002, **C58**, i47.

Sulfides and Sulfates

Name	*Formula*	*MW*	*Color*	*Crystal structure*	*Ref.*
	α-HgS	232.656			See Appendix I
	β-HgS	232.656			See Appendix I
Mercury chloride sulfide	α-$Hg_3Cl_2S_2$	736.808	White	Cubic	1
	β-$Hg_3Cl_2S_2$	736.808	White		
Mercury(II) sulfate	$HgSO_4$	296.652		Orthorhombic	2
				Orthorhombic	3
	$HgSO_4 \cdot H_2O$	314.667		Orthorhombic	
				Orthorhombic	4
	$Hg_3(SO_4)O_2$	729.83			5, 6
	$HgSO_4 \cdot 2HgO$	1121.225			7
Trimercury(II) dihydroxide disulfate monohydrate	$Hg_3(OH)_2(SO_4)_2 \cdot H_2O$	848.923		Monoclinic	8
Mercury manganese sulfide	HgMnS	287.594		Cubic	9
Mercury iron sulfide	HgFeS	288.503		Cubic	9
Mercury cobalt sulfide	HgCoS	291.589		Cubic	9

1. H. Puff and J. Küster, *Naturwissenschaften*, 1962, **49**, 299.
2. P. A. Kokkoros and P. J. Rentzeperis, *Z. Kristallogr.*, 1963, **119**, 234.
3. C. Stålhandske, *Acta Crystallogr.*, 1980, **B36**, 23.
4. L. K. Templeton, D. H. Templeton and A. Zalkin, *Acta Crystallogr.*, 1964, **17**, 933.
5. M. Weil, *Acta Crystallogr.*, 2001, **E57**, i98.
6. M. A. K. Ahmed, H. Fjellvåg and A. Kjekshus, *Thermochim. Acta*, 2002, **390**, 113.
7. G. Nagorsen, S. Lyng, A. Weiss and A. Weiss, *Angew. Chem.*, 1962, **74**, 119.
8. K. Aurivillius and C. Stålhandske, *Z. Kristallogr.*, 1976, **144**, 1.
9. W. Paszkowicz, *Powder Diffract.*, 2000, **15**, 116.

Tellurides and Tellurates

Name	*Formula*	*MW*	*Color*	*Crystal structure*	*Ref.*
Mercury(II) bromide telluride	$Hg_3Br_2Te_2$	1016.778	Yellow		1, 2
Mercury chloride telluride	$Hg_3Cl_2Te_2$	927.875	Yellow		1
Mercury telluride bromide iodide	Hg_3Te_2BrI	1063.778		Monoclinic	3
	Hg_2TeO_4	592.776		Monoclinic	4
	$HgTeO_4 \cdot 2H_2O$	1020.330		Orthorhombic	
Mercury(II) tellurite	$HgTeO_3$	376.187		Triclinic	5
Mercury(II) tellurite(IV) tellurate(VI)	α-$Hg_2Te_2O_7$	768.373		Monoclinic	6
	β-$Hg_2Te_2O_7$	768.373		Orthorhombic	6
Mercury(II) orthotellurate(VI)	Hg_3TeO_6	825.364	Amber	Cubic	7
Basic mercury(II) tetraoxotellurate(VI)	Hg_2TeO_5	608.775	Dark red	Orthorhombic	7
Disilver(I) dimercury(II) tris[tetraoxotellurate(VI)]	$Ag_2Hg_2(TeO_4)_3$	1191.722	Red		8

1. H. Puff, *et al.*, *Naturwissenschaften*, 1962, **49**, 299.
2. P. Khodadad, *et al.*, *Ann. Chim. (Paris)*, 1965, **10**, 83.
3. Yu. V Minets, Yu. V Voroshilov and V. V Pan'ko, *J. Alloys Compd.*, 2004, **367**, 109.
4. G. Brandt and R. Moritz, *Mater. Res. Bull.*, 1985, **20**, 49.
5. V. Krämer and G. Brandt, *Acta Crystallogr.*, 1986, **C42**, 917.
6. M. Weil, *Z. Kristallogr.*, 2003, **218,** 691.
7. M.Weil, *Z. Anorg. Allg. Chem.*, 2003, **629**, 653.
8. M. Weil, *Acta Crystallogr.*, 2005, **C61**, i103.

Tantalates

Name	*Formula*	*MW*	*Color*	*Crystal structure*	*Ref.*
	$Hg_2Ta_20_7$	875.075		Cubic	1

1. A. W. Sleight, *Inorg. Chem.*, 1968, **7**, 1704.

Tungstates

Name	*Formula*	*MW*	*Color*	*Crystal structure*	*Ref.*
	Hg_2WO_4	649.026		Monoclinic	1, 2

1. A. L. Wessels, R. Czekalla and W. Jeitschko, *Mater. Res. Bull.*, 1998, **33**, 95.
2. T. J. Mormann and W. Jeitschko, *Inorg. Chem.*, 2000, **39**, 4219.

Vanadates

Name	*Formula*	*MW*	*Color*	*Crystal structure*	*Ref.*
Mercury vanadate(IV,V)	$Hg_2V_8O_{20}$	1128.696			1
Mercury(I) vanadate(V)	$HgVO_3$	299.529		Triclinic	2
Mercury(I,II) vanadate	Hg_2VO_4	516.118			2
Mercury(II) vanadate(V)	α-HgV_2O_6	398.468			
	β-HgV_2O_6	398.468			3, 4
	α-$Hg_2V_2O_7$	615.057			5
	β-$Hg_2V_2O_7$	615.057			6

1. M. Weil, B. Stoeger, A. L. Wessels and W. Jeitschko, *Z. Naturforsch.*, 2007, **62B**, 1390.
2. S. V. Borisov, S. A. Magarill, N. V. Pervukhina and N. A. Kryuchkova, *J. Struct. Chem.*, 2002, **43**, 293.
3. J. Angenault and A. Rimsky, *C. R. Acad. Sci. Ser. C*, 1968, **266**, 978.
4. T. J. Mormann and W. Jeitschko, *Z. Kristallogr. New Cryst. Struct.*, 2000, **216**, 3.
5. M. Quarton, J. Angenault and A. Rimsky, *Acta Crystallogr.*, 1973, **B29**, 567.
6. A. W. Sleight, *Mater. Res. Bull.*, 1972, **7**, 827.

Mixed Halides

Name	*Formula*	*MW*	*Color*	*Crystal structure*	*Ref.*
Mercury bromide chloride	α-HgBrCl	315.947		Orthorhombic	1
Mercury bromide chloride	β-HgBrCl	315.947		Orthorhombic	1
Mercury bromide fluoride	HgBrF	299.492			2
Mercury bromide iodide	HgBrI	407.398			3
Mercury chloride iodide	HgClI	362.947			3

1. S. Mehdi and S. Mumtaz Ansari, *J. Solid State Chem.*, 1981, **40**, 122.
2. R. P. Rastogi, *et al.*, *J. Inorg. Nucl. Chem.*, 1975, **37**, 1167.
3. R. P. Rastogi and B. L. Dubey, J. Am. Chem. Soc., 1967, **89**, 200.

Intercalation Compounds

Name	*Formula*	*MW*	*Color*	*Crystal structure*	*Ref.*
	$Mg(NH_3)_6Hg_{22}$				1
	Hg–TiS_2				2
	$Hg_{1.24}TiS_2$				3
	K–Hg–graphite				4
	Graphite–Hg–alkalis				5
	Hg_xTaS_2 ($x = 0.58$, 1.19 and 1.3)				6

1. I.-C. Hwang, T. Drews and K. Seppelt, *J. Am. Chem. Soc.*, 2000, **122**, 8486.
2. E. W. Ong, M. J. McKelvy, G. Ouvrard and W. S. Glaunsinger, *Chem. Mater.*, 1992, **4**, 14.

3. P. Ganal, P. Moreau, G. Ouvrard, M. Sidorov, M. McKelvy and W. Glaunsinger, *Chem. Mater.*, 1995, **7**, 1132.
4. G. Roth, A. Chaiken, T. Enoki, N. C. Yeh, G. Dresselhaus and P. M. Tedrow, *Phys. Rev. B*, 1985, **32**, 533.
5. Y. Iye and S.-i. Tanuma, *Phys. Rev. B*, 1982, **25**, 4583.
6. P. Ganal, P. Moreau, G. Ouvrard, W. Olberding and T. Butz, *Phys. Rev. B*, 1995, **52**, 11359.

APPENDIX IV

Selected Organometallic Compounds of Mercury

The physical properties of approximately 800 organomercury compounds have been summarized in Wardell [1]. Physical properties of selected organometallic compounds are reviewed here. Most of these compounds are either highly toxic or extremely toxic and extreme care must be exercised in handling and using any of these compounds.

Ethylmercury Compounds

Name	*Formula*	*MW*	*Density (g cm^{-3})*	*CAS No.*	*M.P. (°C)*	*B.P. (°C)*	*Ref.*
Ethylmercury(II) acetate	$C_4H_8HgO_2$	288.71		109-62-6	69.9	117	1, 2
Ethylmercury(II) bromide	C_2H_5HgBr	309.56		107-26-6	198		3, 4
Ethylmercury(II) chloride	C_2H_5HgCl	265.10		107-27-7	196–198		1, 4
Ethylmercury(II) hydroxide	C_2H_6HgO	246.66		107-28-8	37		1
Ethylmercury(II) iodide	C_2H_5HgI	356.56		2440-42-8	186		5, 6

Methylmercury Compounds

Name	*Formula*	*MW*	*Density (g cm^{-3})*	*CAS No.*	*M.P. (°C)*	*B.P. (°C)*	*Ref.*
Methylmercury(II) acetate	$C_3H_6HgO_2$	274.67		108-07-6	125.5–127.5		1, 7
Methylmercury(II) bromide	CH_3HgBr	295.53		506-83-2	161–172		7
Methylmercury(II) chloride	CH_3HgCl	251.08		115-09-3	167		4, 7

Mercury Handbook: Chemistry, Applications and Environmental Impact
By Leonid F Kozin and Steve Hansen

Published by the Royal Society of Chemistry, www.rsc.org

Name	*Formula*	*MW*	*Density (g cm^{-3})*	*CAS No.*	*M.P. (°C)*	*B.P. (°C)*	*Ref.*
Methylmercury(II) cyanide	CH_3HgCN	241.64	3.97	2597-97-9	93		1, 7, 8
Methylmercury(II) hydroxide	CH_4HgO	232.63		1184-57-2	137		1, 7
Methylmercury(II) iodide	CH_3HgI	342.53		143-36-2	152		9

Phenylmercury Compounds

Name	*Formula*	*MW*	*Density (g cm^{-3})*	*CAS No.*	*M.P. (°C)*	*B.P. (°C)*	*Ref.*
Diphenylmercury	$C_{12}H_{10}Hg$	354.80	2.25^{meas} $2.38^{X\text{-}ray}$	587-85-9	124.5–125		1, 7, 10
Diphenylethynyl mercury	$C_{16}H_{10}Hg$	402.9	~2.0	6077-10-7	125		11
Phenylmercury(II) acetate	$C_6H_8HgO_2$	336.74	2.4	62-38-4	149.5		1, 12
Phenylmercury(II) benzoate	$C_{13}H_{10}HgO_2$	398.81		25358-71-8	97–98	220–240 dec.	1
Phenylmercury(II) borate	$C_6H_7BHgO_3$	338.52		102-98-7	112–113		13, 14
Phenylmercury(II) chloride	C_6H_5HgCl	313.18		100-56-1	249		14–16
Phenylmercury(II) bromide	C_6H_5HgBr	357.60		1192-89-8	283		15
Phenylmercury(II) hydroxide	C_6H_6HgO	294.70		100-57-2	197–205		14
Phenylmercury(II) iodide	C_6H_5HgI	404.60		823-04-1	269		15
Phenylmercury(II) nitrate	$C_6H_5HgNO_3$	339.70		55-68-5	114.5–116.5		1
Phenylmercury nitrate, basic	$C_6H_5HgOH\text{-}C_6H_5HgNO_3$	634.45		8003-05-2		175–185 dec.	14
Phenylmercury oleate	$C_6H_5HgO\text{-}COC_{17}H_{33}$	559.17		104-68-9	45		14

Other R_2Hg Molecules

Name	*Formula*	*MW*	*Density (g cm^{-3})*	*CAS No.*	*M.P. (°C)*	*B.P. (°C)*	*Ref.*
Dibenzylmercury	$C_{14}H_{14}Hg$	382.86	$2.17^{X\text{-}ray}$	780-24-5	111		17
Diethylmercury	$(C_2H_5)_2Hg$	258.71	2.45	627-44-1	–45	159	7
Dimethylmercury	$Hg(CH_3)_2$	230.66	3.07	593-74-8	–43	96	14
Dipropylmercury	$C_6H_{14}Hg$	286.77	2.02	628-85-3		189–191	1
Divinylmercury	C_4H_6Hg	254.68	2.76	1119-20-6		157	1

References

1. J. L. Wardell (ed.), *Organometallic Compounds of Zn, Cd and Hg*, Chapman and Hall, London, 1985.

2. F. C. Whitmore, *Organic Compounds of Mercury*, Chemical Catalog Co., New York, 1921.
3. C. S. Marvel, C. G. Gauerke and E. L. Hill, *J. Am. Chem. Soc.*, 1925, **47**, 3009.
4. D. R. Grdenic and A. I. Kitaigorodskii, *Zh. Fiz. Khim.*, 1949, **23**, 1161.
5. M. M. Baig, MSc thesis, University of British Columbia, 1961.
6. E. Krause and A. Von Grosse, *Die Chemie der Metal-Organischen Verbindung*, Borntraeger, Berlin, 1937.
7. H. L. Roberts, *Adv. Inorg. Chem. Radiochem*, 1968, **11**, 309–338.
8. J. C. Mills, H. S. Preston and H. L. Kennard, *J. Organomet. Chem.*, 1968, **14**, 33.
9. A. J. Brown, O. W. Howarth and P. J. Moore, *J. Chem. Soc., Dalton Trans.*, 1976, 1589.
10. D. Grdenic, B. Kamenar and A. Nagl, *Acta Crystallogr*, 1977, **B33**, 587.
11. J. Trotter, *Can. J. Chem.*, 1962, **40**, 1218.
12. B. Kamenar, *et al.*, *Inorg. Chim. Acta*, 1972, **6**, 191.
13. G. W. A. Milne, *Drugs: Synonyms and Properties*, Ashgate Publishing, Brookfield, VT, 2000, p. 1280.
14. M. Simon, P. Jönk, G. Wühl-Couturier and S. Halbach, *Mercury, Mercury Alloys and Mercury Compounds*, Wiley-VCH, Weinheim, 2006.
15. V. I. Pakhomov, *J. Struct. Chem*, 1963, **4**, 540.
16. J. Fayos, G. Artioli and R. Torres, *J. Chem. Crystallogr*, 1993, **23**, 595.
17. P. B. Hitchcock, *Acta Crystallogr*, 1979, **B35**, 746.

APPENDIX V

Solubility of Common Metals in Mercury

Ag

T *(°C)*	T *(K)*	*Mole fraction Ag*	*Ref.*
16.2	289.4	0.000558	1
20.0	293.2	0.00071	2
50.0	323.2	0.0016	2
99.6	372.8	0.004121	1
100.0	373.2	0.0041	2
150.0	423.2	0.0091	2
184.4	457.6	0.034192	1
200.0	473.2	0.018	2
250.0	523.2	0.031	2
260.0	533.2	0.0345	1
275.0	548.2	0.044	2
300.0	573.2	0.051	2
306.0	579.2	0.0525	1
338.0	611.2	0.0687	1
350.0	623.2	0.120	2
356.7	629.9	0.0929	1
400	673	0.200	2
405	678	0.1805	1
450	723	0.290	2
500	773	0.390	2
550	823	0.440	2

1. D. R. Hudson, *Metallurgia*, 1943, **28**, 203.
2. G. Jangg and H. Palman, Z. *Metallkd.*, 1963, **54**, 364.

Mercury Handbook: Chemistry, Applications and Environmental Impact
By Leonid F Kozin and Steve Hansen

Published by the Royal Society of Chemistry, www.rsc.org

Al

T (°C)	T (K)	Mole fraction Al	Ref.
76	349	0.001	1
101	374	0.001	1
103	376	0.001	1
125	398	0.002	1
160	433	0.003	1
260	533	0.008	1
312	585	0.013	1
370	643	0.046	2
460	733	0.100	2
480	753	0.124	2
510	783	0.204	2
524	797	0.274	2
542	815	0.355	2
550	823	0.402	2
558	831	0.447	2
561	834	0.465	2
566	839	0.500	2
576	849	0.606	2
582	855	0.667	2
590	863	0.746	2
595	868	0.805	2
600	873	0.812	2
604	877	0.854	2
610	883	0.880	2
643	916	0.882	2
650	923	0.959	2
652	925	0.986	2

1. H. A. Leibhafsky, *J. Am. Chem. Soc.*, 1949, **27**, 1468.
2. C. J. De Gruyer, *Recl. Trav. Chim. Pays-Bas*, 1925, **44**, 937.

Au

T (°C)	T (K)	Mole fraction Au	Ref.
80.8	353.95	0.00467	1
101.2	374.35	0.00697	1
121.7	394.85	0.01211	1
142.1	415.25	0.01482	1
159.2	432.35	0.01847	1
182.3	455.45	0.02434	1
200.0	473.15	0.030	1
219.6	492.75	0.037	1
239.2	512.35	0.051	1
260.2	533.35	0.065	1
269.6	542.75	0.078	1
279.6	552.75	0.081	1
292.6	565.75	0.126	1
299.5	572.65	0.140	1
308	581	0.203	2
328	601	0.300	2

T (°C)	T (K)	Mole fraction Au	Ref.
351	624	0.351	2
375	648	0.401	2
418	691	0.449	2

1. A. A. Sunier and E. B. Gramkee, *J. Am. Chem. Soc.*, 1929, **51**, 1703.
2. C. Rolfe and W. Hume-Rothery, *J. Less-Common Met.*, 1967, **13**, 1.

Bi

T (°C)	T (K)	Mole fraction Bi	Ref.
–35.4	237.75	0.001	1, 2
–30.3	242.85	0.0015	1, 2
–22.1	251.05	0.0022	1, 2
–9.85	263.3	0.0036	1, 2
–2.6	270.55	0.0046	1, 2
17.6	290.75	0.0097	1, 2
22.5	295.65	0.0112	1, 2
32.4	305.55	0.0175	1, 2
37.0	310.15	0.02	2
42.2	315.35	0.0275	1, 2
47.0	320.15	0.03	2
50.85	324.00	0.04	1, 2
54.0	327.15	0.04	2
61.6	334.75	0.058	1, 2
62.0	335.15	0.05	2
69.5	342.65	0.077	1, 2
71	344.15	0.08	2
79	352.15	0.11	2
81	354.15	0.13	2
86	359.15	0.15	2
90	363.15	0.17	2
96	369.15	0.2	2
108	381.15	0.25	2
118	391.15	0.3	2
120	393.15	0.3	2
135	408.15	0.4	2
155	428.15	0.5	2
170	443.15	0.6	2
200	473.15	0.7	2

1. G. Petot-Ervas, P. Desre and E. Bonnier, *C. R. Acad. Sci.*, 1965, **261**, 3406–3409.
2. G. Petot-Ervas, M. Allibert, C. Petot, P. Desre and E. Bonnier, *Bull. Soc. Chim. Fr.*, 1969, 1477–1481.

Cd

T (°C)	T (K)	Mole fraction Cd	Ref.
–36.4	236.8	0.0047	1
–35.0	238.2	0.0080	2

T (°C)	T (K)	Mole fraction Cd	Ref.
–34.6	238.6	0.0094	1
–34.0	239.2	0.0130	2
–25.0	248.2	0.0200	2
–19.0	254.2	0.0250	2
–13.0	260.2	0.0300	2
–10.0	263.2	0.0350	2
–2.0	271.2	0.0500	2
–1.6	271.6	0.0552	1
6.5	279.7	0.0650	2
12.5	285.7	0.0750	2
17.5	290.7	0.0850	2
25.0	298.2	0.0900	1
28.5	301.7	0.1050	2
34.0	307.2	0.1244	1
48.0	321.2	0.1550	2
50.0	323.2	0.1600	1
54.4	327.6	0.1839	1
57.0	330.2	0.1850	2
57.0	330.2	0.1840	1
65.5	338.7	0.2000	2
68.8	342.0	0.2221	1
74.0	347.2	0.2200	2
74.0	347.2	0.2221	1
75.0	348.2	0.2800	1
76.0	349.2	0.2400	2
84.6	357.8	0.2722	1
85.5	358.7	0.2750	2
86.0	359.2	0.2722	1
88.0	361.2	0.2800	2
117.0	390.2	0.3839	1
121.8	395.0	0.4004	1
149.6	422.8	0.5028	1
150.0	423.2	0.5028	1
163.6	436.8	0.5510	1
190.8	464.0	0.6433	1
214.6	487.8	0.7090	1
221.0	494.2	0.7090	1
234.0	507.2	0.7450	1
237.3	510.5	0.7458	1
273.4	546.6	0.8496	1

1. H. C. Bijl, *Z. Phys. Chem.*, 1902, **41**, 641.
2. R. E. Mehl and C. S. Barrett, *Trans. AIME*, 1930, **89**, 575.

Co

T (°C)	T (K)	Mole fraction Co	Ref.
160	433.15	2.00×10^{-8}	1
500	773.15	6.80×10^{-7}	1
525	798.15	6.50×10^{-7}	3–5
550	823.15	8.20×10^{-7}	2

T (°C)	T (K)	Mole fraction Co	Ref.
550	823.15	1.80×10^{-6}	3–5
575	848.15	2.70×10^{-6}	3–5
600	873.15	1.40×10^{-6}	3–5
625	898.15	1.40×10^{-6}	3–5
650	923.15	4.10×10^{-6}	3–5
675	948.15	2.40×10^{-6}	3–5
700	973.15	7.10×10^{-6}	3–5
725	998.15	6.10×10^{-6}	3–5
750	1023.15	1.10×10^{-5}	3–5

1. J. Borodzinski, University of Warsaw, unpublished data, personal communicaton to C. Guminski, 1987.
2. G. Jangg and H. Palman, *Z. Metallkd.*, 1963, **54**, 364.
3. J. R. Weeks, A. Minardi and S. Fink, *USAEC Report*, BNL-841, 1963, p. 76.
4. J. R. Weeks and S. Fink, *USAEC Report*, BNL-900, 1964, pp. 136–138.
5. J. R. Weeks, *Corrosion*, 1967, **23**, 98.

Cu

T (°C)	T (K)	Mole fraction Cu	Ref.
20	293.15	0.0001	1
50	323.15	0.00023	1
60	333.15	0.00025	2
80	353.15	0.00037	2
100	373.15	0.00048	2
150	423.15	0.0011	3
250	523.15	0.0034	3
350	623.15	0.0076	3
450	723.15	0.0153	3
550	823.15	0.0360	3

1. S. A. Levitskaya and A. I. Zebreva, *Trans. Inst. Khim. Akad. Nauk Kazakh. SSR*, 1967, **15**, 66.
2. A. A. Lange, S. P. Bukhman and A. A. Kairbaeva, *Izv. Akad. Nauk Kazakh. SSR, Ser. Khim.*, 1974, **24**, 37.
3. G. Jangg and H. Palman, *Z. Metallkd.*, 1964, **54**, 364.

Cr

T (°C)	T (K)	Mole fraction Cr	Ref.
500	773	0.0120	1
500	773	0.0083	2
505	778	0.0118	2

1. G. Jangg and H. Palman, *Z. Metalld.*, 1963, **54**, 364.
2. J. R. Weeks, *Corrosion*, 1967, **23**, 98.

Fe

T (°C)	T (K)	*Mole fraction Fe*	*Ref.*
25	298	5.4×10^{-6}	1
100	373	6.8×10^{-6}	1
200	473	1.1×10^{-5}	1
300	573	1.9×10^{-5}	1
400	673	4.0×10^{-5}	1
500	773	7.5×10^{-5}	1
600	873	1.6×10^{-4}	1
700	973	3.4×10^{-4}	1

1. A. L. Marshall, L. F. Epstein and F. J. Norton, *J. Am. Chem. Soc.*, 1950, **72**, 3514.

Gd

T (°C)	T (K)	*Mole fraction Gd*	*Ref.*
25	298	9.8×10^{-5}	1
25	298	5.3×10^{-5}	2
92.5	365.5	0.000377	3
132.5	405.5	0.00081	3
147.5	420.5	0.00121	3
150	423	0.0013	4
207.5	480.5	0.0027	3
215.0	488	0.00274	3
282.5	555.5	0.00664	3
295.0	568	0.00533	3
340.0	613	0.00967	3
450	723	0.013	4

1. V. A. Bulina, A. I. Zebreva and R. Sh. Enikeev, *Izv. V. U. Z. Khim. Khim. Tekhno*l., 1977, **20**, 959.
2. V. A. Bulina, L. V. Guminichenko, A. I. Zebreva and R. Sh. Enikeev, *Radiokhimiya*, 1977, **19**, 89.
3. A. E. Messing and O. C. Dean, *USAEC Rep.*, ORNL-2871, 1960.
4. H. Kirchmayr and W. Lugscheider, *Z. Metallkd.*, 1966, **57**, 725.

In

T (°C)	T (K)	*Mole fraction In*	*Ref.*
25	298	0.700	1
37	310	0.725	1
53	326	0.750	1
66	339	0.775	1
80	353	0.8025	1
90	363	0.825	1
101	374	0.850	1
103	376	0.855	1
106	379	0.860	1

T (°C)	T (K)	Mole fraction In	Ref.
108	381	0.875	1
114	387	0.880	1
123	396	0.900	1
134	407	0.936	1
150	423	0.975	1

1. L. F. Kozin and N. N. Tananaeva, *Russ. J. Inorg. Chem.*, 1961, **6**, 463.

Mn

T (°C)	T (K)	Mole fraction Mn	Ref.
20	293	4.6×10^{-5}	1
20	293	4.6×10^{-5}	2
25	298	4.4×10^{-5}	3
25	298	4.4×10^{-5}	4
30	303	6.2×10^{-5}	5
86	359	0.00087	6
100	373	0.0010	6
114	387	0.0012	6
125	398	0.0017	6
148	421	0.0026	6
166	439	0.0031	6
198	471	0.0036	6
225	498	0.0051	6
246	519	0.0069	6
270	543	0.0087	6
300	573	0.013	6
330	603	0.019	6
350	623	0.022	6
370	643	0.026	6
400	673	0.031	6
418	691	0.036	6
450	723	0.046	6
470	743	0.056	6
500	773	0.063	6
552	825	0.076	6
565	838	0.082	6

1. N. M. Irvin and A. S. Russell, *J. Chem. Soc.*, 1932, 891.
2. W. Kemula and Z. Galus, *Roczn. Chem.*, 1962, **36**, 1223.
3. I. E. Krasnova and A. I. Zebreva, *Elektrokhimiya.*, 1966, **2**, 96.
4. T. Hurlen and R. Smaaberg, *J. Electroanal. Chem.*, 1976, **71**, 157.
5. J. F. deWet and R. A. W. Haul, *Z. Anorg. Allg. Chem.*, 1954, **277**, 96.
6. G. Jangg and H. Palman, *Z. Metallkd.*, 1963, **54**, 364.

Pb

T (°C)	T (K)	Mole fraction Pb	Ref.
–36	237.00	0.0044	1
–15	258.00	0.0075	1
0	273.00	0.0096	1

T (°C)	T (K)	Mole fraction Pb	Ref.
15	288.00	0.0131	1
19.7	292.70	0.01469	2
24.0	297.00	0.015	3
24.9	297.90	0.0153	3
25	298.00	0.0165	4
25	298.00	0.0162	1
25.4	298.40	0.0154	3
26.3	299.30	0.0157	3
28.0	301.00	0.0161	3
30.7	303.70	0.01811	2
39.9	312.90	0.02203	2
47.4	320.40	0.02588	2
48.2	321.20	0.02631	2
50	323.00	0.0269	1
60.6	333.60	0.03438	2
69.2	342.20	0.04279	2
115	388.00	0.25	5
120	393.00	0.275	5
137	410.00	0.35	5
145	418.00	0.40	5
164	437.00	0.475	5
172	445.00	0.50	5
184	457.00	0.55	5
198	471.00	0.60	5
202	475.00	0.625	5
264	537.00	0.85	5
273	546.00	0.875	5
278	551.00	0.90	5
293	566.00	0.95	5
308	581.00	0.975	5

1. A. S. Moshkevich and A. A. Ravdel, *Zh. Prikl. Khim.*, 1970, **43**, 71.
2. H. E. Thompson, *J. Phys. Chem.*, 1935, **39**, 655.
3. P. Dumas, L. Bougarfa and J. Bensaid, *J. Phys.*, 1984, **45**, 1543.
4. M. M. Haring, M. R. Hatfield and P. T. Zapponi, *Trans. Electrochem. Soc.*, 1939, **75**, 473.
5. G. V. Yan-Sho-Syan, M. V. Nosek, N. M. Semibratova and A. E. Shalamov, *Tr. Inst. Khim. Nauk Akad. Nauk Kazakh. SSR*, 1967, **15**, 139.

Pd

T (°C)	T (K)	Mole fraction Pd	Ref.
25	298	0.000055	1
25	298	0.000050	2
41	314	0.000053	2
57	330	0.000058	2
81	354	0.000074	2
90	363	0.000089	2
95	368	0.000089	2
98	371	0.000094	2
120	393	0.00016	2

T (°C)	T (K)	Mole fraction Pd	Ref.
135	408	0.00021	2
161	434	0.00032	2
162	435	0.00036	2
175	448	0.00047	2
200	473	0.00068	2
214	487	0.00081	2
226	499	0.00117	2
234	507	0.0014	2
240	513	0.0017	2
253	526	0.0019	2
260	533	0.0020	2
286	559	0.0031	2
305	578	0.0042	2

1. J. N. Butler and A. C. Makrides, *Trans. Faraday Soc.*, 1964, **60**, 938.
2. G. Jangg and W. Gröll, *Z. Metallkd.*, 1965, **56**, 232.

Pu

T (°C)	T (K)	Mole fraction Pu	Ref.
19	292	0.000 136	1
21	294	0.000 131	2
24	297	0.000 161	2
50	323	0.000 255	2
100	373	0.000 625	2
150	423	0.00126	2
190	463	0.00182	2
200	473	0.00190	2
225	498	0.00275	2
260	533	0.00380	2
280	553	0.00421	2
300	573	0.00496	2
325	598	0.00561	2

1. A. G. White, *The Preparation of Plutonium Amalgam and Its Reaction with Dilute Hydrochloric Acid*, Tecnical Report AERE-C/R-1468, Atomic Energy Research Establishment, Harwell, 1955.
2. D. F. Bowersox and J. A. Leary, *J. Inorg. Nucl. Chem.*, 1959, **9**, 108.

Rh

T (°C)	T (K)	Mole fraction Rh	Ref.
500	773	1×10^{-6}	1

1. G. Jangg and T. Dörtbudak, *Z. Metallkd.*, 1973, **64**, 715.

Sn

T (°C)	T (K)	Mole fraction Sn	Ref.
–35.4	237.75	0.0016	1
–28.4	244.75	0.0029	1

T (°C)	T (K)	*Mole fraction Sn*	*Ref.*
–17.9	255.25	0.0041	1
–8.4	264.75	0.0052	1
1.1	274.25	0.0065	1
16.5	289.65	0.0097	1
26	299.15	0.0127	1
30	303.15	0.0140	1
40	313.15	0.0188	1
50	323.15	0.0259	1
60	333.15	0.0334	1
72	345.15	0.0560	1
54	327.15	0.025	1
61	334.15	0.030	1
67.5	340.65	0.040	1
70	343.15	0.050	1
78	351.15	0.080	1
81.5	354.65	0.126	2
85	358.15	0.15	1
88.75	361.90	0.020	2
92	365.15	0.20	1
93.5	366.65	0.254	2
97	370.15	0.267	2
98	371.15	0.285	2
101.5	374.65	0.308	2
102	375.15	0.318	2
103	376.15	0.30	1
105	378.15	0.333	2
108	381.15	0.362	2
108.5	381.65	0.35	1
113.5	386.65	0.40	1
114	387.15	0.399	2
117.5	390.65	0.418	2
122.75	395.90	0.454	2
123	396.15	0.45	1
129	402.15	0.50	1
132.5	405.65	0.500	2
140.5	413.65	0.543	2
142.5	415.65	0.55	1
152	425.15	0600	2
159.25	432.40	0.638	2
166	439.15	0.668	2
170.5	443.65	0.691	2
180	453.15	0.736	2
185.25	458.40	0.765	2
192.5	465.65	0.800	2
199.75	472.90	0.838	2
207.5	480.65	0.879	2
211.7	484.85	0.900	2
215.5	488.65	0.922	2
218.2	491.35	0.937	2
221	494.15	0.952	2
224	497.15	0.970	2
227	500.15	0.983	2
229.4	502.55	0.993	2

1. G. Petot-Ervas, M. Gaillet and P. Desre, *C. R. Acad. Sci.*, 1967, **264**, 490.
2. N. A. Puschin, *Z. Anorg. Allg. Chem.*, 1903, **36**, 210.

Tl

T (°C)	T (K)	Mole fraction Tl	Ref.
0.5	273.5	0.4050	1
184	457	0.7252	1
218	491	0.7959	1
231	504	0.8316	1
244	517	0.8685	1
261	534	0.9083	1
278	551	0.9462	1
283	556	0.9682	1

1. Y. Claire and J. Rey, *J. Less-Common Met.*, 1980, **70**, 33.

Tm

T (°C)	T (K)	Mole fraction Tm	Ref.
25	298	4×10^{-6}	1

1. V. A. Bulina, A. I. Zebreva and R. Sh. Enikeev, *Izv. V. U. Z. Khim. Khim. Tekhnol.*, 1977, **20**, 959.

Zn

T (°C)	T (K)	Mole fraction Zn	Ref.
–41.50	231.65	0.0260	1
0.30	273.45	0.0409	2
13.00	286.15	0.0570	1
19.90	293.05	0.0586	2
30.00	303.15	0.0696	2
36.00	309.15	0.0840	1
39.95	313.10	0.0828	2
50.00	323.15	0.0966	2
51.50	324.65	0.1060	1
64.75	337.90	0.1206	2
72.00	345.15	0.1420	1
80.10	353.25	0.1480	2
88.25	361.40	0.1800	1
89.50	362.65	0.1662	2
94.80	367.95	0.1779	2
99.60	372.75	0.1885	2
103.50	376.65	0.2150	1
120.00	393.15	0.2510	1
134.75	407.90	0.2860	1
155.00	428.15	0.3340	1
172.25	445.40	0.3710	1
184.00	457.15	0.4000	1
196.75	469.90	0.4320	1

T (°C)	T (K)	Mole fraction Zn	Ref.
209.75	482.90	0.4640	1
223.75	496.90	0.5000	1
233.50	506.65	0.5270	1
246.75	519.90	0.5610	1
262.25	535.40	0.6000	1
274.50	547.65	0.6320	1
285.00	558.15	0.6670	1
300.00	573.15	0.7050	1
317.00	590.15	0.7500	1
325.75	598.90	0.7720	1
334.00	607.15	0.7960	1
342.50	615.65	0.8250	1
354.00	627.15	0.8490	1
372.00	645.15	0.8940	1
396.00	669.15	0.9490	1

1. N. A. Pushin, *Z. Anorg. Chem.*, 1903, **34**, 201.
2. E. Cohen and K. Inouye, *Z. Phys. Chem.*, 1910, **71**, 625

Zr

T (°C)	T (K)	Mole fraction Zr	Ref.
350	623	1.1×10^{-3}	1
350	623	1.6×10^{-3}	2, 3
482	755	2.2×10^{-3}	4
500	773	6.6×10^{-3}	5
525	798	5.5×10^{-3}	5–7
545	818	3.4×10^{-3}	5–7
550	823	0.016	5
572	845	3.7×10^{-3}	5–7
600	873	9.9×10^{-3}	5–7
600	873	0.012	5–7
625	898	0.033	5–7
650	923	0.043	5–7
700	973	0.145	5–7
760	1033	0.40	8

1. A. J. Nerad, General Electric Co., unpublished work, cited by L. R. Kelman, W. D. Wilkinson and F. L. Yagee, Resistance of Materials to Attack by Liquid Metals, *USAEC Rep.*, ANL-4417, 1950, p. 68.
2. J. A. Leary, *USAEC Rep.*, LA-2218, 1958.
3. J. A. Leary, R. Benz, D. F. Bowersox, C. W. Bjørklund, K. W. H. Johnson, W. J. Maraman, L. J. Mullins and J. G. Reavis, Pyrometallurgical purification of Pu reactor fuels, in *Peaceful Uses of Atomic Energy*, United Nations, Geneva, 1958, vol. 17, pp. 376–382.
4. J. F. Nejedlik, *USAEC Rep.*, NAA-SR-6306, 1961.
5. J. R. Weeks, *Corrosion*, 1967, **23**, 98.
6. J. R. Weeks and S. Fink, *USAEC Rep.*, BNL-782, 1962, pp. 73–75.
7. J. R. Weeks and S. Fink, *USAEC Rep.*, BNL-900, 1964, pp. 136–38.
8. A. Fleitman and J. Brandon, *USAEC Rep.*, BNL-799, 1963, pp. 75–76.

Subject Index